Lecture Notes in Computer Science 2866

Edited by G. Goos, J. Hartmanis, and J. van Leeuwen

Springer
Berlin
Heidelberg
New York
Hong Kong
London
Milan
Paris
Tokyo

Jin Akiyama   Mikio Kano (Eds.)

Discrete and Computational Geometry

Japanese Conference, JCDCG 2002
Tokyo, Japan, December 6-9, 2002
Revised Papers

Springer

Series Editors

Gerhard Goos, Karlsruhe University, Germany
Juris Hartmanis, Cornell University, NY, USA
Jan van Leeuwen, Utrecht University, The Netherlands

Volume Editors

Jin Akiyama
Tokai University, Research Institute of Educational Development
2-28-4 Tomigaya, Shibuya-ku, Tokyo 151-0063, Japan
E-mail: fwjb5117@mb.infoweb.ne.jp

Mikio Kano
Ibaraki University, Department of Computer and Information Sciences
Nakanarusawa, Hitachi, Ibaraki 316-8511, Japan
E-mail: kano@cis.ibaraki.ac.jp

Cataloging-in-Publication Data applied for

A catalog record for this book is available from the Library of Congress.

Bibliographic information published by Die Deutsche Bibliothek
Die Deutsche Bibliothek lists this publication in the Deutsche Nationalbibliografie;
detailed bibliographic data is available in the Internet at <http://dnb.ddb.de>.

CR Subject Classification (1998): I.3.5, G.2, F.2.2, E.1

ISSN 0302-9743
ISBN 3-540-20776-7 Springer-Verlag Berlin Heidelberg New York

Springer-Verlag is a part of Springer Science+Business Media

springeronline.com

© Springer-Verlag Berlin Heidelberg 2003

Typesetting: Camera-ready by author, data conversion by Olgun Computergrafik
Printed on acid-free paper   SPIN: 10920909   06/3142   5 4 3 2 1 0

Preface

Since it was first organized in 1997, the Japan Conference on Discrete and Computational Geometry (JCDCG) continues to attract an international audience. The first five conferences of the series were held in Tokyo, the sixth in Manila, Philippines. This volume consists of the refereed papers presented at the seventh conference, JCDCG 2002, held in Tokai University, Tokyo, December 6–9, 2002. An eighth conference is planned to be held in Bandung, Indonesia.

The proceedings of JCDCG 1998 and JCDCG 2000 were published by Springer-Verlag as part of the series Lecture Notes in Computer Science: LNCS volumes 1763 and 2098, respectively. The proceedings of JCDCG 2001 were also published by Springer-Verlag as a special issue of the journal *Graphs and Combinatorics*, Vol. 18, No. 4, 2002.

The organizers are grateful to Tokai University for sponsoring the conference. They wish to thank all the people who contributed to the success of the conference, in particular, Chie Nara, who headed the conference secretariat, and the principal speakers: Takao Asano, David Avis, Greg N. Frederickson, Ferran Hurtado, Joseph O'Rourke, János Pach, Rom Pinchasi, and Jorge Urrutia.

October 2003

Jin Akiyama
Mikio Kano

Organizing Committee

Co-chairs: Jin Akiyama and Mikio Kano

Members: Tetsuo Asano, David Avis, Vašek Chvátal, Hiroshi Imai, Hiro Ito, Naoki Katoh, Midori Kobayashi, Chie Nara, Joseph O'Rourke, János Pach, Kokichi Sugihara, Xuehou Tan, Takeshi Tokuyama, Masatsugu Urabe, and Jorge Urrutia

Members of the Executive Committee: Takako Kodate, Haruhide Matsuda, Mari-Jo Ruiz, and Toshinori Sakai

Table of Contents

Universal Measuring Devices
with Rectangular Base

Jin Akiyama[1], Hiroshi Fukuda[2], and Gisaku Nakamura[1]

[1] Research Institute of Educational Development, Tokai University
Tokyo, 151-0063 Japan
[2] School of Administration and Informatics, University of Shizuoka
Shizuoka, 422-8526 Japan

Abstract. We consider a device with rectangular base having no gradations. We show that the number of directly measurable amounts of liquid
using the device with its vertices as markers is always 13, independent
of its shape. Then we show how the device can measure any integral
amount of liquid between 1 and 858 liters.

1   Universal Measuring Devices

A common device used to measure liquid in many Japanese stores some years
back, was a measuring box with a square base of area 6 and height 1. By tilting
the box, and using its vertices as markers, it is possible to keep 6, 3, and 1 liters.
The box would have no extra gradations to measure 2, 4, and 5 liters.

A store would keep a container holding large amounts of a certain liquid. If
a customer wanted to buy a certain integral amount of liquid between 1 and 6,
the store owner would proceed as follows:

1. He would immerse his measuring box into the store's container, and fill it
   just once.
2. Then he would alternately pour out a certain amount of liquid into the
   store's container and the customer's container.

We are interested in studying measuring devices without gradations which
nevertheless can be used for measuring any integral amount of liquid up to their
full capacity by using the procedure described above. We call such measuring
devices *universal measuring devices*. In the previous works [1,2], we have determined that the largest capacity of a universal measuring device *with triangular*
base is 41.

In this work, we consider devices *with rectangular* bases. Although we include
some discussion of this case in the previous work [1], we used the horizontal plane
to keep liquid in the device. In this present work, we refer back to the Japanese
store owners practice of using only vertices of the device as markers. In particular,
we use at least three vertices of the device to specify the plane formed by the
surface of the water.

J. Akiyama and M. Kano (Eds.): JCDCG 2002, LNCS 2866, pp. 1–8, 2003.

2   Devices with a Rectangular Base

We consider a universal measuring device with rectangular base of area 6. We first find all the amounts of liquid that can be directly measured with the device. We assume that the vertices of these devices are labeled a_i, a'_i for $i = 1, 2, 3, 4$, as shown in Figure 1.

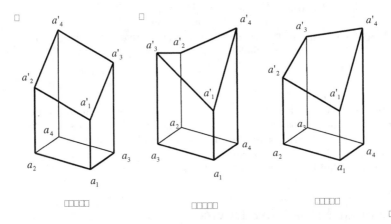

Fig. 1. Possible shapes of measuring devices with rectangular base.

Consider any device of heights $h_1 \leq h_2 \leq h_3 \leq h_4$, where h_i is the distance between a_i and a'_i. Starting at a'_1, record the labels found while visiting the vertices on the top of the device in a clockwise fashion. We can code the shape of any device by a 4–bit 0–1 vector $b_1 b_2 b_3 b_4$, in such a way that $b_i = 1$ if we move upwards from the ith visited vertex to the next. We call this the *shape vector* of the device. We can see by inspection that from the 16 possible shapes of devices only 3, up to symmetries, are meaningful. The corresponding shape vectors are 1100, 1010 and 1110.

In order to measure liquid, we use three vertices of the device as reference points to the surface plane of liquid. The following five choices of three vertices from the device correspond to the possible surface of liquid.

1. The surface of liquid is defined by one upper vertex and two vertices on the base in diagonal position as shown in Figure 2 (a). The amount of liquid is

$$h_4.$$

2. The surface of liquid is defined by one upper vertex and two vertices on the same edge of the base as shown in Figure 2 (b). The amount of liquid is

$$3h_3.$$

3. The surface of liquid is defined by two upper vertices in diagonal position and one vertex on the base as shown in Figure 2 (c). The amount of liquid is

$$3(h_2 + h_3).$$

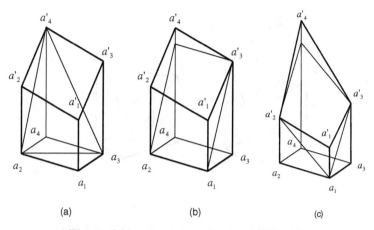

Fig. 2. Measuring integral amount of liquid.

4. The surface of liquid is defined by two upper vertices on the same edge and one vertex on the base. The amount of liquid is given by

$$3h_4$$

or by fractional expression

$$h_2 + h_3 + \frac{h_2^2}{h_3}$$

depending on the shape of the devices as shown in Figure 3 (a) and Figure 3 (b), respectively.

5. The surface of liquid is defined by three upper vertices. The amount of liquid is given by

$$3(h_2 + h_3)$$

or by the fractional expression

$$3(h_2 + h_3) + \frac{(h_4 - h_2 - h_3)^3}{(h_4 - h_2)(h_4 - h_3)}$$

depending on the shape of the devices as shown in Figure 4 (a) and Figure 4 (b), respectively.

We can summarize these directly measurable amounts in the following theorem.

Theorem 1. *The device with rectangular base of area 6 and of heights $h_1 \leq h_2 \leq h_3 \leq h_4$, shown in Figure 1, can measure the amounts of liquid given below after each shape vector. We use the notation*

$$v_{ij} = h_i + h_j + \frac{h_i^2}{h_j},$$

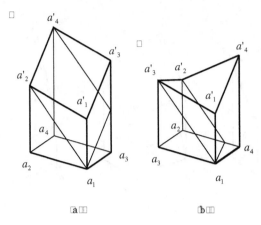

Fig. 3. Measuring liquid using two upper vertices on the same edge.

and

$$v_{ijk} = 3(h_j + h_k) + \frac{(h_i - h_j - h_k)^3}{(h_i - h_j)(h_i - h_k)}.$$

1. *The device with the shape vector 1100:*
   If $h_4 > h_2 + h_3$, the measurable amounts are

   $$h_1, h_2, h_3, h_4, 3h_1, 3h_2, 3h_3, 3(h_2 + h_3), v_{13}, v_{12}, v_{24}, v_{34}, v_{432}.$$

   If $h_2 + h_3 - h_1 < h_4 \leq h_2 + h_3$, the measurable amounts are

   $$h_1, h_2, h_3, h_4, 3h_1, 3h_2, 3h_3, 3h_4, 3(h_2 + h_3), v_{13}, v_{12}, v_{24}, v_{34}.$$

   If $h_4 \leq h_2 + h_3 - h_1$, the measurable amounts are

   $$h_1, h_2, h_3, h_4, 3h_1, 3h_2, 3h_3, 3h_4, 3(h_1 + h_4), v_{13}, v_{12}, v_{24}, v_{34}.$$

2. *The device with the shape vector 1010:*
   If $h_3 > h_1 + h_2$, the measurable amounts are

   $$h_1, h_2, h_3, h_4, 3h_1, 3h_2, 3(h_1 + h_2), v_{13}, v_{14}, v_{23}, v_{24}, v_{412}, v_{312}.$$

   If $h_3 \leq h_1 + h_2 < h_4$, the measurable amounts are

   $$h_1, h_2, h_3, h_4, 3h_1, 3h_2, 3h_3, 3(h_1 + h_2), v_{13}, v_{14}, v_{23}, v_{24}, v_{412}.$$

   If $h_4 \leq h_1 + h_2$, the measurable amounts are

   $$h_1, h_2, h_3, h_4, 3h_1, 3h_2, 3h_3, 3h_4, 3(h_1 + h_2), v_{13}, v_{14}, v_{23}, v_{24}.$$

3. *The device with the shape vector 1110:*
   If $h_4 > h_1 + h_3$, the measurable amounts are

   $$h_1, h_2, h_3, h_4, 3h_1, 3h_2, 3h_3, 3(h_1 + h_3), v_{23}, v_{34}, v_{14}, v_{12}, v_{413}.$$

   Otherwise if $h_4 \leq h_1 + h_3$, the measurable amounts are

   $$h_1, h_2, h_3, h_4, 3h_1, 3h_2, 3h_3, 3h_4, 3(h_1 + h_3), v_{23}, v_{34}, v_{14}, v_{12}.$$

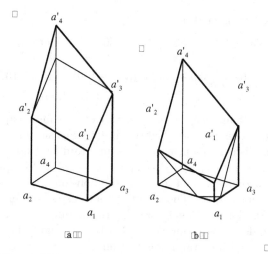

Fig. 4. Measuring liquid using three upper vertices.

The next result follows immediately from Theorem 1:

Corollary 1. *The number of directly measurable amounts using the device with rectangular base, given by Theorem 1, is always 13 independent of the shape of the device.*

Example.

Consider a measuring device with heights $h_1 = 130$, $h_2 = 132$, $h_3 = 156$, $h_4 = 169$. This device can measure eleven integral amounts of liquid : 130, 132, 156, 169, 390, 396, 399, 468, 469, 507 and 858 by Theorem 1. Suppose we want to sell 730 liters of liquid to a customer. We measure 730 liters in the following manner while pouring out liquid from the device eleven times:

0. First fill the device with 858 liters from the store's container.
1. Pour out 351 liters to the customer's container while retaining 507 liters in the device.
2. From the remaining 507 liters, pour back 38 liters into the store's container while retaining 469 liters.
3. From the remaining 469 liters, pour 1 liter into the customer's container while retaining 468 liters.
4. Pour back 69 liters into the store's container while retaining 399 liters.
5. Pour 3 liters into the customer's container while retaining 396 liters.
6. Pour back 6 liters into the store's container while retaining 390 liters.
7. Pour 221 liters into the customer's container while retaining 169 liters.
8. Pour back 13 liters into the store's container while retaining 156 liters.
9. Pour 24 liters into the customer's container while retaining 132 liters.
10. Pour back 2 liters into the store's container while retaining 130 liters.
11. Finally, pour the remaining 130 liters into the customer's container.

In summary, the customer receives

$$351 + 1 + 3 + 221 + 24 + 130 = 730 \text{ liters}$$

or

$$(858 - 507) + (469 - 468) + (399 - 396) + (390 - 169) + (156 - 132) + (130 - 0)$$
$$= 730 \text{ liters.}$$

Having these results at hand, we can find a device of maximum capacity by brute force as we did in the previous work [1] if the directly measurable amounts are all integrals. We accomplished this using a C++ program that does the following: For any set of heights $1 \leq h_1 \leq h_2 \leq h_3 \leq h_4 \leq N$, we compute the set of amounts given in Theorem 1. We then omit the non-integral amounts from the set and apply Theorem 2 in [1]. The set includes the sporadic integral amounts obtained from the fractional forms v_{ij} or v_{ijk},. In the range $N = 2^{12}$, we found a universal measuring device of shape vector 1110 with heights 130, 132, 156 and 169 that can measure all integral amounts from 1 to 858. We are restricted to the range $N = 2^{12}$ by computer limitations. Thus we have:

Theorem 2. *The capacity of a universal measuring device with a rectangular base is* ≥ 858.

Remarks.

1. The device with shape vector 1110 is the only one in the given range which can measure up to 858 liters.
2. The maximum of the minimum number of times that liquid has to be poured out of this device to measure varying integral amounts of liquid is 11, while the average number is 4.17. Liquid has to be poured out of the device 11times only in one case, that is, when the amount of liquid to be measured is 730 liters as shown in the example. The minimum number of times liquid has to be poured out to measure other integral amounts is shown in the Appendix.
3. The lower bound 858 obtained in this paper is better than 691 which was the bound obtained in a previous paper [1]. The improvement is due to the fact the fractional expressions v_{ij} and v_{ijk} were included among the numbers considered and that they actually yielded integral values in a few cases.
4. It may be possible to improve on the lower bound 858 with further work and a more powerful computer .
5. It would be interesting to investigate measuring devices with bases which are convex polygons.

Acknowledgements

The authors thank Mari-Jo Ruiz and referees for their helpful suggestions.

References

1. Akiyama, J., Fukuda, H., Nakamura, G., Sakai, T., Urrutia, J., Zamora-Cura, C.: Universal measuring devices without gradations, Discrete and Computational Geometry, JCDCG 2000, (LNCS 2098) 31–40 (Springer, 2001)
2. Akiyama, J., Fukuda, H., Nakamura, G., Nara, C., Sakai, T., Urrutia, J.: Universal measuring devices with triangular bases, submitted

Appendix

Let l denote the number of liters to be measured and let $d(l)$ denote the minimum number of times liquid has to be poured out of the measuring device in order to measure l liters.

l	$d(l)$	l	$d(l)$	l	$d(l)$	l	$d(l)$	l	$d(l)$	l	$d(l)$	l	$d(l)$	l	$d(l)$	l	$d(l)$
1	1	51	3	101	5	151	5	201	3	251	3	301	3	351	1	401	5
2	1	52	3	102	3	152	5	202	3	252	5	302	3	352	3	402	3
3	1	53	5	103	3	153	5	203	3	253	3	303	3	353	3	403	3
4	3	54	5	104	3	154	3	204	3	254	3	304	3	354	3	404	5
5	3	55	5	105	3	155	5	205	3	255	5	305	3	355	5	405	5
6	1	56	7	106	3	156	1	206	5	256	3	306	3	356	5	406	5
7	3	57	5	107	3	157	3	207	3	257	5	307	3	357	3	407	7
8	3	58	5	108	1	158	7	208	3	258	1	308	3	358	5	408	5
9	1	59	7	109	3	159	3	209	3	259	3	309	3	359	3	409	5
10	3	60	5	110	3	160	5	210	3	260	3	310	3	360	3	410	5
11	3	61	5	111	1	161	7	211	3	261	3	311	5	361	5	411	5
12	5	62	3	112	3	162	3	212	5	262	5	312	1	362	3	412	5
13	1	63	3	113	3	163	5	213	5	263	3	313	1	363	5	413	3
14	3	64	3	114	3	164	7	214	5	264	1	314	3	364	3	414	3
15	3	65	3	115	3	165	3	215	5	265	3	315	3	365	5	415	3
16	3	66	5	116	3	166	5	216	5	266	1	316	5	366	3	416	3
17	5	67	5	117	5	167	3	217	5	267	1	317	5	367	5	417	5
18	5	68	5	118	3	168	3	218	5	268	3	318	5	368	3	418	5
19	3	69	1	119	3	169	1	219	7	269	1	319	5	369	5	419	5
20	5	70	1	120	5	170	3	220	7	270	3	320	5	370	3	420	3
21	5	71	3	121	3	171	3	221	1	271	5	321	5	371	3	421	3
22	3	72	1	122	7	172	3	222	3	272	3	322	5	372	3	422	5
23	5	73	1	123	5	173	5	223	3	273	3	323	3	373	3	423	3
24	1	74	3	124	3	174	5	224	3	274	3	324	3	374	3	424	3
25	3	75	3	125	7	175	3	225	3	275	5	325	3	375	1	425	5
26	1	76	3	126	5	176	5	226	3	276	5	326	5	376	3	426	3
27	3	77	3	127	5	177	5	227	1	277	3	327	3	377	1	427	3
28	5	78	1	128	9	178	3	228	3	278	3	328	3	378	3	428	3
29	3	79	1	129	5	179	5	229	3	279	3	329	3	379	3	429	3
30	3	80	3	130	1	180	5	230	1	280	1	330	3	380	3	430	3
31	5	81	3	131	5	181	5	231	3	281	3	331	3	381	5	431	3
32	3	82	3	132	1	182	5	232	3	282	3	332	3	382	7	432	3
33	3	83	3	133	3	183	5	233	5	283	5	333	3	383	3	433	5
34	5	84	5	134	3	184	5	234	1	284	5	334	5	384	5	434	3
35	3	85	3	135	3	185	7	235	3	285	5	335	3	385	7	435	3
36	5	86	3	136	3	186	7	236	3	286	3	336	1	386	3	436	5
37	1	87	5	137	3	187	7	237	3	287	7	337	1	387	5	437	3
38	1	88	5	138	3	188	7	238	3	288	3	338	1	388	3	438	3
39	1	89	5	139	3	189	7	239	3	289	5	339	1	389	1	439	5
40	3	90	7	140	5	190	7	240	1	290	3	340	3	390	1	440	5
41	3	91	3	141	3	191	7	241	3	291	3	341	5	391	3	441	5
42	3	92	3	142	5	192	5	242	3	292	5	342	3	392	3	442	3
43	3	93	3	143	3	193	5	243	1	293	3	343	7	393	3	443	3
44	3	94	3	144	5	194	3	244	3	294	3	344	5	394	5	444	3
45	3	95	3	145	3	195	3	245	3	295	5	345	3	395	3	445	3
46	3	96	3	146	5	196	7	246	5	296	3	346	7	396	1	446	5
47	3	97	3	147	3	197	5	247	3	297	3	347	5	397	3	447	5
48	3	98	3	148	3	198	5	248	3	298	3	348	3	398	3	448	5
49	5	99	3	149	5	199	3	249	3	299	1	349	7	399	1	449	5
50	5	100	5	150	3	200	3	250	5	300	1	350	3	400	3	450	5

l	$d(l)$	l	$d(l)$	l	$d(l)$	l	$d(l)$	l	$d(l)$	l	$d(l)$	l	$d(l)$	l	$d(l)$	l	$d(l)$
451	7	501	3	551	5	601	7	651	3	701	3	751	5	801	7	851	5
452	5	502	5	552	5	602	5	652	5	702	1	752	5	802	9	852	3
453	5	503	5	553	5	603	5	653	3	703	5	753	5	803	7	853	5
454	5	504	3	554	5	604	5	654	3	704	3	754	5	804	7	854	5
455	3	505	3	555	5	605	5	655	3	705	7	755	5	805	7	855	3
456	3	506	3	556	5	606	7	656	3	706	5	756	5	806	5	856	3
457	5	507	1	557	5	607	5	657	3	707	5	757	7	807	5	857	3
458	3	508	5	558	3	608	7	658	3	708	5	758	7	808	7	858	1
459	1	509	9	559	3	609	3	659	3	709	5	759	5	809	7		
460	3	510	5	560	5	610	3	660	5	710	5	760	5	810	5		
461	3	511	7	561	5	611	3	661	5	711	5	761	5	811	5		
462	1	512	9	562	5	612	5	662	7	712	5	762	5	812	5		
463	3	513	5	563	7	613	5	663	3	713	3	763	5	813	5		
464	3	514	7	564	5	614	5	664	3	714	5	764	5	814	5		
465	3	515	9	565	5	615	3	665	5	715	3	765	5	815	5		
466	3	516	5	566	7	616	5	666	3	716	5	766	5	816	5		
467	3	517	7	567	5	617	3	667	7	717	3	767	5	817	5		
468	1	518	5	568	5	618	3	668	7	718	5	768	9	818	5		
469	1	519	3	569	7	619	3	669	7	719	3	769	7	819	3		
470	3	520	3	570	9	620	3	670	7	720	3	770	7	820	3		
471	5	521	3	571	9	621	5	671	7	721	5	771	7	821	3		
472	3	522	3	572	3	622	5	672	7	722	3	772	5	822	7		
473	7	523	5	573	5	623	7	673	7	723	3	773	5	823	5		
474	5	524	5	574	5	624	3	674	5	724	5	774	7	824	7		
475	3	525	5	575	5	625	5	675	5	725	3	775	5	825	5		
476	7	526	5	576	5	626	5	676	5	726	1	776	5	826	5		
477	5	527	5	577	5	627	5	677	5	727	3	777	5	827	7		
478	5	528	5	578	3	628	3	678	5	728	1	778	5	828	5		
479	7	529	5	579	5	629	3	679	5	729	5	779	3	829	5		
480	5	530	5	580	5	630	3	680	5	730	11	780	3	830	7		
481	3	531	5	581	3	631	3	681	5	731	7	781	5	831	5		
482	5	532	5	582	5	632	3	682	5	732	7	782	5	832	3		
483	3	533	5	583	5	633	3	683	3	733	9	783	5	833	5		
484	5	534	5	584	7	634	5	684	5	734	5	784	5	834	3		
485	3	535	5	585	3	635	5	685	5	735	7	785	3	835	7		
486	3	536	7	586	5	636	5	686	3	736	9	786	3	836	5		
487	5	537	7	587	5	637	3	687	3	737	5	787	5	837	7		
488	3	538	7	588	5	638	7	688	3	738	7	788	3	838	7		
489	5	539	7	589	3	639	7	689	1	739	5	789	3	839	5		
490	5	540	7	590	5	640	5	690	3	740	5	790	7	840	7		
491	5	541	7	591	3	641	5	691	3	741	3	791	7	841	7		
492	3	542	7	592	3	642	5	692	5	742	5	792	7	842	5		
493	5	543	5	593	5	643	5	693	3	743	5	793	5	843	5		
494	3	544	5	594	3	644	5	694	7	744	5	794	5	844	5		
495	5	545	3	595	5	645	5	695	5	745	5	795	5	845	3		
496	3	546	3	596	5	646	5	696	3	746	5	796	5	846	7		
497	5	547	7	597	5	647	3	697	7	747	3	797	7	847	5		
498	3	548	5	598	3	648	3	698	5	748	5	798	7	848	5		
499	3	549	5	599	5	649	3	699	3	749	5	799	9	849	3		
500	5	550	5	600	3	650	3	700	7	750	3	800	7	850	5		

Maximin Distance for n Points
in a Unit Square or a Unit Circle

Jin Akiyama[1], Rika Mochizuki[2], Nobuaki Mutoh[2], and Gisaku Nakamura[1]

[1] Research Institute of Educational Development, Tokai University
Tokyo, 151-0063 Japan
[2] School of Administration and Informatics, University of Shizuoka
Shizuoka, 422-8526 Japan

Abstract. Given n points inside a unit square (circle), let $d_n(c_n)$ denote the maximum value of the minimum distance between any two of the n points. The problem of determining $d_n(c_n)$ and identifying the configuration of that yields $d_n(c_n)$ has been investigated using geometric methods and computer-aided methods in a number of papers. We investigate the problem using a computer-aided search and arrive at some approximations which improve on earlier results for $n = 59$, 73 and 108 for the unit square, and also for $n = 70$, 73, 75 and $77, \cdots, 80$ for the unit circle. The associated configurations are identified for all the above-mentioned improved results.

1 Introduction

Given n points inside a unit square (circle), determine the maximum value of the minimum distance (*maximin* distance) between any two of the n points. This is the problem we address in this paper together with the closely related problem of identifying the configuration which yields the maximin distance. Let d_n and c_n denote the maximin distance for a unit square and a unit circle, respectively.

Quite a number of papers have been written on this problem. In particular, Donovan and Specht independently present up dated results in their web sites [1] and [3], respectively. While many results have been obtained, we are able to improve on earlier results for the unit square for $n = 59$, 73 and 108; we also obtain improvements on earlier results for the unit circle for $n = 70$, 73, 75 and $77, \cdots, 80$. We are able to improve on configurations suggested in earlier papers. We also identify all the configurations associated with the above-mentioned improved results.

A computer program was devised for this investigation. The program assigns a random initial placement for the n points and uses a greedy algorithm to determine d_n or c_n. For each n, the program is executed repeatedly until the values d_n and c_n converge.

We note that the problem of determining the configuration of n points in a unit square (circle) which yields the maximin distance, expressed below as Problem 1, is equivalent to the problem of determining the densest packing of n circles of equal diameter in a unit square (circle), expressed below as Problem 2.

J. Akiyama and M. Kano (Eds.): JCDCG 2002, LNCS 2866, pp. 9–13, 2003.

Problem 1. *How should n points be arranged in a unit square (circle) so that the minimum distance between them is maximum?*

Problem 2. *What is the maximum diameter of n equal circles that can be packed into a unit square (circle)?*

If a collection of points in a unit square (circle) are a distance of at least d from each other, the points can serve as the centers of a collection of circles of diameter d that will pack a square (circle) of side (diameter) $1 + d$ (Fig. 1).

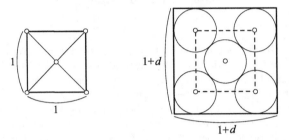

Fig. 1.

The improvements we have made are given in Section 2 and we discuss our computer algorithm in Section 3.

The problem of determining the maximin distance of n points in a square was first studied by Leo Moser in 1960. In his paper [2], we find the following conjecture for $n = 8$:

Conjecture 1. 8 points in or on a unit square determine a distance $\leq \frac{1}{2} \sec 15°$.

The reader who is interested in a historical account of the solution to the problem, for various values of n, is referred to [4].

2   Improvements on Earlier Results

(a) Packing Equal Circles in a Unit Square

We calculated d_n and obtained three improvements on previously obtained values for $n = 59$, 73 and 108. Table 1 compares the values we obtained with earlier results. The associated configurations are identified in Fig. 2.

Table 1. Best results of the packing circles in a unit square for $n = 59$, 73 and 108.

n	Our results	The best result in the literature	Ref.
59	0.15156206560892	0.151561918317	[1]
73	0.13471560564007	0.134709827696	[1]
108	0.10921914670419	0.109176694811	[1]

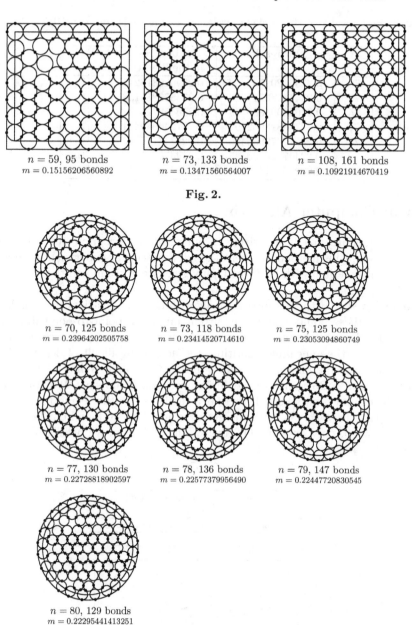

$n = 59$, 95 bonds
$m = 0.15156206560892$

$n = 73$, 133 bonds
$m = 0.13471560564007$

$n = 108$, 161 bonds
$m = 0.10921914670419$

Fig. 2.

$n = 70$, 125 bonds
$m = 0.23964202505758$

$n = 73$, 118 bonds
$m = 0.23414520714610$

$n = 75$, 125 bonds
$m = 0.23053094860749$

$n = 77$, 130 bonds
$m = 0.22728818902597$

$n = 78$, 136 bonds
$m = 0.22577379956490$

$n = 79$, 147 bonds
$m = 0.22447720830545$

$n = 80$, 129 bonds
$m = 0.22295441413251$

Fig. 3.

(b) Packing Equal Circles in a Unit Circle

We calculated c_n and obtained seven better results than the previous values for $n = 70, 73, 75$ and $77, \cdots, 80$.

Table 2 compares our results with those obtained previously. The associated configurations are identified in Fig. 3.

Table 2. Best results of the packing circles in a unit circle for $n = 70, 73, 75$ and $77, \cdots, 80$.

n	Our results	The best result in the literature	Ref.
70	0.23964202505758	0.239616818213	[1]
73	0.23414520714610	0.233873855138	[1]
75	0.23053094860749	0.230458713438	[1]
77	0.22728818902597	0.227244949481	[1]
78	0.22577379956490	0.225744765603	[1]
79	0.22447720830545	0.224476757605	[1]
80	0.22295441413251	0.222950838488	[1]

3   Our Computer Algorithm

We use a standard greedy algorithm in our computer program but before we decided to adopt his method, we went though a process of experimentation and compared this method with both the highly greedy algorithm and the annealing algorithm.

We observed that the highly greedy algorithm, which restricts the search by moving only those points giving the current minimum point-to-point distance, confines the choice of points too much and tends to be trapped early by local solutions. With the annealing algorithm, we faced the difficult of deciding how much to decrease the minimum distance during the search. When a little flexibility is allowed, convergence to the desired solution is too slow. We also noted that with the annealing method, there is a tendency for the search to arrive at repeated patterns; whereas, with the greedy algorithm, we found our best solution, for $n = 50$, only once in 1000 trials. The algorithm we used is shown in Fig. 4:

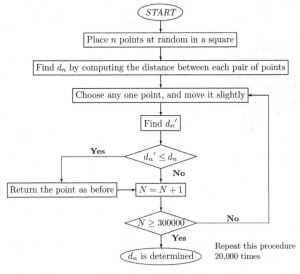

Fig. 4.

Acknowledgements

The authors thank Ronald L. Graham, Mari-Jo Ruiz and Toshinori Sakai for their helpful suggestions.

References

1. Donovan, J.: Packing Circles in Squares and Circles Page,
   http://home.att.net/~donovanhse/Packing/index.html
2. Moser, L.: Problem 24 (corrected), Canadian Mathematical Bulletin 3 (1960) 78
3. Specht, E.: www.packomania.com,
   http://www.packomania.com/, http://hydra.nat.uni-magdeburg.de/packing/
4. Szabó, P. G., Csendes, T., Casado, L. G., García, I.: Equal circles packing in a square
   I – Problem setting and bounds for optimal solutions, New Trends in Equilibrium
   Systrems (2000) 1-15

Congruent Dudeney Dissections of Polygons
All the Hinge Points on Vertices of the Polygon

Jin Akiyama and Gisaku Nakamura

Research Institute of Educational Development, Tokai University,
Tokyo, 151-8677 Japan
fwjb5117@mb.infoweb.ne.jp

Abstract. Let α and β be polygons with the same area. A Dudeney
dissection of α to β is a partition of α into parts which can be reassembled
to produce β as follows: Hinge the parts of α like a string along the
perimeter of α, then fix one of the parts to form β with the perimeter of
α going into its interior and with its perimeter consisting of the dissection
lines in the interior of α, without turning the surfaces over. In this paper
we discuss a special case of Dudeney dissection where α is congruent to
β, in particular, when all hinge points are on the vertices of the polygon
α. We determine necessary and sufficient conditions under which such
dissections exist.

1   Introduction

Given an equilateral triangle α and a square β of the same area, Henry E. Dudeney introduced in [6] a partition of α into parts that can be reassembled in
some way, without turning over the surfaces, to form β (Fig. 1.1). An examina-
tion of Dudeney's method of partition motivated us to introduce the notion of
Dudeney dissection of a polygon.

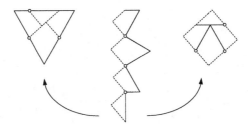

Fig. 1.1

Let α and β be polygons with the same area. A *Dudeney dissection of α to β*
is a partition of α into parts which can be reassembled to produce β as follows:
Hinge the parts of α like a string along the perimeter of α, then fix one of the
parts and rotate the remaining parts about the fixed part to form β with the
perimeter of α in its interior, and with its perimeter consisting of the dissection
lines in the interior of α, without turning the surfaces over.

J. Akiyama and M. Kano (Eds.): JCDCG 2002, LNCS 2866, pp. 14–21, 2003.

Dudeney dissections will be denoted simply by the notation DD and β is said to be a *Dudeney partner of* α. Throughout this paper β will denote the Dudeney partner of α. In the paper [1], we discussed procedures for obtaining Dudeney dissections of quadrilaterals to other quadrilaterals, quadrilaterals to parallelograms, triangles to parallelograms, parallel hexagons to trapezoids, parallel hexagons to triangles, and trapezoidal pentagons to trapezoids. Frederickson's book [7] surveys results involving a more general procedure called *fully hinged dissection*. Some results are related to those in [1] but are obtained by different methods.

In this paper, we consider a special case of Dudeney dissections in which polygon β is congruent to polygon α i.e., α itself is a Dudeney partner of α. We refer to such a dissection as a *congruent Dudeney dissection* of polygon α and denote it simply by CDD of α (Fig. 1.2(a)). We divide the discussion into three cases depending on the location of hinge points. This paper deals only with the case where all the hinge points are on vertices of polygon α and so also of β (Fig. 1.2(a)). The papers [4,5] deal with the case where all the hinge points are interior points of the sides of α and so also of β (Fig. 1.2(b)). The paper [3] deals with the case where some hinge points are on vertices of α while others are interior to the sides of α (Fig. 1.2(c)).

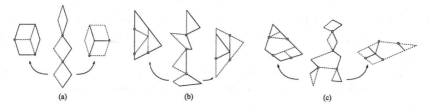

Fig. 1.2

We determine necessary and sufficient conditions under which such dissections exist. The discussion is confined only to convex polygons.

2 Simple Observations on Dudeney Dissection of Polygons

It is clear that a congruent Dudeney dissection is first of all a Dudeney dissection, so we begin by considering conditions under which a polygon α has a Dudeney dissection to some polygon β.

Throughout this paper, the sides of polygon α will be drawn using solid lines while the lines of the dissection will be drawn using dotted lines (Fig. 2.1(a)). We refer to the dotted lines as *dissection lines* and the resulting parts of the polygon as *components of the dissection*.

Attach hinges to all the vertices of α from which at least one dissection line emanates and call the resulting polygon *the hinged polygon of* α. The hinge points are denoted by small circles (Fig. 2.1(b)). All the vertices of polygon α

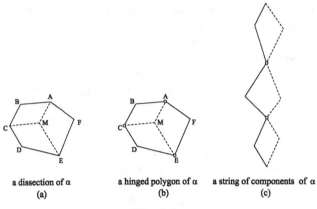

a dissection of α   a hinged polygon of α   a string of components of α
(a)   (b)   (c)

Fig. 2.1

which are not hinge points are referred to as *non-hinge points*. Hinge points can be suppressed from a hinged polygon, as appropriate, to obtain a *string of components of α* (Fig. 2.1(c)).

Suppose β is a Dudeney partner of α. The following are immediate consequences of the fact that both α and β are convex.

Proposition 2.1. *In a DD, every component of a polygon is convex.*

Proposition 2.2. *In a DD, every component is bounded by both sides of α and dissection lines.*

Proposition 2.3. *In the hinged polygon of α, each hinge point has at least two lines emanating from it, one is a dissection line and the other is a side of α (Fig. 2.2(a)).*

If there is a hinge point H at which more than two components meet (Fig. 2.2(b)), we choose to let H connect two components both of which contain a side of α emanating from H.

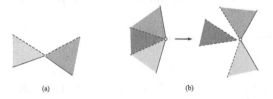

(a)   (b)

Fig. 2.2

3   Dudeney Dissections with All Hinge Points on Vertices of the Polygon

Choose any hinge point, say H, of the hinged polygon of α and suppress appropriate hinge points other than H so that the result is a string of components.

For two components C_1 and C_2 hinged on H, the solid line of C_1 emanating from H overlaps exactly with the solid line of C_2 emanating from H when C_1 is rotated toward C_2 about H (Fig. 3.1), where the solid lines represent sides of α. Consider these two sides of α, both adjacent to H, as a pair. We obtain the following:

Proposition 3.1. *If a polygon α has a DD, then α is an even polygon, i.e. α has an even number of sides, and hinged points appear alternately among the vertices of the hinged polygon of α.*

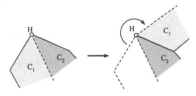

Fig. 3.1

Proposition 3.2. *If a polygon α has a DD, α is a hexagon.*

Proof. Let β be a Dudeney partner of α. Then every non-hinge point of α is in the interior of β. Therefore at least three non-hinge points of α meet at an interior point of β and the sum of their vertex angles must be 2π. We divide the proof into two cases depending on the number of non-hinge points of α meeting at an interior point of β.

Case 1. Three non-hinge points of α meet at an interior point of β.

If α is a 12-gon, 6 non-hinge points meet at two interior points of β. Since the sum of the vertex angles of α is 10π, the average size of the angles at the 6 hinge points of α would be $(10\pi - 4\pi)/6 = \pi$, which contradicts the convexity of α. Similarly, we arrive at a contradiction if we assume α has more than 12 sides. Thus we may assume that α is a polygon with either 10, 8, 6 or 4 sides containing 5,4,3,2 non-hinge points, respectively. However, for any polygon other than a hexagon, no matter how non-hinge points are combined when they meet, they will not end up in the interior of β. Thus the only possible case is that α is a hexagon.

Case 2. Four non-hinge points of α meet at an interior point of β.

If α is an octagon, 4 non-hinge points meet at an interior point of β. Since the sum of the vertex angles of α is 6π, the average size of the vertex angles at the 4 hinge points of α would be $(6\pi - 2\pi)/4 = \pi$, which contradicts the convexity of α. Similarly, we arrive at a contradiction if we assume that α is a polygon with more than 8 sides. Therefore it is sufficient to consider those cases when α is either a hexagon or a quadrilateral with 3 or 2 non-hinge points, respectively. However, neither of these have 4 non-hinge points meeting at an interior point of β, which contradicts the assumption. □

From the observation above, it is sufficient to consider a hexagon α with three non-hinge points of α meeting at an interior point of β.

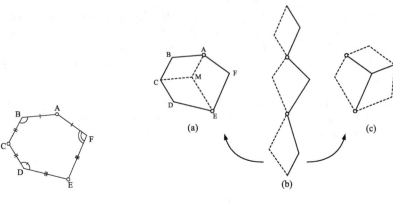

Fig. 3.2   Fig. 3.3

Theorem 3.1. *A hexagon* $\alpha = ABCDEF$ *has a Dudeney dissection, where* A, C, E *are hinge points and* B, D, F *are not, if and only if* α *satisfies the following conditions (Fig. 3.2):*

$$FA = AB, \ BC = CD, \ DE = EF \tag{3.1}$$
$$\angle ABC + \angle CDE + \angle EFA = 2\pi \tag{3.2}$$

Proof. Since necessity follows immediately from the previous discussion, we show only sufficiency. Consider any hexagon α satisfying the conditions (3.1) and (3.2), and choose any interior point M of α.

Dissect α into 3 parts along the segments AM, CM, and EM and hinge 3 quadrilaterals at A, C, M. The resulting hinged hexagon is illustrated in Fig. 3.3(a). The string of components obtained by suppressing the hinge at A is illustrated in Fig. 3.3(b).   □

4 Congruent Dudeney Dissections with All Hinge Points on the Vertices of a Polygon

If a polygon α has a CDD, then it must be a hexagon satisfying the conditions (3.1) and (3.2) by Theorem 3.1. In this section, we determine all hexagons which have a CDD.

Theorem 4.1. *Let* $\alpha = ABCDEF$ *be a hexagon and let* M *be an interior point of* α. α *has a congruent Dudeney dissection with hinge points* A, C *and* E *if and only if* α *satisfies either the conditions (4.1) or (4.2).*

$$\left. \begin{array}{l} CDEM \ is \ a \ parallelogram \\ ABCM \equiv AMEF \\ AB = AM = AF \end{array} \right\} \tag{4.1}$$

$$AB = BC = CD = DE = EF = FA = MA = MC = ME \tag{4.2}$$

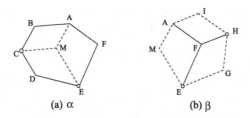

Fig. 4.1

Proof. Without loss of generality, we adopt the notation in Fig. 4.1(a) and Fig. 4.1(b), $\alpha = ABCDEF$ and its Dudeney partner $\beta = IAMEGH$.

Since β is the partner of α, $AMEF$ in Fig. 4.1(a) is congruent to $AMEF$ in Fig. 4.1(b) and

$$ABCM \equiv AFHI \tag{4.3}$$
$$CDEM \equiv HFEG. \tag{4.4}$$

Suppose that α has a CDD. There are additional conditions. In order to obtain them, we consider two cases depending on whether the hinge points of α correspond to the hinge points of β (Case 1) or not (Case 2).

Case 1. Although there are three possibilities, i.e.

$$\alpha \equiv AMEGHI \text{ or } \alpha \equiv EGHIAM \text{ or } \alpha \equiv HIAMEG, \tag{4.5}$$

we may assume, without loss of generality, that

$$\alpha \equiv AMEGHI \tag{4.6}$$

since we can interchange the roles of the three vertices A, C, E.

Consider $CDEM$. Then it follows from (4.6) that $CD = EG$ and $DE = GH$, and from (4.4) that $EG = EM, GH = MC$ and thus $CDEM$ is a parallelogram. Next consider $ABCM$ and $AMEF$. Then it follows from (4.6) and (3.1) that $AM = AB = AF$ and from (4.6) that $BC = ME$, and from (4.6) and (4.3) that $EF = HI = CM$. Thus all corresponding sides of the quadrilaterals $ABCM$ and $AMEF$ have the same lengths.

Consider the two triangles $\triangle ABC$ and $\triangle AME$. Then it follows from (4.6) that $\angle ABC = \angle AME$, thus $\triangle ABC \equiv \triangle AMC$. Consider $\triangle ACM$ and $\triangle AEF$. Then it follows from (4.3) and (4.6) that $\angle AMC = \angle AIH = \angle AFE$, and thus $\triangle AMC \equiv \triangle AFE$. From these facts, we obtain $ABCM \equiv AMEF$. In summary, we arrive at the necessary condition (4.1) for a hexagon $ABCDEF$ to have a CDD.

On the other hand, since every hexagon with (4.1) has a CDD as illustrated in Fig. 4.2, (4.1) is also a sufficient condition.

Case 2. Although there are three possibilities

$$\alpha \equiv IAMEGH, \ \alpha \equiv MEGHIA, \ \alpha \equiv GHIAME, \tag{4.7}$$

we may assume, without loss of generality, that

$$\alpha \equiv IAMEGH \tag{4.8}$$

since we can interchange the roles of the three vertices A, C, E.

In this case, it follows from (3.1) that $FA = AB, BC = CD, DE = EF$ and from (4.8) that $AB = IA, BC = AM, CD = ME, DE = EG, EF = GH$ and

$FA = HI$. Also from (4.3) and (4.4) it follows that $IA = MA, HI = CM, GH = MC$ and $EG = EM$. From these facts, we obtain (4.2), i.e.

$$AB = BC = CD = DE = EF = FA = MA = MC = ME,$$

which imply that hexagon $ABCDEF$ consists of 3 rhombic components.

For every hexagon $\alpha = ABCDEF$ with interior point M satisfying (4.2), dissect α into 3 components along AM, CM and EM and hinge them as illustrated in Fig. 4.3.

It is easy to see that the Dudeney partner of α is congruent to α, thus (4.2) is also a sufficient condition for α to have a CDD.   □

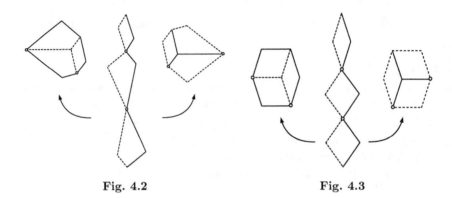

Fig. 4.2   **Fig. 4.3**

Remarks. Dudeney dissections and congruent Dudeney dissections for polyhedrons have also been discussed in [2].

Acknowledgments

The authors thank Mari-Jo Ruiz, Yuji Ito and Toshinori Sakai for their assistance in the preparation of this paper.

References

1. Akiyama, J., Nakamura, G.: Dudeney dissection of polygons, *Discrete and Computational Geometry* (J. Akiyama et al (eds.)), 14-29, Springer Lecture Notes in Computer Science, Vol.1763, 2000
2. Akiyama, J., Nakamura, G.: Dudeney dissections of polyhedrons I–VI (in Japanese), *The Rep. of Res. Inst. Educ., Tokai Univ.*, Vol.8, 1-66, 2000
3. Akiyama, J., Nakamura, G.: Congruent Dudeney dissections of polygons – Some hinge points on vertices, others interior to the sides of the polygon –, to appear
4. Akiyama, J., Nakamura, G.: Congruent Dudeney dissections of triangles and convex quadrilaterals – All hinge points interior to the sides of the polygon –, *Discrete and Computational Geometry* (B. Aronov et al (eds.)), 43-63, *Algorithms and Combinatorics*, Vol.25, 2003

5. Akiyama, J., Nakamura, G.: Determination of all convex polygons which are chameleons – Congruent Dudeney dissections of polygons –, *IEICE TRANS. FUNDAMENTALS*, Vol.E86-A, No.5, 978-986, 2003
6. Dudeney, H.E.: *The Canterbury Puzzles and Other Curious Problems*, W. Heinemann, 1907
7. Frederickson, G.N.: *Dissections: Plane & Fancy*, Cambridge University Press, 1997

Playing with Triangulations

Oswin Aichholzer[1], David Bremner[2], Erik D. Demaine[3],
Ferran Hurtado[4], Evangelos Kranakis[5], Hannes Krasser[6],
Suneeta Ramaswami[7], Saurabh Sethia[8], and Jorge Urrutia[9]

[1] IICM - Softwaretechnology, Graz University of Technology
Inffeldgasse 16b/I, A-8010 Graz, Austria
oaich@igi.tu-graz.ac.at
[2] Faculty of Computer Science, University of New Brunswick
P.O. Box 4400 Fredericton, N,B E3B 5A3, Canada
bremner@unb.ca
[3] Laboratory for Computer Science, Massachusetts Institute of Technology
200 Technology Square, Cambridge, MA 02139, USA
edemaine@mit.edu
[4] Departament de Matemàtica Aplicada II, Universitat Politècnica de Catalunya
Pau Gargallo,5, 08028 Barcelona, Spain
hurtado@ma2.upc.es
[5] School of Computer Science, Carleton University
1125 Colonel By Drive, Ottawa, Ontario K1S 5B6, Canada
kranakis@scs.carleton.ca
[6] Institute for Theoretical Computer Science, Graz University of Technology
Inffeldgasse 16b/I, A-8010 Graz, Austria
hkrasser@igi.tu-graz.ac.at
[7] Computer Science Department, Rutgers University
321 Business & Science Building, Camden, NJ 08102, USA
rsuneeta@crab.rutgers.edu
[8] Department of Computer Science, Oregon State University
Corvallis, OR 97331-3202, USA
saurabh@cs.orst.edu
[9] Instituto de Matemáticas, Universidad Nacional Autónoma de México
Área inv. científ., Circ. Ext., Ciudad Univ., Coyoacán 04510, México D.F., México
urrutia@math.unam.mx

Abstract. We analyze several perfect-information combinatorial games played on planar triangulations. We introduce three broad categories of such games: constructing, transforming, and marking triangulations. In various situations, we develop polynomial-time algorithms to determine who wins a given game under optimal play, and to find a winning strategy. Along the way, we show connections to existing combinatorial games such as Kayles.

1 Introduction

Let S be a set of n points in the plane, which we assume to be in general position, i.e., no three points of S lie on the same line. A *triangulation* of S is a simplicial decomposition of its convex hull having S as vertex set.

J. Akiyama and M. Kano (Eds.): JCDCG 2002, LNCS 2866, pp. 22–37, 2003.

In this work we consider several games involving the vertices, edges (straight-line segments) and faces (triangles) of some triangulation. We present games where two players $\mathcal{R}$(ed) and $\mathcal{B}$(lue) play in turns, as well as *solitaire* games for one player. In some *bichromatic* versions, player $\mathcal{R}$ will use red and player $\mathcal{B}$ will use blue, respectively, to color some element of the triangulation. In *monochromatic* variations, all players (maybe the single one) use the same color, green.

Games on triangulations come in three main flavors:

- *Constructing (a triangulation).* The players construct a triangulation $T(S)$ on a given point set S. Starting from no edges, players $\mathcal{R}$ and $\mathcal{B}$ play in turn by drawing one or more edges in each move. In some variations, the game stops as soon as some structure is achieved. In other cases, the game stops when the triangulation is complete, the last move or possibly some counting decides then who is the winner.
- *Transforming (a triangulation).* A triangulation $T(S)$ on top of S is initially given, all edges originally colored black. In each turn, a player applies some local transformation to the current triangulation, resulting in a new triangulation with some edges possibly recolored. The game stops when a specific configuration is achieved or no more moves are possible.
- *Marking (a triangulation).* A triangulation $T(S)$ on top of S is initially given, all edges and nodes originally colored black. In each turn, some of its elements are marked (e.g., colored) in a game-specific way. The game stops when some configuration of marked elements is achieved (possibly the whole triangulation) or no more moves are possible.

For each of the variety of games described in Section 3, we are interested in characterizing who wins the game, and designing efficient algorithms to determine the winner and compute a winning strategy. We present several such results in Sections 4–6 to give a taste of the area, and leave further details to the companion paper [1].

Besides beauty and entertainment, games keep attracting the interest of mathematicians and computer scientists because they also have applications to modeling several areas and because they often reveal deep mathematical properties of the underlying structures, in our case the combinatorics of planar triangulations.

2 Combinatorial Games

Games on triangulations belong to the more general area of *combinatorial games* which typically involve two players, $\mathcal{R}$ and $\mathcal{B}$. We define next a few more terms from combinatorial game theory that we will use in this paper. For more information, refer to the books [3, 5] and the survey [7]. The paper [8] contains a list of more than 900 references.

We consider games with *perfect information* (no hidden information as in many card games) and without chance moves (like rolling dice). In such a game, a *game position* consists of a set of options for $\mathcal{R}$'s moves and a set of options for $\mathcal{B}$'s moves. Each option is itself a game position, representing the result of the move.

Most of the games we consider in this paper (the monochromatic games) are also *impartial* in the sense that the options for $\mathcal{R}$ are the same as the options for $\mathcal{B}$. In this case, a game position is simply a set of game positions, and can thus be viewed as a tree. The leaves of this tree correspond to the empty set, meaning that no options can be played; this game is called the *zero game*, denoted 0.

In general, each leaf might be assigned a label of whether the current player reaching that node is a winner or loser, or the players tied. However, a common and natural assumption is that the zero game is a losing position, because the next player to move has no move to make. We usually make this assumption, called *normal play*, so that the goal is to make the last move. In contrast, *misère play* is just the opposite: the last player able to move loses. In more complicated games, the winner is determined by comparing scores.

Any impartial perfect-information combinatorial game without ties has one of two *outcomes* under optimal play (when the players do their best to win): a *first-player win* or a *second-player win*. In other words, whoever moves first can force herself to reach a winning leaf, or else whoever moves second can force herself to reach a winning leaf, no matter how the other player moves throughout the game. Such forcing procedures are called *winning strategies*. For example, under normal play, the game 0 is a second-player win, and the game having a single move to 0 is a first-player win, in both cases no matter how the players move. More generally, impartial games may have a third outcome: that one player can force a *tie*.

The Sprague-Grundy theory of impartial games (see e.g. [3, Chapter 3]) says that, under normal play, every impartial perfect-information combinatorial game is equivalent to the classic game of Nim. In (single-pile) Nim, there is a pile of $i \geq 0$ beans, denoted $*i$, and players alternate removing any positive number of beans from the pile. Only the empty pile $*0$ results in a second-player win (because the first player has no move); for any other pile, the first player can force a win by removing all the beans. If a game is equivalent to $*i$, then i is called the *Nim value* of the game.

Given two or more games, their *sum* is the game in which, at each move, a player chooses one subgame to move in, and makes a single move in that subgame. In this sense, sums are *disjunctive*: a player makes exactly one move at each turn. Games often split into sums of independent games in this way, and combinatorial game theory explains how the sum relates to its parts. In particular, if we sum two games with Nim values i and j, then the resulting game has Nim value equal to the bitwise XOR of i and j.

3   Examples of Games

We describe next the rules of several specific games that we have studied.

3.1   Constructing

3.1.1 Monochromatic Complete Triangulation. The players construct a triangulation $T(S)$ on a given point set S. Starting from no edges, players $\mathcal{R}$ and $\mathcal{B}$ play in turn by drawing one edge in each move. Each time a player completes one or more empty triangle(s), it is (they are) given to this player and it is again her turn (an "extra move"). Once the triangulation is complete, the game stops and the player who owns more triangles is the winner.

3.1.2 Monochromatic Triangle. Starts as in 3.1.1, but has a different stopping condition: the first player who completes one empty triangle is the winner.

3.1.3 Bichromatic Complete Triangulation. As in 3.1.1, but the two players use red and blue edges. Only monochromatic triangles count.

3.1.4 Bichromatic Triangle. As in 3.1.2, but with red and blue edges. The first empty triangle must be monochromatic.

3.2 Transforming

3.2.1 Monochromatic Flipping. Two players start with a triangulation whose edges are initially black. Each move consists of choosing a black edge, flipping it, and coloring the new edge green. The winner is determined by normal play.

3.2.2 Monochromatic Flipping to Triangle. Same rules as for 3.2.1, except now the winner is who completes the first empty green triangle.

3.2.3 Bichromatic Flipping. Two players play in turn, selecting a flippable black edge e of $T(S)$ and flipping it. Then e as well as any still-black boundary edges of the enclosing quadrilateral become red if it was player $\mathcal{R}$'s turn, and blue if it was player $\mathcal{B}$'s move. The game stops if no more flips are possible. The player who owns more edges of her color wins.

3.2.4 All-Green Solitaire. In each move, the player flips a flippable black edge e of $T(S)$; then e becomes green, as do the four boundary edges of the enclosing quadrilateral. The goal of the game is to color all edges green.

3.2.5 Green-Wins Solitaire. As in 3.2.4, but the goal of the game is to obtain more green edges than black edges.

3.3 Marking

3.3.1 Triangulation Coloring Game. Two players move in turn by coloring a black edge of $T(S)$ green. The first player who completes an empty green triangle wins.

3.3.2 Bichromatic Coloring Game. Two players $\mathcal{R}$ and $\mathcal{B}$ move in turn by coloring red respectively blue a black edge of $T(S)$. The first player who completes an empty monochromatic triangle wins.

3.3.3 Four-Cycle Game. Same as 3.3.1 but the goal is to get an empty quadrilateral.

3.3.4 Nimstring Game. Nimstring is a game defined in Winning Ways [3] as a special case of the classic children's (but nonetheless deep) combinatorial game Dots and Boxes [2,3]. In the context of triangulations, players in Nimstring alternate *marking* one-by-one the edges of a given triangulation (i.e., coloring green an edge, initially black), and whenever a triangle has all three of its edges marked, the completing player is awarded an extra move and must move again. The winner is determined by normal play, meaning that the goal is to make the last complete move. Thus, the player marking the last edge of the triangulation actually loses, because that last edge completes one or two triangles, and the player is forced to move again, which is impossible.

4 Metamorphosis of Some Games

In this section we show that some of the above games are equivalent to famous combinatorial games and describe their solutions.

4.1 Triangulation Coloring Game

Obviously the Triangulation Coloring Game (3.3.1) terminates after a linear number of moves and there are no ties. For point sets S in convex position and several classes of triangulations $T(S)$ of S we will show a one-to-one relation to seemingly unrelated games on piles of beans, which will provide us with an optimal winning strategy for these settings. To this end, consider the dual of the triangulation $T(S)$, i.e., the graph G_T with a vertex per triangle of $T(S)$ and an edge between each pair of vertices corresponding to triangles of $T(S)$ that share a diagonal. An *inner* triangle of $T(S)$ consists entirely of diagonals of $T(S)$, and therefore it does not use an edge of the convex hull of S. Thus, exactly those vertices of G_T corresponding to inner triangles have degree three, whereas all other vertices have degree one (ears of the triangulation) or two.

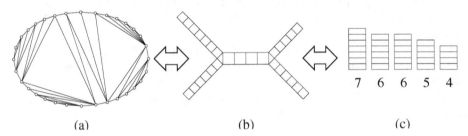

Fig. 1. Different incarnations of the triangulation coloring game

Motivation for considering the dual graph of $T(S)$ stems from the following observation. Coloring an edge of a triangle Δ for which one edge has already been colored leads immediately to a winning move for the opponent: she just has to color the third edge of Δ. Thus we call any triangle Δ with one colored edge 'taken', because we can never color another edge of Δ unless we are ready to lose. Thus, coloring an edge e of $T(S)$ means, in the dual setting, marking either a single vertex of G (if e was on the convex hull) or two adjacent vertices (if e was a diagonal) as taken. Vertices already marked cannot be marked again and whoever marks the last vertex will win. Figure 1(b) shows G as a set of connected arrays of boxes, where marking a vertex of G might be seen as drawing a cross inside the corresponding box.

If the triangulation is *serpentine*, i.e., its dual is a simple path (equivalently, a single array of boxes without branches), we can show that the first player has a winning strategy by applying a symmetry principle (called the *Tweedledum-Tweedledee Argument* in [3]). For an odd number of triangles, she first takes the central triangle by coloring the edge of this triangle that belongs to the convex hull. For an even the number of triangles, she first takes both triangles adjacent to the central diagonal by coloring this diagonal. In both cases, the remainders are two combinatorially identical triangulations (two equal-sized box arrays), in which all possible moves can be played independently. Thus the winning strategy of player one is just to mimic any of her opponent's moves by simply coloring the corresponding edge in the other triangulation. This strategy ensures that she always can make a valid move, forcing the second player finally to color a second edge of an already taken triangle, leading to a winning position for the first player.

If $T(S)$ contains inner triangles, the dual is a tree and the problem of finding optimal strategies is more involved. We consider this situation with the only restriction that no two inner triangles share a common diagonal; see for example Figure 1(a). We say that in this case the corresponding triangulation is a *simple-branching* triangulation. The main observation for our game is that all inner triangles can be ignored: Consider an inner triangle Δ and observe first that it cannot be taken on its own, because Δ does not have an edge from the convex hull of S. Thus the situation after the three neighbors of Δ have been taken is the same, regardless of whether Δ was taken together with one of them: Δ is blocked in any case. In the dual setting, this observation means that we can remove the vertex of G corresponding to Δ (plus adjacent edges) without changing the game. Drawing G with blocks as in Figure 1(b), we can thus remove the 'triangular' blocks and consider the remaining block arrays independently. Instead of playing with these arrays, we might as well deal with integers reflecting the length of each array; see Figure 1(c).

Surprisingly, this setting turns out to be an incarnation of a well-known taking-and-breaking game played on heaps of beans or sets of coins, called *Kayles* [3]. This game was introduced by Dudeney and independently also by Sam Loyd, who originally called it 'Rip Van Winkle's Game'. The following description is taken from [3, Chapter 4]: *Each player, when it is his turn to move, may take 1 or 2 beans from a heap, and, if he likes, split what is left of that heap into two smaller heaps.*

Any triangulation of a convex set S without inner triangles sharing a common diagonal can be represented by Kayles, while the reverse transformation is less general. The number of heaps which can be represented by a single legal triangulation has to be odd because any inner triangle has degree three. During a game of course any number of heaps may occur, as the triangulation may split into several independent parts. A generalization is thus to play the game on more than one point set from the very beginning.

Because Kayles is impartial, the Sprague-Grundy theory described in Section 2 applies, so the game is completely described by its sequence of Nim values for a single pile of size n. It has been shown that this Nim sequence has a periodicity of length 12, with 14 irregularities occurring, the last for $n = 70$; see Table 1. To compute the Nim value for a game with several heaps, we can xor-add up the Nim values (given by Table 1) for the individual heaps. Moreover, in this case, we can xor-add up just a four-bit vector, corresponding to the four 'magic'

Table 1. Nim values for Kayles: the Nim sequence has periodicity 12 and there are 14 exceptional numbers

---

$\mathcal{G}(n) = K[n \text{ modulo } 12]$, $K[0, \ldots, 11] = (4, 1, 2, 8, 1, 4, 7, 2, 1, 8, 2, 7)$
exceptional values:
$\mathcal{G}(0) = 0$ $\mathcal{G}(3) = 3$ $\mathcal{G}(6) = 3$ $\mathcal{G}(9) = 4$ $\mathcal{G}(11) = 6$ $\mathcal{G}(15) = 7$ $\mathcal{G}(18) = 3$
$\mathcal{G}(21) = 4$ $\mathcal{G}(22) = 6$ $\mathcal{G}(28) = 5$ $\mathcal{G}(34) = 6$ $\mathcal{G}(39) = 3$ $\mathcal{G}(57) = 4$ $\mathcal{G}(70) = 6$

---

heap sizes 1, 2, 5, and 27, respectively, where powers of two in the Nim sequence appear for the first time. For example, a single heap of size 42 (Nim value 7) is equivalent to the situation of three heaps with sizes 1, 2, and 5, respectively, reflecting the 'ones' in the 4-bit representation of 7.

It follows that, in time linear in the number of heaps, a position can be determined to be either a first-player win (nonzero Nim value) or a second-player win (zero Nim value). Any move from a second-player-winning position leads to a first-player-winning position; and for any first-player-winning position there is always at least one move that leads to a second-player-winning position. A winning strategy just needs to follow such moves, because after one move, the players effectively reverse roles. Because any position has at most a linear number of possible moves, we conclude that for the triangulation-coloring game a winning move (if it exists) can be found in time linear in the size of the triangulation. It is interesting to note that there are no zeros in the Nim sequence of Kayles. This reflects the fact that when starting with a single integer number the first player can always win, as has been pointed out above for triangulations without inner triangles.

From the previous discussion we obtain the following result:

Theorem 1. *Deciding whether the Triangulation Coloring Game on a simple-branching triangulation on n points in convex position is a first-player win or a second-player win, as well as finding moves leading to an optimal strategy, can be solved in time linear in the size of the triangulation.*

At this point it is worth mentioning that there is a version of Kayles played on graphs: two players play in turn by selecting a vertex of a given graph G that must be nonadjacent to (and different from) any previously chosen vertex. The last player that can select a vertex, completing a maximal independent set, is the winner. Deciding which player has a winning strategy is known to be PSPACE-complete [10].

Now, given a triangulation T on a point set S, let us define a graph $EG(T)$ having a vertex per each edge in T and an adjacency between any two nodes whose corresponding edges in T belong to the same triangle; an example is shown in Figure 2. From the preceding paragraphs, it is clear that playing the Triangulation Coloring Game on T is equivalent to playing Kayles on $EG(T)$.

While such a reduction does not prove hardness of the Triangulation Coloring Game, it does transfer any solutions to special cases of Kayles. In [4] it is shown

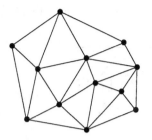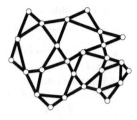

Fig. 2. A triangulation T of a point set (left) and the graph $EG(T)$ associated to adjacent edges (right)

that there are polynomial-time algorithms to determine the winner in Kayles on graphs with bounded asteroidal number, on cocomparability graphs, and on circular-arc graphs. Theorem 1 can be rephrased as a similar (and computationally efficient) result for outerplanar graphs in which every block is a triangle and blocks that contain three articulation vertices do not share any of them.

4.2   Monochromatic Triangle

We present an optimal strategy for Monochromatic Triangle (3.1.2) provided S is a convex set. First observe that an edge should be drawn only if it connects two vertices that have not been used before. Otherwise, a vertex p of degree at least two occurs, leading to a winning move for the opponent: she just has to close the triangle formed by two neighboring edges of p. (Note that it is important here that we consider point sets in convex position.) In other words, when drawing a diagonal pq in S, the two vertices p and q are taken for the rest of the game. Moreover, pq splits S into two independent subsets with cardinality n_1 and n_2, respectively, such that $n_1 + n_2 = n - 2$. The player who draws the last edge according to these observations will win the game with her succeeding move.

Because no edge can be drawn in sets of cardinality of at most one, we have just shown that our game is an incarnation of a known game called *Dawson's Kayles*, a cousin of Kayles [3]. In terms of bowling, the game reads as follows: A row of n pins is given and the only legal move is to knock down two adjacent pins. Afterwards, one or two shorter rows of pins remain, and single pins are removed immediately. Whoever makes the last strike wins.

In more mathematical terms, the game is defined by a set of k integers $n_1, \ldots, n_k$. A move consists of choosing one n_i, $1 \le i \le k$, reducing it by two to $\hat{n}_i$ and eventually replacing it afterwards by two numbers n'_i and n''_i, $n'_i + n''_i = \hat{n}_i$. Any $n_i \le 1$ can be removed from the set, because it cannot be used for further moves. Whoever can make the last legal move wins. Note the case that n_i is not split after reduction corresponds to drawing an edge on the convex hull of S (or the respective subset).

Sprague-Grundy theory also applies to Dawson's Kayles. It has been shown that its Nim sequence has a periodicity of length 34, with 8 irregularities occurring, the last for $n = 52$; see Table 2. As with Kayles, to compute the Nim value

Table 2. Nim values for Dawson's Kayles: the Nim sequence has periodicity 34 and there are 8 exceptional numbers

---

$\mathcal{G}(n) = K[n \bmod 34]$, $K[0, \ldots, 33] =$
$(4, 8, 1, 1, 2, 0, 3, 1, 1, 0, 3, 3, 2, 2, 4, 4, 5, 5, 9, 3, 3, 0, 1, 1, 3, 0, 2, 1, 1, 0, 4, 5, 3, 7)$
exceptional values:
$\mathcal{G}(0) = 0$ $\mathcal{G}(1) = 0$ $\mathcal{G}(15) = 0$ $\mathcal{G}(17) = 2$ $\mathcal{G}(18) = 2$
$\mathcal{G}(32) = 2$ $\mathcal{G}(35) = 0$ $\mathcal{G}(52) = 2$

---

for a position consisting of k heaps, i.e., to xor-add up the Nim values given by Table 2 for the k heap sizes, a vector with four bits is sufficient, corresponding to the heap sizes 2, 4, 14, and 69.

Theorem 2. *The Monochromatic Triangle game on n points in convex position is a second-player win when $n \equiv 5, 9, 21, 25, 29$ (mod 34) and for the special cases $n = 15$ and $n = 35$; otherwise it is a first-player win. Each move in a winning strategy can be computed in time linear in the size of the triangulation.*

For n even, this result was clear from the very beginning, as in this case the first player, say $\mathcal{R}$, may start by drawing a diagonal d leaving $(n - 2)/2$ points on each side and apply the symmetry principle: for every move of $\mathcal{B}$, player $\mathcal{R}$ either makes a winning move, if available, or mimics her opponent's last move on the opposite side of d.

5   Monochromatic Complete Triangulation

In this section we consider the triangulation-construction game 3.1.1. In this context, we show by direct arguments that for a set S in convex position a greedy strategy is optimal for this game where, depending on the parity of n, the first player can always win (odd n) or either player can force a tie (even n).

Theorem 3. *The outcome of the Monochromatic Complete Triangulation Game on n points in convex position is a first-player win for n odd, and a tie for n even.*

Let us call two edges sharing a common point p an *open triangle* if we can build a valid triangle (no intersections with other edges occur) by connecting the two endpoints not adjacent to p by inserting a third edge, called *closing* edge. Obviously closing edges are drawn between vertices of the same connected component. When drawing an edge connecting two formerly different components we call it an *i-edge*, $i \in \{0, 1, 2\}$, if i of its endpoints already have at least one other incident edge. Thus, 0-edges connect isolated points while, by convexity, 1-edges produce one additional open triangle, and 2-edges give rise to two additional open triangles. Because we have n points overall, the total number of i-edges throughout a game is $n - 1$.

Note that in addition to these two types of edges there exist so-called *re-dundant* edges: connecting two points from the same connected component, but not closing a triangle. This happens if a cycle of length at least 4 occurs, containing several open triangles. We first argue that any optimal strategy uses no redundant edges, i.e., open triangles will be closed immediately. Otherwise the opponent might close the triangles, getting the points, and continue afterwards with the same number of possible i-edges. Here it is crucial to observe that when an edge connects two different connected components, it is not important for the strategy which points of these components are used, since when closing all open triangles of the new connected component everything within its convex hull is triangulated. Thus for analyzing strategies not the exact shape but only the number of connected components counts.

The greedy strategy works as follows. As long as there are closing edges, draw them. Recall that after closing a triangle, it remains the same player's turn. Then draw an i-edge for the smallest possible i.

To analyze our strategy, let e_i denote the number of i-edges drawn during an entire game. The first time a point of S is used, it is either part of a 0-edge or a 1-edge. Also, a 0-edge uses two previously unused points, whereas a 1-edge uses one previously unused point. Thus, $2e_0 + e_1 = n$. Moreover, $e_2 = e_0 - 1$ because $e_0 + e_1 + e_2 = n - 1$. Further observe that if there are no open triangles left and a player plays an i-edge then her opponent can, and will by the observations above, close exactly i open triangles in her next move. (Note that only i triangles can be closed only because S is in convex position.) Thus, the goal of a player is to globally minimize the sum of i over all i-moves she makes.

We split the remaining proof of the theorem into three parts:

Lemma 1. *For n odd, player $\mathcal{R}$ can win by playing greedily.*

Proof. We have $e_0 + e_1 + e_2 = n - 1$ which is even. Thus, there will be $(n - 1)/2$ rounds of both players picking i-edges. In each round, player $\mathcal{R}$ picks first and greedily, and hence in each round $\mathcal{R}$ wins or ties with $\mathcal{B}$. For a tie to occur, $\mathcal{B}$ must tie with $\mathcal{R}$ in all rounds, but that requires that e_0, e_1, e_2 all be even, which is not possible because $e_2 = e_0 - 1$.   □

Lemma 2. *For n even, player $\mathcal{B}$ can force a tie by playing greedily.*

Proof. We have $e_0 + e_1 + e_2 = n - 1$ which is odd. If the first move for player $\mathcal{R}$ is an i-edge, then player $\mathcal{B}$ wins i points. After this first move, there are $n - 2$ i-edges remaining. Players $\mathcal{B}$ and $\mathcal{R}$ will pick these i-edges alternately in $(n - 2)/2$ rounds. In each round, player $\mathcal{B}$ picks first and greedily, and hence in each round $\mathcal{B}$ either wins or ties with $\mathcal{R}$. Thus, $\mathcal{B}$ either wins or ties overall by playing greedily.   □

Lemma 3. *For n even, player $\mathcal{R}$ can force a tie.*

Proof. Here we diverge from the greedy strategy, because if e_0 ended up even, then player $\mathcal{B}$ would win by two triangles by playing greedily (only e_2 is odd). Instead, $\mathcal{R}$ employs a symmetry strategy to ensure that e_0 ends up odd, so that both e_1 and e_2 are even, leading to a tie. Player $\mathcal{R}$ begins by playing a diagonal splitting S into two equal sets (recall that n is even). Then as long as $\mathcal{B}$ does not unnecessarily leave triangles open, she plays symmetrically: close open triangles and mimic whatever $\mathcal{B}$ has done in the opposite part of S. In this way, it is guaranteed that e_0 will be odd: the first diagonal plus two times the number of 0-edges $\mathcal{B}$ has drawn. If at some point $\mathcal{B}$ does not close an open triangle, then $\mathcal{R}$ closes it and starts playing according to the ordinary greedy strategy. Because $\mathcal{R}$ now won a triangle from $\mathcal{B}$, the scoring difference changed by two and thus with the greedy strategy $\mathcal{R}$ will win for e_0 odd and get at least a tie for e_0 even. $\qquad\square$

6   Solitaire Games

In this section we consider games in which there is only one player.

6.1   All-Green Solitaire

In each move, the player flips a flippable black edge e of $T(S)$; then e becomes green, as do the four boundary edges of the enclosing quadrilateral. The goal of the game is to color all edges green.

Notice that this is not always possible, as can be seen from the example of a triangulated regular pentagon. Two questions are of interest: (1) Characterize classes of triangulations for which it is possible; (2) give an efficient algorithm to find an appropriate flip sequence for such triangulations. Our next result settles the second question in the convex case:

Theorem 4. *Whether the player can win the All-Green Solitaire Game for a given triangulation of n points in convex position can be decided in time $O(n)$. When the player can win, a winning sequence of moves can be found within the same time bound.*

Proof. Let S be the triangulated subpolygon to the right of a given oriented diagonal d. There are two diagonals d_1 and d_2 in S, that form a triangle together with d, which we orient leaving d to their left, as shown in Figure 3. Let us denote by S^1 and S^2 the subpolygons these diagonals define (we follow the counterclockwise order). When the notation is iterated we write simply $S^{i,j}$ instead of $(S^i)^j$.

Let us define the logic value $b(S)$ to be true if, and only if, it is possible to color green all the edges of the polygon S, by flipping black diagonals strictly interior to S, assuming that all edges in S, including d, are initially black. Let the logic value $g(S)$ to be true if, and only if, it is possible to color green all the edges of the polygon S, by flipping black diagonals strictly interior to S, assuming that all edges in S are initially black with the only exception of d, which is initially green. Notice that $b(S)$ implies $g(S)$. In the simplest case S consists of merely a single boundary edge (d), then $b(S)$ is false and $g(S)$ is true.

We use the symbols "$\wedge$" and "$\vee$" for the logic operators "and" and "or", respectively. If d is initially black, at some point we have to flip either d_1 or d_2

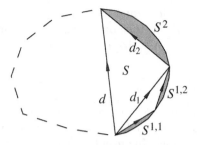

Fig. 3. Illustrating the technique for the All-Green Solitaire Game in convex position

to give the color green to d. Say it is d_1, then none of the boundary edges of the quadrilateral that has d_1 as diagonal may have been flipped before, therefore in order to complete a green coloration we must have $g(S^{1,1}) \wedge g(S^{1,2}) \wedge g(S^2)$. Hence

$$b(S) = \left[g(S^{1,1}) \wedge g(S^{1,2}) \wedge g(S^2)\right] \vee \left[g(S^{2,1}) \wedge g(S^{2,2}) \wedge g(S^1)\right].$$

For $g(S)$ to be true it is obviously sufficient that $b(S)$ is true; if $b(S)$ is false we must have both $b(S_1)$ and $b(S_2)$ to be true, and use the additional fact that d is initially green. Therefore we have

$$g(S) = [b(S_1) \wedge b(S_2)] \vee b(S) =$$
$$= [b(S_1) \wedge b(S_2)] \vee \left[g(S^{1,1}) \wedge g(S^{1,2}) \wedge g(S^2)\right] \vee \left[g(S^{2,1}) \wedge g(S^{2,2}) \wedge g(S^1)\right].$$

Let us call *jump* of an oriented diagonal d the number of convex hull edges that d has on its right side. For example, if d is an edge at the convex hull the jump is 0 or $n-1$ when it is oriented counterclockwise and clockwise, respectively.

Notice that in the above formulas for $b(S)$ and $g(S)$ these values depend on similar values associated to diagonals with strictly smaller jump. Therefore if we process the polygons to the right of the oriented diagonals ordered by increasing jump, the computation of $g(S)$ and $b(S)$ is carried out in constant time for a fixed S, and in time $O(n)$ for the whole triangulation.

Once the computation is complete, we can consider in turn each diagonal e of the polygon as a tentative candidate for starting the All-Green Solitaire Game. This would color green e and the four boundary edges e_1, e_2, e_3 and e_4, of the enclosing quadrilateral, oriented counterclockwise. If we call P_1, P_2, P_3 and P_4 the polygons to the right of these diagonals, we would be able to complete a green coloring if, and only if, the four values $g(S_i)$ are true. As this is checked in constant time after our precomputation, the overall exploration can be done in time $O(n)$.

If a first move in a winning sequence is found, the same technique can easily be adapted to explore down the recurrence of $b(S)$ and $g(S)$ and find the entire sequence of winning moves in $O(n)$ additional time. Basically, depending on the case that caused either $b(S)$ or $g(S)$ to be true, you flip the appropriate edge to reduce to the appropriate subcases and recurse down the dual tree. In this way,

$O(1)$ time is spent per triangle / dual node. For example, if $g(S)$ was set true because $g(S^{1,1}) \wedge g(S^{1,2}) \wedge g(S^2)$, then one should flip d_1 and then recurse in the subproblems $g(S^{1,1})$, $g(S^{1,2})$, and $g(S^2)$.   □

6.2 Green-Wins Solitaire

Suppose the rules are the same as All-Green Solitaire, but the goal of the game is to obtain at the end more green edges than black edges. It is an open question whether this can always be done. In our next result we give bounds on how many green edges we can always guarantee.

Theorem 5. *The player of the Green-Wins Solitaire Game can obtain from any given triangulation on n points at least $1/6$ of the edges to be green at the end of the game. There are triangulated point sets such that no sequence of flips of black edges provides more than $5/9$ of the edges to be green at the end. (In the above fractions we don't pay attention to additive constants).*

Proof. For the lower bound, it is known that any triangulation of n points contains at least $\frac{n-4}{6}$ independently flippable edges, in the sense that no two of them are sides of the same triangle [9]. Each one of these edges will color 5 edges by its flip. A green non-flipped edge might get counted twice this way, thus we get at least $\frac{n-4}{6} + \frac{4}{2} \times \frac{n-4}{6} = \frac{n-4}{2}$ colored edges. As there are at most $3n$ edges, and we have colored at least $n/2$ edges (we disregard additive constants for both numbers), we have got at least $1/6$ of the edges to be green, as claimed.

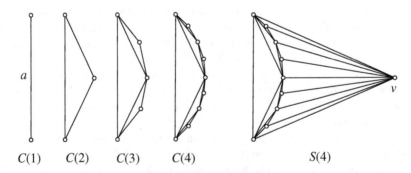

Fig. 4. Recursive construction of triangulated point sets $S(t)$

As for the upper bound, we define a triangulated convex polygon $C(t)$ as follows (see Figure 4). The vertices of $C(t)$ are placed on an arc of circle with central angle below π. Take $C(1)$ equal to the chord a associated with the arc, add a triangle with the third vertex in the arc in order to get $C(2)$. Attach externally triangles to the two outer chords of $C(2)$ for constructing $C(3)$, and iterate this process in order to obtain $C(t)$. The number of vertices of $C(t)$ is $2^{t-1} + 1$.

Now let v be a point that sees completely the circular arc from outside the circle; a triangulated point set $S(t)$ is defined by connecting v to all the vertices of $C(t)$. The edges in $S(t)$ are those in the boundary of $C(t)$, plus its diagonals, plus the edges incident to v, therefore their total number is

$$e(t) = (2^{t-1} + 1) + (2^{t-1} + 1 - 3) + (2^{t-1} + 1) = 3 \cdot 2^{t-1}.$$

Notice that no boundary edge of $C(t)$ and no edge incident to v can ever be flipped, therefore the edges in $S(t)$ incident to v are never colored green. On the other hand observe that if we suppress a from $S(t)$ we obtain two instances of $S(t-1)$; despite the fact that the two copies share an edge, the coloring process behaves independently.

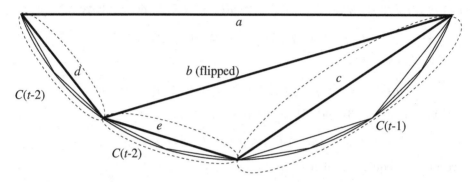

Fig. 5. Bounding the value $g_2(t)$. The five solid lines are green once b has been flipped

Let $g(t)$ be the maximum number of green edges that can be obtained from $S(t)$ after any flip sequence of black edges; we show next that $g(t)/e(t)$ approaches $5/9$ for large values of t. The numbers $g(1) = g(2) = 0$ and $g(3) = 5$ are directly computable. Let b and c be the edges that together with a form a triangle in $S(t)$. As $S(t)$ contains two copies of $S(t-1)$ which we can color independently, we have

$$2 \cdot g(t-1) \leq g(t). \tag{1}$$

Let $g_1(t)$ be the maximum number of green edges achievable from $S(t)$ when neither b nor c are flipped. We have

$$g_1(t) = 2 \cdot g(t-1). \tag{2}$$

Let $g_2(t)$ be the maximum number of green edges achievable from $S(t)$ when either b or c are flipped (refer to Figure 5). Assume it is b, and notice that in this case none of the edges d, e and c has been flipped before b is, as otherwise b would be green and disallowed to be flipped. Therefore the maximum number of green edges we can achieve this way is

$$g_2(t) \leq g(t-1) + 2 \cdot g(t-2) + 5. \tag{3}$$

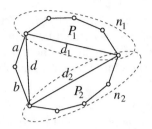

Fig. 6. Illustrating a winning strategy for the Green-Wins Solitaire in convex position

Combining the above equation with the fact that $2 \cdot g(t-2) \leq g(t-1)$ that we know from (1), we obtain $g_2(t) \leq 2 \cdot g(t-1) + 5$. On the other hand the equality (2) directly gives that $g_1(t) \leq 2 \cdot g(t-1) + 5$. Hence we have

$$g(t) = \max(g_1(t), g_2(t)) \leq 2 \cdot g(t-1) + 5, \tag{4}$$

and from this we get

$$g_1(t) \leq g(t-1) + g(t-1) \leq g(t-1) + 2 \cdot g(t-2) + 5. \tag{5}$$

Using equations (3) and (5) we arrive to

$$g(t) = \max(g_1(t), g_2(t)) \leq g(t-1) + 2 \cdot g(t-2) + 5,$$

a recursive inequality which solves to

$$g(t) \leq \frac{5}{6} \cdot (2^t - (-1)^t - 3).$$

Therefore

$$\frac{g(t)}{e(t)} \leq \frac{5}{6} \cdot \frac{2^t - (-1)^t - 3}{3 \cdot 2^{t-1}} \xrightarrow[t \to \infty]{} \frac{5}{9}$$

as claimed.   □

Finally, it is quite easy to prove that the game is not very exciting in convex position:

Theorem 6. *The player of the Green-Wins Solitaire Game can always win for any given triangulation on $n \geq 4$ points in convex position.*

Proof. The number of edges in any triangulation is $2n - 3$, therefore we have to prove that we can always achieve at least $n - 1$ green edges after a suitable sequence of flips.

We proceed by induction. The cases $n = 4, 5, 6$ are easily checked directly, hence we can assume $n \geq 7$. Let a and b be consecutive boundary edges of an ear of the triangulation, and let d be the diagonal which completes a triangle with a and b (refer to Figure 6). Let d_1 and d_2 be the edges of the other triangle which shares the diagonal d. Consider the polygons P_1 and P_2 respectively separated by these diagonals from the whole polygon, and let n_1 and n_2 be their respective number of vertices, where $n_1 + n_2 = n$. We can assume that $n_1 \leq n_2$.

If $n_1 = 2$, we flip d and apply induction to P_2. In this way we obtain at least $4 + (n_2 - 1)$ green edges, and

$$4 + (n_2 - 1) = 4 + (n - 2) - 1 = n + 1 > n - 1$$

as desired. If $n_1 = 3$, we proceed in the same way and obtain at least $4 + (n_2 - 1)$ green edges. Now

$$4 + (n_2 - 1) = 4 + (n - 3) - 1 = n > n - 1.$$

Finally, if $n_2 \geq n_1 \geq 4$, we flip d and apply induction both to P_1 and P_2. In this way we obtain at least $3 + (n_1 - 1) + (n_2 - 1)$ green edges, where

$$3 + (n_1 - 1) + (n_2 - 1) = n_1 + n_2 + 1 = n + 1 > n - 1. \qquad \square$$

Acknowledgments

Ferran Hurtado is partially supported by Projects MEC-DGES-SEUID PB98-0933, MCYT-FEDER BFM2002-0557 and Gen. Cat 2001SGR00224. Research of Evangelos Kranakis is supported in part by NSERC and MITACS grants. Research of Hannes Krasser is supported by the FWF (Austrian Fonds zur Förderung der Wissenschaftlichen Forschung). Suneeta Ramaswami is partially supported by a Rutgers University ISATC pilot project grant. Research of Jorge Urrutia is supported in part by CONACYT grant 37540-A and grant PAPIIT.

References

1. O. Aichholzer, D. Bremner, E. D. Demaine, F. Hurtado, E. Kranakis, H. Krasser, S. Ramaswami, S. Sethia, J. Urrutia: Games on Triangulations: Several Variations. Manuscript in preparation
2. E. R. Berlekamp: The Dots and Boxes Game: Sophisticated Child's Play. Academic Press (1982)
3. E. R. Berlekamp, J. H. Conway, R. K. Guy: Winning Ways for your Mathematical Plays. Academic Press (1982). Second edition in print, A K Peters Ltd., (2001)
4. H. L. Boedlander, D. Kratsch: Kayles and Nimbers. J. of Algorithms **43** (2002) 106–119
5. J. H. Conway: On Numbers and Games. Academic Press (1976). Second edition, A K Peters Ltd. (2002)
6. D. G. Corneil, S. Olariu, L. Stewart: Asteroidal triple-free graphs. SIAM J. Discrete Math. **10** (1997) 399–430
7. E. D. Demaine: Playing games with algorithms: Algorithmic combinatorial game theory. In: Proc. 26th Symp. on Math Found. in Comp. Sci., Lect. Notes in Comp. Sci., Springer-Verlag (2001) 18–32
8. A. S. Fraenkel: Combinatorial Games: Selected Bibliography with a Succinct Gourmet Introduction. Electronic Journal of Combinatorics, http://www.wisdom.weizmann.ac.il/~fraenkel
9. J. Galtier, F. Hurtado, M. Noy, S. Pérennes, J. Urrutia: Simultaneous edge flipping in triangulations. Submitted
10. T. J. Schaefer: On the complexity of some two-person perfect-information games. J. Comput. Syst. Sci. **16** (1978) 185–225

The Foldings of a Square to Convex Polyhedra

Rebecca Alexander, Heather Dyson, and Joseph O'Rourke

Dept. Comput. Sci., Smith College, Northampton, MA 01063, USA
{ralexand,hdyson,orourke}@cs.smith.edu
http://cs.smith.edu/~orourke/

Abstract. The structure of the set of all convex polyhedra foldable from a square is detailed. It is proved that five combinatorially distinct nondegenerate polyhedra, and four different flat polyhedra, are realizable. All the polyhedra are continuously deformable into each other, with the space of polyhedra having the topology of four connected rings.

1 Introduction

If the perimeter of a polygon is glued to itself in a length-preserving manner, and in such a way that the resulting complex is homeomorphic to a sphere, then a theorem of Aleksandrov [1] establishes that as long as no more than 2π face angle is glued together at any point, the gluing corresponds to a unique convex polyhedron (where "polyhedron" here includes doubly-covered flat polygons). Exploration of the possible foldings of a polygon to convex polyhedra via these *Aleksandrov gluings* was initiated in [2], and further explored in [3,4]. Theorem 1 in [4] established that every convex polygon can fold to a nondenumerably infinite number of incongruent convex polyhedra. Although this set is infinite, it arises from a finite collection of *gluing trees*, which record the combinatorially possible ways to glue up the perimeter. Enumerating the gluing trees leads to an inventory of the possible foldings of a given polygon to polyhedra. These ideas were implemented in two computer programs, developed independently by Anna Lubiw and Koichi Hirata[1]. These programs only list the gluings, not the polyhedra. Even though the polyhedra are uniquely determined by the gluings, there is no known practical algorithm for computing the creases and reconstructing the 3D shape of the polyhedra [5].

The contribution of this paper is to construct the 3D structure of all the polyhedra foldable from one particularly simple polygon: a unit square. Spare remarks on regular n-gons will be ventured in the final Section 4. The polyhedra foldable from a square have from 3 to 6 vertices, and we show they fall into nine distinct combinatorial classes: tetrahedra, two different pentahedra, hexahedra, and octahedra; and a flat triangle, square, rectangle, and pentagon. Each achievable shape can be continuously deformed into any other through intermediate foldings of the square, i.e., no shape is isolated. An illustration of the foldings for

---

[1] Personal communications, Fall 2000. Hirata's program is available at http://weyl.
ed.ehime-u.ac.jp/cgi-bin/WebObjects/Polytope2

J. Akiyama and M. Kano (Eds.): JCDCG 2002, LNCS 2866, pp. 38–50, 2003.

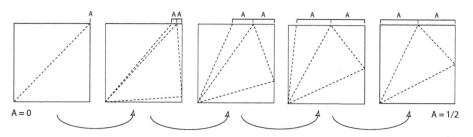

Fig. 1. Creases for a section of a continuum of foldings: as A varies in $[0, \frac{1}{2}]$, the polyhedra vary between a flat triangle and a symmetric tetrahedron. See also Figure 7) below.

a portion of a continuum are shown in Figure 1. In general the crease patterns vary continuously, as in this figure, with discontinuous jumps at coplanarities to different combinatorial types. The continua fall into four distinct rings (A, B, C, D), each corresponding to a single parameter change in the gluing, which join together topologically as depicted in Figure 2. Three of the rings (A, B, D) share and join at the flat $1 \times \frac{1}{2}$ rectangle; rings A and C join at a symmetric tetrahedron.

An animated GIF of the entire set of polyhedra is available at `http://cs.smith.edu/~orourke/Square/animation.html`.

2   Proof

Our goal in this (long) section is to prove that Figure 2 represents a complete inventory of the shapes foldable from a square; but we will not prove formally every detail depicted in the figure. The starting point for the proof is a lemma that limits the combinatorial structure of gluing trees for convex polygons, specialized to $n = 4$:

Lemma 1. *[3,4]; see also [6]. The possible gluing trees for a convex quadrilateral ($n = 4$) are of four combinatorial types:*

1. *'$|$': a tree of two leaves, i.e., a path.*
2. *'Y': a tree of three leaves and one internal degree-3 node.*
3. *'I': a tree of four leaves and two internal degree-3 nodes.*
4. *'+': a tree of four leaves and one internal degree-4 node.*

We will now make an exhaustive list of the possible gluings of a unit square, applying three facts to increasingly restrict the possibilities:

1. Lemma 1.
2. A square has four corners, each of internal angle $\pi/2$.
3. These corners are separated by edges of length 1, and the perimeter is 4.

Let c_0, c_1, c_2, c_3 be the corners of the square. The proof, an extended case analysis, follows the structure provided by Lemma 1.

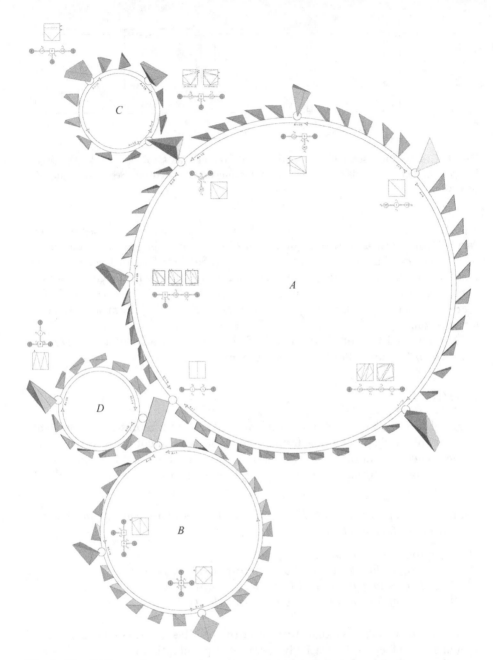

Fig. 2. The foldings of a square to convex polyhedra. Selected polyhedra are shown enlarged (including the four flat shapes), together with their crease patterns and corresponding gluing trees.

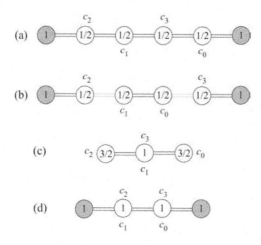

Fig. 3. (a) When all corners are distinct and not leaves; (b) A configuration that forces the lightly shaded arcs to have zero length; (c) A flat right triangle; (d).A flat rectangle.

Case 1. The four corners must be distributed along the path. If no corner is glued to another, and neither leaf is a corner, then we have the structure illustrated in Figure 3(a). Here, and in subsequent figures, nonzero-curvature vertices of the polyhedron are marked by circles, with the curvature (in units of π) indicated inside the circle. (Thus the circled numbers must add to 4 to satisfy the Gauss-Bonnet theorem.) A shaded circle represents a *fold point* – a creased edge – necessarily of curvature π. The corners are labeled c_i, with the position of the labels indicating to which side of the pieces of paper gluing there the corner resides. The staggered distribution of the corners in (a) are necessary, for if, say, c_0 and c_1 were adjacent on the lower side as shown in (b), then two arcs would have to be zero length in order to have a total perimeter of 4, which would force nodes to merge. The gluing tree in (a) is a six-vertex polyhedron, generally an octahedron, although there one spot in the continuum (with parameter $A \approx 0.35$) where it becomes a pentahedron with three quadrilateral faces: a triangular prism, two nearly parallel triangles joined by four near-rectangles. Figure 4 shows an enlargement of the A- (triangle) continuum in Figure 2[2] that includes the octahedra.

If one corner of the square is a leaf of the gluing tree, then both leaves are forced to be opposite corners, and the two other opposite corners mate, resulting in the structure shown in (c) of the figure, which corresponds to a flat right triangle, the right end of the octahedron continuum in Figure 4.

If two adjacent corners mate, the structure shown in (d) of the figure is forced, which folds to a flat $1 \times \frac{1}{2}$ rectangle the left end of the octahedron continuum in Figure 4 (where it joins with two other continuua, B and D).

---

[2]   All subsequent closeups of the continuua are rotated 90° with respect to the orientation used in Fig. 2.

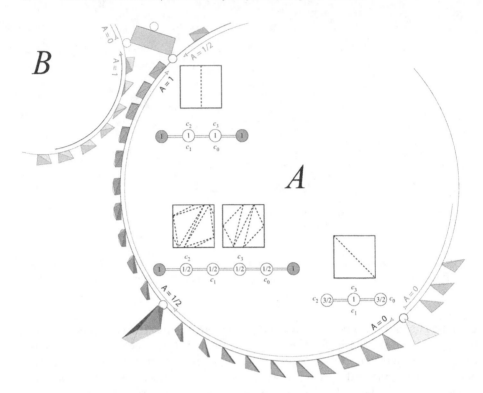

Fig. 4. The octahedra continuum, a subpart of the A-loop.

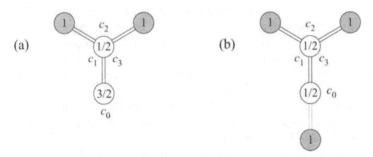

Fig. 5. (a) A fixed tetrahedron; (b) An impossible gluing tree.

Case Y. The single internal node of a 'Y' must consist of two (cc) or three (ccc) corners, for otherwise the paper there would exceed 2π.

1. ccc. The only issue remaining is where the fourth corner lies. If it is a leaf, the structure shown in Figure 5(a) is forced. This represents a "fixed" tetrahedron, fixed in the sense that the two fold-point leaves are forced to be side midpoints and so do not form a "rolling belt" [3,4]. (This tetrahedron is the shape shared by loops A and C in Figure 2). If the fourth corner is a path node, then the length of the shaded arc is forced to be zero, which reduces this case to that in (a).

2. cc. Two corners and an interior point on a square side forms a total angle of 2π, so the internal node of the 'Y' has zero curvature. Thus it is not a polyhedron vertex; we will indicate such nodes with a box enclosing a 0. There are two possibilities: either the two corners at the junction are opposite corners, or adjacent corners.

(a) Opposite. With c_0 and c_2 (without loss of generality) glued to the junction, c_1 is forced to be at a leaf. This leaves only the location of the fourth corner to be determined. It might be a leaf node or a path node. If c_3 is a path node, we have the tree shown in Figure 6(a). This corresponds to an asymmetric tetrahedron, a continuum with the two shaded leaves forming a rolling belt. If c_3 is a leaf node, then the shaded arc in (b) of the figure must be zero length, yielding (c), which is the same as Figure 3(c): a flat right triangle. The corresponding continuum (part of the A-loop) is shown Figure 7; note that its lower end is the flat right triangle and its upper end the fixed tetrahedron.

(b) Adjacent. With c_0 and c_1 (without loss of generality) glued to the junction, the leaf between is a fixed fold midpoint. The two other corners of the square must be placed on the other two branches of the 'Y' (which now looks like a 'T'). We can distinguish four possibilities, determined by whether each corner is a leaf (L), a path node (P), and when both are path nodes, whether to the opposite (o) or same (s) side of the junction.

   i. LL. When both corners are leaves, we have the structure shown in Figure 8(a). Knowing that the distance between consecutive corners is 1, this structure must have perimeter 5. So it is unachievable.

   ii. LP. When one corner is a leaf and the other a path node, the path node could be to the same or the opposite side of the 'Y' as the leaf. If it is to the same side, the total perimeter can be calculated to be 5, ruling out this structure. If it to the opposite side, we have the structure shown in (b) of the figure. Here the shaded arc must be zero length to have a perimeter of 4, which reduces this structure to the three-corner 'Y' of Figure 5(a).

   iii. PPo. This structure (Figure 8(c)) corresponds to a five-vertex polyhedron continuum, which at its midpoint is a flat pentagon, and at either endpoint (when one corner becomes a leaf and the other corner joins the junction) becomes the fixed tetrahedron of Figure 5(a). This forms the C-loop of the continuum, as shown in Figure 9.

   iv. PPs. This (Figure 8(d)) corresponds to another five-vertex continuum, which at one endpoint is the same fixed tetrahedron, and at the other the flat rectangle of Figure 3(d). See Figure 10. Note that c_2 and c_3 cannot be on the same side of the branch, as then the perimeter would be too long, so they must be on opposite sides. And c_3 cannot be on the lower side, adjacent to c_0, for the same reason. Thus the structure illustrated is the only one possible.

   We will not establish this formally, but the hexahedron continuum in Figure 10 partitions into three sections, separated by two pentahedra which occur when two triangles become coplanar and form a quadrilateral face.

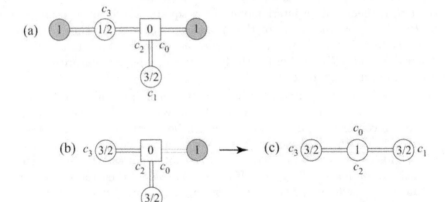

Fig. 6. Opposite corners glued to 'Y'-junction.

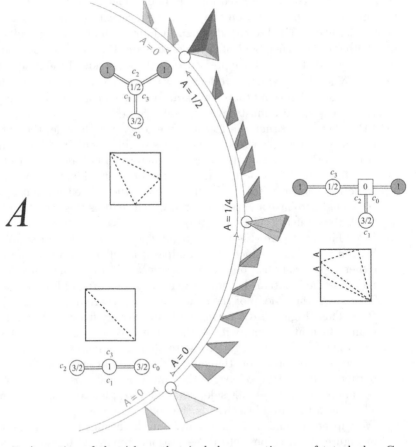

Fig. 7. A portion of the A-loop that includes a continuum of tetrahedra. Compare Figure 1.

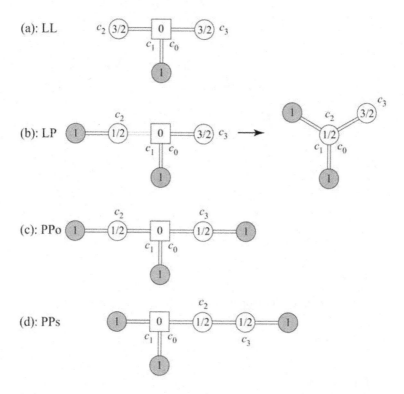

Fig. 8. Adjacent corners glued to 'Y'-junction. Labels: L=leaf; P=Path; o=opposite; s=same.

Case I. Each junction must have two or three corners, which, because there are two junctions, means that both must have two corners. There are three possible patterns for the distribution of the corners, illustrated in (a,b,d) of Figure 11, which we will call mixed, adjacent, and opposite.

1. Mixed. The shaded edge in Figure 11(a) must be zero length to achieve a perimeter of 4, which reduces this case to (c), which we will discuss below.
2. Adjacent. The pattern of corner distribution shown in Figure 11(b) is possible, producing two independent rolling belts. This is a two-dimensional continuum of tetrahedra, all four of whose vertices have curvature π. It is clear that if we roll the upper belt to one extreme and fix it there, thereby gluing two corners together as in Figure 11(c), the shapes produced by the rolling of the lower belt include all the shapes achievable through the rolling of both belts. This is because it is only the relative rolling of the two belts that is significant. Thus, the 2D continuum of shapes can be captured in the 1D loop D, illustrated in Figure 12. Notice that the polyhedra at symmetric positions with respect to the flat rectangle are reflections of one another.

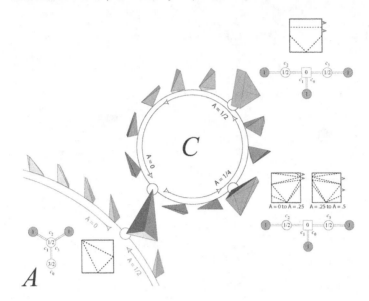

Fig. 9. The C-loop containing the flat pentagon.

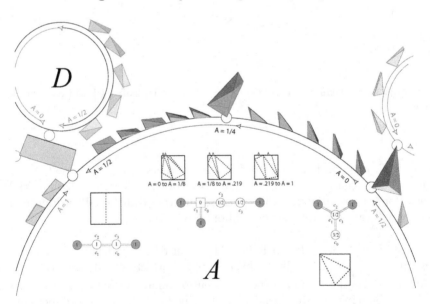

Fig. 10. The portion of loop A consisting of hexahedra.

3. Opposite. The structure forced here, Figure 11(d), is another continuum of tetrahedra with curvature-π vertices. Here the upper and lower fold points form a rolling belt; the two side fold-point leaves are fixed at the midpoint of the folded edge. The continuum forms the B-loop, Figure 13, shapes mirror-symmetric about the flat rectangle and square.

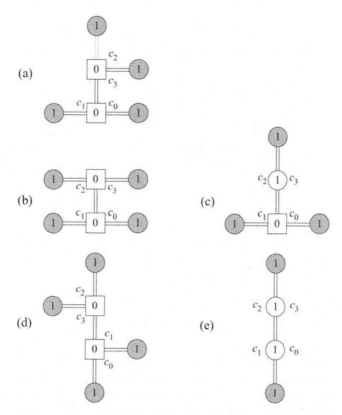

Fig. 11. The gluing trees with two degree-3 nodes: (a) Not possible; (b) Double belt; (c) Equivalent single belt; (d) Another tetrahedron continuum.

The extremes of both continuua, at both ends, lead to the structures shown in Figure 11(e), which can be recognized as the flat rectangle of Figure 3(d).

Case +. Finally, the degree-4 junction of the '+' can only be realized with all four corners glued together, which forces the structure in Figure 14: the flat square of Figure 13.

This completes the inventory of the polyhedra foldable from a square.

3   Reconstructing the 3D Shapes

We mentioned that it is an unsolved algorithmic problem to compute the three-dimensional coordinates of a polyhedron given the face structure determined by a particular gluing. We now describe how it was possible nevertheless to compute the structure of all the polyhedra foldable from a square, as displayed in Figure 2.

There are two issues: (1) Identifying the creases, and therefore the edge lengths; and (2) reconstructing the 3D shape from the edge lengths.

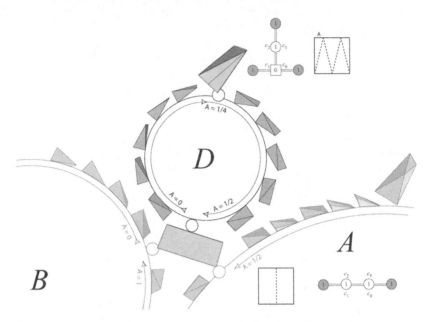

Fig. 12. Twisting tetrahedra: loop D.

(1) Although we have no systematic method for identifying the creases, there are only a finite number of possibilities, as Aleksandrov observed [7]. The vertices are determined as the points of nonzero curvature, and we know that every edge is a shortest path between the vertices that are its endpoints. So, lacking any other information, we could try all $\binom{n}{2}$ possibilities. In practice, some of the creases are obvious from physical models, leaving only a few uncertainties. These were resolved by "trying" each, and relying on Aleksandrov's theorem guaranteeing a unique reconstruction: the wrong choices failed to reconstruct, and the correct choice led to a valid polyhedron.

(2) All of the nonflat polyhedra foldable from a square have 4, 5, or 6 vertices. Aside from isolated special cases caused by coplanar triangles merging into quadrilaterals, all faces are triangles. In fact, only three distinct combinatorial types are realized: tetrahedra, hexahedra equivalent to two tetrahedra glued base-to-base (i.e., a "trigonal dipyramid"), and octahedra combinatorially equivalent to the regular octahedron. Reconstructing tetrahedra from their six edge lengths is not difficult. Reconstructing the hexahedra from their nine edge lengths can be accomplished by reconstructing the two joined tetrahedra.

Reconstructing octahedra is more challenging, for the structure is not determined by the union of tetrahedra all of whose edges are on the surface. Any partitioning into tetrahedra leaves edges of tetrahedra as internal diagonals, whose lengths are not determined by creases of the square. The strategy we used is to partition the octahedron into two hexahedra, each a trigonal bipyramid. Consider one half of the octahedron, for which we know eight lengths: the four edges

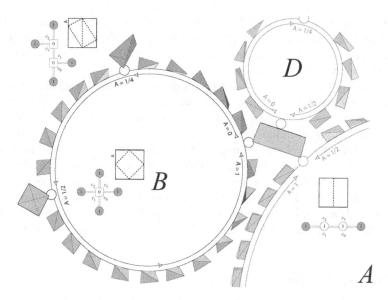

Fig. 13. The B-loop, containing the flat square.

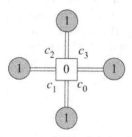

Fig. 14. This gluing tree corresponds to a flat square.

of the quadrilateral "base" (which will not in general be a flat polygon), and the four edges from the base to an apex. These lengths leave the structure with one degree of flexibility. The unknown length x of the "internal" diagonal d splitting the base quadrilateral is one parameter that determines this flex. Two octahedron halves together with the same x may join only if the dihedral angle in each at the edge d sum to 2π. This gives us a method to solve for x: find an x such that the angles sum to 2π. Although one can express x as a polynomial of some large degree (perhaps 16), we found it possible to solve for x via numerical search.

4   Discussion

Constructing the entire set of polyhedra permits us to answer a special case of a question posed by Joseph Malkevitch[3]: What is the maximum volume poly-

---

[3]   Personal communication, Feb. 2002.

hedron foldable from a given polygon? We found (by numerical search) that the maximum is achieved by an octahedron along the A-ring (at about the 4-5 o'clock position), with a volume of approximately 0.055849. We expected to observe symmetry here but did not.

The work reported here is part of a larger project to understand the structure of all the convex polyhedra foldable from any given polygon. Although that goal does not appear close with our current understanding, the foldings of regular n-gons seem more approachable. It was established in [3] that, for $n > 6$, there is only one nonflat folding of regular n-gons, the class of "pita polytopes" produced by perimeter halving. (For the square, this corresponds to the path gluing of Figure 3(a).) One might expect a one-ring continuum of polyhedra for these n-gons. The complexity of the multiple rings manifest with the square may be an artifact of small $n \leq 6$. For arbitrary convex polygons, it would be of interest to establish that the space of folded polyhedra is connected, i.e., that no shape is isolated.

Acknowledgements

The third author thanks Koichi Hirata and Anna Lubiw for the use of their enumeration code, Martin Demaine for folding polyhedra, and Erik Demaine for a conversation that led to a method for reconstructing octahedra. All authors are grateful to Michiko Charley, Beenish Chaudry, Melody Donoso, Monta Lertpachin, Sonya Nikolova, and Emily Zaehring for discussions. The referees were both very helpful. This work was supported by NSF Distinguished Teaching Scholars award DUE-0123154.

References

1. Aleksandrov, A.D.: Konvexe Polyeder. Akademie Verlag, Berlin (1958)
2. Lubiw, A., O'Rourke, J.: When can a polygon fold to a polytope? Technical Report 048, Dept. Comput. Sci., Smith College (1996) Presented at AMS Conf., 5 Oct. 1996.
3. Demaine, E.D., Demaine, M.L., Lubiw, A., O'Rourke, J.: Examples, counterexamples, and enumeration results for foldings and unfoldings between polygons and polytopes. Technical Report 069, Smith College, Northampton, MA (2000) LANL ArXive cs.CG/0007019.
4. Demaine, E.D., Demaine, M.L., Lubiw, A., O'Rourke, J.: Enumerating foldings and unfoldings between polygons and polytopes. Graphs and Combinatorics **18** (2002) 93–104
5. O'Rourke, J.: Folding and unfolding in computational geometry. In: Discrete Comput. Geom. Volume 1763 of Lecture Notes Comput. Sci., Springer-Verlag (2000) 258–266 Papers from the Japan Conf. Discrete Comput. Geom., Tokyo, Dec. 1998.
6. Shephard, G.C.: Convex polytopes with convex nets. Math. Proc. Camb. Phil. Soc. **78** (1975) 389–403
7. Aleksandrov, A.D.: Existence of a convex polyhedron and a convex surface with a given metric. In Reshetnyak, Y.G., Kutateladze, S.S., eds.: A. D. Aleksandrov: Selected Works: Part I. Gordon and Breach, Australia (1996) 169–173. Translation of Doklady Akad. Nauk SSSR, Matematika, Vol. 30, No. 2, 103–106 (1941).

On the Complexity of Testing Hypermetric, Negative Type, k-Gonal and Gap Inequalities

David Avis

Computer Science, Mcgill University and GERAD
3480 University, Montreal, Quebec, Canada H3A 2A7
avis@cs.mcgill.ca

Abstract. Hypermetric inequalities have many applications, most recently in the approximate solution of max-cut problems by linear and semidefinite programming. However, not much is known about the separation problem for these inequalities. Previously Avis and Grishukhin showed that certain special cases of the separation problem for hypermetric inequalities are NP-hard, as evidence that the separation problem is itself hard. In this paper we show that similar results hold for inequalities of negative type, even though the separation problem for negative type inequalities is well known to be solvable in polynomial time. We also show similar results hold for the more general k-gonal and gap inequalities.

1   Introduction

Let $b = (b_1, ..., b_n)$ be an integer vector, let $k = \sum_{1 \leq i \leq n} |b_i|$ and $s = \sum_{1 \leq i \leq n} b_i$. Let $x = (x_{ij}), 1 \leq i < j \leq n$ be a vector in $R^{\binom{n}{2}}$. We say that b defines a k-gonal inequality:

$$Q(b, x) = \sum_{1 \leq i < j \leq n} b_i b_j x_{ij} \leq \lfloor \frac{s^2}{4} \rfloor \tag{1}$$

Since the vector $-b$ generates the same inequality as the vector b, we will assume throughout the paper that $s \geq 0$. The inequality (1) is called *hypermetric* if $s = 1$, in which case k is necessarily odd. It is called *negative type* if $s = 0$, in which case k is even. It is called *pure* if $b_i \in \{\pm 1, 0\}$. Inequalities of type (1) have been well studied and rediscovered many times. The hypermetric inequalities appear in Deza [3], and the negative type inequalities in the work of Cayley. The book of Deza and Laurent [5] collects a wealth of information about them and their applications.

For fixed n, it is easy to show that the cone formed by the negative type inequalities is not polyhedral. However, Deza, Grishukhin and Laurent [4] showed that the hypermetric inequalities do form a polyhedral cone. Furthermore, each $2k+2$-gonal inequality can be obtained by a non-negative combination of $2k+1$-gonal inequalities. This was proved by Deza [3] for negative type inequalities, and for the general k-gonal inequalities by Avis and Umemoto [2].

J. Akiyama and M. Kano (Eds.): JCDCG 2002, LNCS 2866, pp. 51–59, 2003.

The k-gonal inequalities have application in both linear programming (LP) [2] and semidefinite programming (SDP) [8] relaxations of the well known MAX_CUT problem. The hypermetric inequalities form a good approximation of the cut polytope, and in fact many induce facets. The SDP relaxation essentially consists of optimizing over the feasible region of all k-gonal inequalities (1) where the floor function is omitted. The efficiency of the SDP relaxation depends on the fact that separation for this feasible region can be performed in polynomial time (see, e.g. Section 28.4.1 in [5]).

In order to use hypermetric inequalities effectively, it is necessary to be able to test whether a vector x satisfies all of them, and if not, to find a violated inequality. This is the *separation* problem for hypermetric inequalities. Its complexity status is unknown, although it is known that *membership* testing for hypermetric inequalities is in the class co-NP [1]. The paper [1] also contains the following complexity results:

P1. Hypermetric (2m+1)-gonality testing
Instance: A rational vector x of length $\binom{n}{2}$ and an integer m.
Question: Does x satisfy all $(2m + 1)$-gonal hypermetric inequalities?
Complexity: Co-NP complete.
Comments: Remains co-NP complete for testing if x satisfies all pure $(2m+1)$-gon hypermetric inequalities.

P2. Strong hypermetricity
Instance: A rational vector x of length $\binom{n}{2}$.
Question: Is x hypermetric? If not, give smallest k such that x violates a $(2k + 1)$-gonal hypermetric inequality.
Complexity: NP-hard.

These results may seem to give evidence to the claim that the separation problem for hypermetric inequalities is hard (see, eg. [5], p. 454). However, in the next section we prove that the following similar results apply to negative type inequalities.

P1N. Negative type 2m-gonality testing
Instance: A rational vector x of length $\binom{n}{2}$ and an integer m.
Question: Does x satisfy all $2m$-gonal negative type inequalities?
Complexity: co-NP complete.
Comments: Remains co-NP complete for testing if x satisfies all pure $2m$-gonal negative type inequalities.

P2N. Strong negative type
Instance: A rational vector x of length $\binom{n}{2}$.
Question: Is x of negative type? If not, give smallest m such that x violates a $2m$-gonal negative type inequality.
Complexity: NP-hard.

For fixed n, the cone of negative type inequalities is a linear transformation of the cone of positive semidefinite matrices, for which separation can be performed in $O(n^3)$ time (see, eg. [10]). Therefore separation for negative type inequalities can be performed in $O(n^3)$ time also. These results, therefore, cast some doubt on the evidence that hypermetric separation is hard. We can also show similar results for testing k-gonality.

P1K. Testing k-gonality
Instance: A rational vector x of length $\binom{n}{2}$ and an integer k.
Question: Does x satisfy all k-gonal inequalities?
Complexity: co-NP complete.

P2K. Strong k-gonality
Instance: A rational vector x of length $\binom{n}{2}$.
Question: Is x k-gonal? If not, give smallest m such that x violates a k-gonal inequality.
Complexity: NP-hard.

Laurent and Poljak [9] introduced a set of inequalities, called *gapinequalities*, that can be stronger than the k-gonal inequalities when $s \geq 2$. The gap $g = g(b)$ of an integer sequence $b_1, \ldots, b_n$ is defined by

$$g = g(b) = min_{S \subseteq \{1,2,\ldots,n\}} | \sum_{i \in S} b_i - \sum_{i \notin S} b_i|. \qquad (2)$$

It is easy to see that for a negative type inequality, the b vector has gap zero and for a hypermetric inequality it has gap one. A gap inequality if formed by modifying the right hand side of (1):

$$Q(b, x) = \sum_{1 \leq i < j \leq n} b_i b_j x_{ij} \leq \frac{1}{4}(s^2 - g^2). \qquad (3)$$

Since s and g have the same parity, the right hand side is always integral. We call a gap inequality k-gonal If $k = \sum_{i=1}^n |b_i|$, we call (3) a $k - gonal\ gap\ inequality$.

The gap inequalities are also valid for the MAX_CUT problem. Very little is known about gap inequalities. For example, it is not known if they form facets of the cut polytope, if they form a polyhedron for each n, or if separation can be efficiently performed. Here we show that complexity results similar to those above can be obtained.

P1G. Testing k-gonal gap inequalities
Instance: A rational vector x of length $\binom{n}{2}$ and an integer k.
Question: Does x satisfy all k-gonal gap inequalities?
Complexity: co-NP complete.

P2G. Strong gap testing
Instance: A rational vector x of length $\binom{n}{2}$.
Question: Does x satisfy all gap inequalities. If not, give smallest m such that x violates a k-gonal gap inequality.
Complexity: NP-hard.

2   Preliminary Results

The proofs follow similar lines to the proofs of P1 and P2 in [1], see also Section 28.3 of [5]. We make use of the following problems known to be hard.

P3. Complete Bipartite Subgraph
Instance: Graph G on n vertices and an integer m.
Problem: (a) Does G contains an induced complete bipartite subgraph $K_{m,m}$?
(b) Does G contains an induced complete bipartite subgraph $K_{m,m+1}$?
Complexity: Both NP-complete. Reduction from independent set [6].

P4. Largest Complete Bipartite Subgraph
Instance: Graph G on n vertices.
Problem: (a) Find largest m such that G contains an induced $K_{m,m}$ subgraph.
(b) Find largest m such that G contains an induced $K_{m,m+1}$ as an induced subgraph.
Complexity: Both NP-hard. Reduction from independent set [6].

Let G be an undirected graph with vertices $V = V(G)$ and edges $E = E(G)$. Construct the edge weights

$$x^t_{ij}(G) = \begin{cases} 1 & \text{if } (ij) \in E \\ 1 + t & \text{if } (ij) \notin E \end{cases} \tag{4}$$

Let b be an integer vector of length n and define

$$k = k(b) = \sum_{i=1}^{n} |b_i|, \quad s = s(b) = \sum_{i=1}^{n} b_i, \tag{5}$$

which implies that

$$\sum_{i \in V_+(b)} b_i = (s+k)/2, \quad \sum_{i \in V_-(b)} b_i = (s-k)/2. \tag{6}$$

Let $Q_G(b, x^t)$ be the left hand side of the inequality (1) with $x^t = x^t(G)$. We calculate $Q_G(b, x^t)$. We set

$$V_+(b) = \{i : b_i > 0\}, \quad V_-(b) = \{i : b_i < 0\}, \quad V(b) = V_+(b) \cup V_-(b)$$

$$n_+ = |V_+(b)|, \quad n_- = |V_-(b)|, \quad n_b = n_+ + n_-.$$

We denote by $G(b)$ the subgraph of G induced on the set $V(b)$. Let K_{n_+,n_-} be the complete bipartite graph on the set $V(b)$ with the partition $(V_+(b), V_-(b))$. Let

$$E_b(G) = E(G(b)) \triangle E(K_{n_+,n_-}),$$

where $\triangle$ denoted the symmetric difference of 2 sets.

Lemma 1. *Let G be a graph and let b and $E_b(G)$ be defined as above. Then*

$$Q_G(b, x^t) = \frac{s^2}{2} + \frac{t}{4}(s^2 + k^2) - \frac{1+t}{2} \sum_{i=1}^{n} b_i^2 - t \sum_{(ij) \in E_b(G)} |b_i||b_j| \tag{7}$$

Proof. Suppose at first that $E_b(G) = \emptyset$, i.e. the set $V(b)$ induces a complete bipartite graph $K(b)$. From the definitions we have

$$Q_G(b, x^t) = Q_{K_{n_+, n_-}}(b, x^t) = \sum_{i \in V_+, j \in V_-} b_i b_j + (1+t)\left(\sum_{i,j \in V_+, i<j} b_i b_j + \sum_{i,j \in V_-, i<j} b_i b_j\right).$$

Since for any set X

$$\sum_{i,j \in X, i<j} b_i b_j = \frac{1}{2}\left(\left(\sum_{i \in X} b_i\right)^2 - \sum_{i \in X} b_i^2\right),$$

using (6) we obtain

$$Q_{K_{n_+, n_-}}(b, x^t) = \frac{(s+k)(s-k)}{4} + \frac{1+t}{2}\left(\frac{(s+k)^2}{4} + \frac{(s-k)^2}{4} - \sum_{i=1}^{n} b_i^2\right).$$

After simplification, we obtain

$$Q_{K_{n_+, n_-}}(b, x^t) = \frac{s^2}{2} + \frac{t}{4}(s^2 + k^2) - \frac{(1+t)}{2}\sum_{i=1}^{n} b_i^2$$

which are the first 3 terms of (7).

If $E_b(G) \neq \emptyset$, then it may contain two types of edges: those in G between vertices both in either V_+ or V_-, or edges in K_{n_+, n_-} not in G. In both cases, the right hand side of the equality (2) obtains additional negative summand

$$-t \sum_{(i,j) \in E_b(G)} |b_i||b_j|$$

and we are done. ∎

We remark that for a pure b vector, with $k \leq n$, the equality (7) takes the simple form

$$Q_G(b, x^t) = \frac{s^2}{2} + \frac{t}{4}(s^2 + k^2) - \frac{1+t}{2}k - t|E_b(G)|. \tag{8}$$

We also have the immediate:

Lemma 2. *Let b be an integer vector of length n, and let k and s be defined as in (5). Then*

$$Q_G(b, x^t) \leq \frac{s^2}{2} + \frac{t}{4}(s^2 + k^2) - \frac{1+t}{2}k \tag{9}$$

with equality if and only if $G(b) = K_{n_+, n_-}$.

We first consider the k-gonal inequalities of negative type, ie. the situation when $s = 0$, k is even, and the right hand side of (1) is zero.

Lemma 3. *Let $n \geq 2m \geq 6$. Set $x^t = x^t(G)$, where $t = \frac{m^2+1}{m^2(m-1)}$. Then*

(a) x^t satisfies all k-gonal negative type inequalities with $k < 2m$.

(b) x^t satisfies all $2m$-gonal negative type inequalities, except when G contains $K_{m,m}$ as an induced subgraph. In this case only the pure $2m$-gonal negative type inequality is violated.

(c) $\frac{x^t}{n^2}$ satisfies all k-gonal inequalities with $k \leq n$ and $s \geq 2$.

Proof. If $k < 2m$, then it is easy to verify that $k \leq \frac{2}{t}$, hence $\frac{k^2 t}{4} \leq \frac{k}{2}$. Using Lemma 2 with $s = 0$, we see that $Q_G(b, x^t) \leq 0$, proving (a).

For part (b), first consider the case where $G = G(b) = K_{m,m}$, $k = 2m$, $s = 0$, and b is a pure $2m$-gonal inequality. From Lemma 2 we have that

$$Q_G(b, x^t) = m^2 t - mt - m = (m^2 - m)\frac{m^2 + 1}{m^2(m-1)} - m = \frac{1}{m} > 0. \qquad (10)$$

Therefore this pure negative type inequality is violated by x^t. It is easy to check that $t > \frac{1}{m}$. If G is not $K_{m,m}$, then $|E(b)| \geq 1$, and as we saw in the proof of Lemma 1, $Q_G(b, x^t)$ will be reduced by at least t so it becomes negative. Therefore the inequality holds in this case.

Now let $G = K_{m,m}$, and assume b is $2m$-gonal but not pure. In this case $|b_i| \geq 2$ for some i, and so $\sum_{i=1}^n b_i^2 \geq 2m + 2$. Now from (7) with $s = 0$ and $k = 2m$ we obtain

$$Q_G(b, x^t) \leq m^2 t - \frac{1+t}{2}(2m+2) = m^2 t - mt - m - t - 1 < 0.$$

Therefore x^t satisfies all $2m$-gonal negative type inequalities that are not pure.

For part (c), we observe that when $k \leq n$

$$\sum_{1 \leq i < j \leq n} |b_i||b_j| \leq \frac{1}{2}(\sum_{i=1}^n |b_i|)^2 = \frac{k^2}{2} \leq \frac{n^2}{2},$$

Now $s \geq 2$, and since $m \geq 3$, $x^t \leq 2$, so

$$Q_G(b, \frac{x^t}{n^2}) = \sum_{1 \leq i < j \leq n} b_i b_j \frac{x_{ij}^t}{n^2} \leq \sum_{1 \leq i < j \leq n} |b_i||b_j|\frac{2}{n^2} \leq 1 \leq \lfloor \frac{s^2}{4} \rfloor.$$

This completes the proof of the lemma. ∎

The following similar lemma applies to hypermetric inequalities and parts (a) and (b) appeared in [1] (see also [5], Section 28.3.) Part (c) can be proved as in Lemma 3 (c).

Lemma 4. *Let* $n \geq 2m + 1 \geq 5$. *Set* $x^t = x^t(G)$, *where* $t = \frac{m^2+1}{m^3}$. *Then*
  (a) x^t *satisfies all k-gonal hypermetric inequalities with* $k < 2m + 1$.
  (b) x^t *satisfies all $(2m + 1)$-gonal hypermetric inequalities, except when G contains $K_{m,m+1}$ as an induced subgraph. In this case only the pure $(2m + 1)$-gonal negative type inequality is violated.*
  (c) $\frac{x^t}{n^2}$ *satisfies all k-gonal inequalities with* $k \leq n$ *and* $s \geq 2$. ∎

We now turn to gap inequalities. Recall $g = g(b)$ defined in (2) is the gap of an integer vector b.

Lemma 5. *Let* b, k *and* s *be defined as in (5), such that* $g = s \geq 2$. *Then*

$$\sum_{i=1}^n b_i^2 \geq k + s^2$$

Proof. Assume $b_1 \geq b_2$ are the two largest integers in b. We may also assume without loss in generality that no integer b_i is zero. Since $g = s$ we observe that the set $S = \{1, 2, ..., n\}$ realizes the minimum gap. Assume first that $b_2 > 0$. It is easily seen that $b_1 \geq b_2 \geq g = s \geq 2$, since otherwise a smaller gap could be formed by removing b_2 from S. Therefore

$$\sum_{i=1}^{n} b_i^2 \geq b_1^2 + b_2^2 + \sum_{i=3}^{n} |b_i| = b_1^2 + b_2^2 + k - b_1 - b_2 \geq 2s^2 - 2s + k \geq k + s^2.$$

Otherwise, suppose $b_2 < 0$. From (1), $b_1 = (k + s)/2$. Furthermore, since b has positive and negative components, $k = s + u$, for some $u \geq 2$. Therefore

$$\sum_{i=1}^{n} b_i^2 \geq \frac{(k+s)^2}{4} + \frac{k-s}{2} \geq \frac{(2s+u)^2}{4} + \frac{u}{2} = s^2 + su + \frac{u^2}{4} + \frac{u}{2} \geq s^2 + s + u = k + s^2.$$

∎

3 Proofs of Main Results

In the proofs that follow, we can and do assume that $m \geq 3$.

Proof of P1N.
For a graph G on n vertices, set t as in Lemma 3, and define $x^t(G)$ as in (4). $x^t(G)$ satisfies all $2m$-gonal inequalities if and only if G does not contain $K_{m,m}$ as an induced subgraph. Hence P1N is co-NP complete. Part (b) of Lemma 3 shows that the problem remains co-NP complete even when restricted to testing whether x^t satisfies all pure $2m$-gonal negative type inequalities. ∎

Proof of P2N.
Let G be a graph with n vertices, and let m be the largest integer such that G contains $K_{m,m}$ as an induced subgraph. For each $u = \lfloor n/2 \rfloor, \lfloor n/2 \rfloor - 1, ..., 2, 1$ we ask a question of type P2N. Set $t = \frac{u^2+1}{u^2(u-1)}$ and let $x^u(G)$ be defined as in (4). Consider the answers given to question P2N. If $u > m$, G does not contain a $K_{u,u}$, so by the lemma either the answer to P2N is x^u is of negative type, or x^u is not of negative type but violates a $2p$-gonal negative type inequality for $p > u$. When $u = m$, G contains an induced $K_{u,u}$. By Lemma 3 the answer to P2N must be that x^u is not hypermetric and the minimum hypermetric inequality violated is $2u$-gonal. This answer to P2N gives us the value $m = u$ that answers P4. Therefore P2N is NP-hard. ∎

Proof of P1K
Suppose first $k = 2m$ for some integer m. For a graph G on n vertices, set t as in Lemma 3, and define $x = \frac{x^t(G)}{n^2}$. Since the negative type inequalities form a cone, parts (a) and (b) of the lemma apply to x also. Therefore x satisfies all $2m$-gonal negative type inequalities with except if G contains an induced $K_{m,m}$, in which case it violates only the pure $2m$-gonal inequality.

Now consider any k-gonal inequality that is not of negative type. Since k is even, s is even also, and so we can apply part (c) of Lemma 3. This shows that

x satisfies this k-gonal inequality also. Hence the answer to P1K can be used to determine if G contains an induced $K_{m,m}$.

Next suppose $k = 2m + 1$ for some integer m. We now make use of Lemma 4. Arguing as above, we show that P1K can be used to determine if G contains an induced $K_{m,m+1}$. It follows that P1K is co-NP complete. ∎

Proof of P2K

Let G be a graph with n vertices, and let m be the largest integer such that G has $K_{\lfloor \frac{m}{2} \rfloor, \lceil \frac{m}{2} \rceil}$ as an induced subgraph. For $u = n, n-1, ..$, we ask a question of type P1K for the vector $x = \frac{x^t(G)}{n^2}$, where $x^t(G)$ is defined in Lemma 3 if u is even or Lemma 4 if u is odd. If $u > m$, G does not contain an induced $K_{\lfloor \frac{m}{2} \rfloor, \lceil \frac{m}{2} \rceil}$ and so, using the appropriate lemma, we see that the answer is either that x satisfies all k-gonal inequalities or that x violates a k-gonal inequality for $k > u$. When $u = m$, we get the answer that the smallest k-gonal inequality violated by x is u-gonal. Therefore we have determined m and P2K is NP-hard. ∎

Proof of P1G

We proceed as in the proof of P1K. Suppose first $k = 2m$ for some integer m. For a graph G on n vertices, set t as in Lemma 3, and define $x = \frac{x^t(G)}{n^2}$. As before, we see that x satisfies all $2m$-gonal negative type inequalities, except if G contains an induced $K_{m,m}$, in which case it violates only the pure $2m$-gonal inequality.

We show that x satisfies all other $2m$-gonal gap inequalities. First assume the gap $g = 0$. In this case the gap inequality coincides with the $2m$-gonal inequality (1). We proved in P1K that x satisfies all such inequalities. Next we assume that $g = s \geq 2$. From Lemmas 1 and 5 we have

$$Q_G(b, x^t) \leq \frac{s^2}{2} + \frac{t}{4}(s^2 + k^2) - \frac{1+t}{2}(k + s^2) = \frac{k^2 t}{4} - \frac{k}{2} - \frac{tk}{2} - \frac{s^2 t}{4}.$$

$$= m^2 t - mt - m - \frac{s^2 t}{4} = \frac{1}{m} - \frac{s^2 t}{4} < 0.$$

In the second line we used (10), $s \geq 2$ and $t > 1/m$. By homogeneity it follows that $x = \frac{x^t(G)}{n^2}$ also satisfies all gap inequalities with $g = s \geq 2$.

It remains to check that x satisfies all gap inequalities with $s > g$. Since s and g have the same parity, we have $s \geq g + 2$ and so $\frac{1}{4}(s^2 - g^2) \geq 1$. But the proof of Lemma 3(c) showed that

$$Q_G(b, \frac{x^t}{n^2}) \leq 1,$$

so x satisfies these inequalities also.

The proof for $k = 2m + 1$ is similar and omitted. ∎

Proof of P2G

Let G be a graph with n vertices, and let m be the largest integer such that G has $K_{\lfloor \frac{m}{2} \rfloor, \lceil \frac{m}{2} \rceil}$ as an induced subgraph. We proceed as in the proof of P2K to use answers to questions of type P2G to find m. ∎

Acknowledgements

I would like to thank an anonymous referee for remarks leading to several clarifications of the original manuscript.

References

1. Avis, D., and Grishukhin, V.P.: *A Bound on the k-gonality of Facets of the Hypermetric Cone and Related Complexity Problems*, Computational Geometry: Theory and Applications 2 (1993) 241-254.
2. Avis, D., and Umemoto, J.: *Stronger Linear Programming Relaxations of Max-Cut*, Les cahiers du GERAD G-2002-48, September, 2002.
3. Deza, M. (Tylkin, M.E.): *Realizablility of Distance Matrices in Unit Cubes(in Russian)*, Problemy Kybernetiki 7 (1962) 31-42.
4. Deza, M., Grishukhin, V.P., and Laurent, M.: *The Hypermetric Cone is Polyhedral*, Combinatorica 13 (1993) 397-411.
5. Deza, M., and Laurent, M.: *Geometry of Cuts and Metrics*, Springer, 1997.
6. Garey M.R., Johnson D.S., *Computers and Intractability*, W.H.Freeman and C^o 1979.
7. Goemans, M., and Williamson, D.: *0.878-Approximation Algorithms for MAX CUT and MAX 2SAT*, Proc 26th STOC, (1994) 422-431.
8. Helmberg, C., and Rendl, F.: *Solving Quadratic (0,1)-Problems by Semidefinite Programs and Cutting Planes*, Math. Prog. 82 (1998) 291-315.
9. Laurent, M., and Poljak, S.: *Gap Inequalities for the Gap Polytope*, Europ. J. Combinatorics 17 (1996) 233-254.
10. Reed, B.: *A Gentle Introduction to Semi-Definite Programming*, in Pefect Graphs, ed. J. Alfonsín and B. Reed, Wiley, (2001).

On Partitioning a Cake

Sergei Bespamyatnikh

Department of Computer Science, University of Texas at Dallas
Box 830688, Richardson, TX 75083, USA
besp@utdallas.edu

Abstract. We investigate balanced α-partitions of a generalized cake which has two masses, one distributed in a convex set and the other distributed on its boundary. We characterize all ratio vectors α such that any cake has a perfect α-partition. We also consider convex α-partitions of a cake into three pieces. It is known [4,5,9] that any two masses have convex α-partition for $\alpha = (1/3, 1/3, 1/3)$. We prove the existence of convex α-partitions for any $\alpha = (a, a, 1 - 2a), 0 < a < 1/3$. We also provide an infinite family of α each of which does not guarantee a convex α-partition of a cake.

1 Introduction

A generalization of the Ham Sandwich Theorem in the plane states that, for any kn red and km blue points ($k, n, m > 0$ are integers) in the plane in general position, there are k disjoint convex polygons with n red and m blue points in each of them [3,4,5,6,9]. Kaneko and Kano [7] considered $(n_1, n_2, \ldots, n_k)$-balanced partitions.

Conjecture 1 ([7,8]). Let $a \geq 1, b \geq 1, k \geq 3$ be integers and $n = n_1 + \cdots + n_k$ be an integer-partition such that $1 \leq n_i \leq n/3$ for every $1 \leq i \leq k$. For any an red points and bn blue points in the plane in general position, there are k disjoint convex polygons such that i-th polygon contains exactly an_i red and bn_i blue points.

They proved Conjecture 1 for two cases: (i) if $a = 1$ and $n \leq 8$, or (ii) $a = 1, n_1 = \cdots = n_{k-1} = 2, n_k = 1$.

Akiyama *et al.* [2] first introduced the problem of dividing a cake: "how to divide a cake among the children attending a birthday party in such a way that each child gets the same amount of cake and (perhaps more important to them) the same amount of icing". Assuming that the height of the cake is constant the problem can be stated as follows. Let $\delta(A)$ denote the boundary of a convex set A and $\mathrm{Int}(A)$ denote its interior. Let S be a bounded convex set in the plane which corresponds to the base of the cake. For a convex set $A \subseteq S$, let $area(A)$ and $length(A)$ denote the area of A and the length of $\delta(S) \cap A$. Cutting the cake can be viewed as the partitioning S into n parts $S_1, S_2, \ldots, S_n$. The amount of cake in ith piece can be measured by its area $area(S_i)$. A partition of S into n convex subsets $S_1, \ldots, S_n$ is *perfect* if each subset has the same area

J. Akiyama and M. Kano (Eds.): JCDCG 2002, LNCS 2866, pp. 60–71, 2003.

$area(S_i) = area(S)/n$ and has the exactly one connected part of $\delta(S)$ with the same length $length(S_i) = length(S)/n$.

Akiyama *et al.* [1] proved that every convex set S has a perfect n-partition for any $n \geq 3$. Very recently Kaneko and Kano [8] generalized perfect partitions and proved their existence.

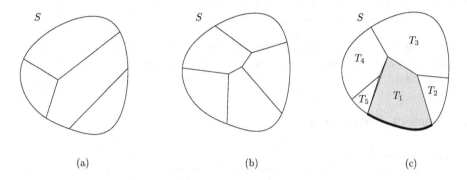

Fig. 1. (a) non-perfect partition, (b) perfect 5-partition, and (c) generalized perfect partition.

Theorem 1 ([8]). *Let S be a convex set in the plane, let $n \geq 2$ be an integer, and let $\alpha_1, \ldots, \alpha_n$ be positive real numbers such that $\alpha_1 + \cdots + \alpha_n = 1$ and $0 < \alpha_i \leq 1/2$ for all $1 \leq i \leq n$. Then S can be partitioned into n convex sets $T_1, \ldots, T_n$ so that each T_i satisfies the following three conditions: (i) $area(T_i) = \alpha_i \times area(S)$; (ii) $length(T_i) = \alpha_i \times length(S)$; and (iii) $T_i \cap \delta(S)$ consists of exactly one continuous curve (see Fig. 1 (b)).*

Theorem 1 can be viewed as affirmative answer to a special case of Conjecture 1[1] where the red points are uniformly distributed on the boundary of a convex set S and the blue points are uniformly distributed on S (by a unifom distribution on S we mean a set B of points satisfying property that (i) $d_{\max}/d_{\min} = O(1)$ where $d_{\max}$ ($d_{\min}$) is the largest (smallest) distance from a point of B to its nearest neighbor in B, and (ii) $|B|$ is sufficiently large). In this paper we further generalize the perfect partitions and define a *cake* as a convex bounded set S with two mass distributions μ_1, μ_2 with domains S and $\delta(S)$ with $\mu_1(S) = \mu_2(S) = 1$. Consider a partition of S into n subsets $P = \{S_1, \ldots, S_n\}$ such that (i) $\cup_{1 \leq i \leq n} S_i = S$ and (ii) $Int(S_i) \cap Int(S_j) = \emptyset$ for any $i \neq j$. P is called *convex* if every S_i is convex. Let $\alpha = (\alpha_1, \ldots, \alpha_n)$ and $\alpha_i \geq 0$, $\sum_{1 \leq i \leq n} \alpha_i = 1$. P is called α-partition if it is convex and $\mu_1(S_i) = \mu_2(S_i) = \alpha_i$. P is called *perfect* if $S_i \cap \delta(S)$ is a continuous curve. We characterize all tuples α such that all masses admit perfect α-partitions.

---

[1] A continuous version of Conjecture 1.

Lemma 1. *Let $n \geq 2$ be an integer, and let $\alpha_1, \ldots, \alpha_n$ be positive real numbers such that $\alpha_1 + \cdots + \alpha_n = 1$ and $0 < \alpha_i \leq 1$ for all $1 \leq i \leq n$. Every cake has a perfect α-partition if and only if $n = 2$ and $\alpha = (1/2, 1/2)$.*

This can be viewed as a negative result since a $(1/2, 1/2)$-partition is the well-known ham sandwich cut. We prove a positive result for convex partitions.

Theorem 2. *Let α_1 be a real number in $[1/4, 1/3)$ and $\alpha_2 = \alpha_1$ and $\alpha_3 = 1 - 2\alpha_1$. Every cake has a convex α-partition.*

Theorem 2 does not cover all possible triples α. We show that for some triples α convex α-partition cannot be guaranteed. For example, there is a cake that does not admit a convex $(0.2532, 0.3734, 0.3734)$-partition.

2   Perfect Partitions

Proof of Lemma 1.

(i) $n = 2$. If $\alpha = (1/2, 1/2)$ one can apply ham sandwich cut to get a perfect α-partition. Suppose that $\alpha \neq (1/2, 1/2)$. Let S be a unit disk. The mass μ_2 is uniform distribution on $\delta(S)$. The mass μ_1 is concentrated in the center of S, see Fig. 2 (a).

(ii) $n \geq 3$. Let

$$A = \{\alpha_{i_1} + \cdots + \alpha_{i_m} \mid \{i_1, i_2, \ldots\} \subseteq \{1, 2, \ldots, n\}\}$$

be a set of all partial sums of elements of α. Let $m \geq 3$ be an integer such that $A \cap \frac{1}{m}\mathbb{Z} = \emptyset$. One can take a large prime number as m since A is finite. Let S be the unit disk. μ_2 is uniform distribution on $\delta(S)$. μ_1 is concentrated at the vertices of a regular m-gon Q inscribed in S such that the weight of each vertex is $1/m$.

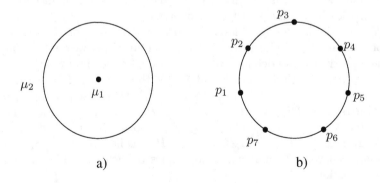

a)   b)

Fig. 2. a) No perfect $(\alpha_1, 1 - \alpha_1)$-partition for $\alpha_1 \in (0, 1/2)$. b) $m = 7$. No perfect α-partition if, say $\alpha = (0.1, 0.3, 0.6)$.

We show that there is no perfect α-partition of S by contradiction. Suppose that $P = \{S_1, \ldots, S_n\}$ is a perfect partition. The unit circle $\delta(S)$ is divided into n arcs of lengths $2\pi\alpha_1, \ldots, 2\pi\alpha_n$. Let P' be the set of endpoints of the arcs. By the choice of m, P' can contain at most one vertex of Q. Since $n \geq 3$ there is an arc a whose endpoints are not vertices of Q. Thus the weight $\mu_1(a) \in \mathbb{Z}/m$. This contradicts the choice of m.

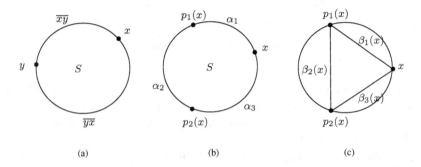

(a)   (b)   (c)

Fig. 3. Notation (a) path $\overline{xy}$, (b) points $p_i(x)$, (c) triple $\beta = (\beta_1, \beta_2, \beta_3)$.

3   Convex Partitions

Let $\mathcal{C} = \delta(S)$. For $x, y \in \mathcal{C}$, let $\overline{xy}$ denote the path along $\mathcal{C}$ from x to y in counterclockwise order, see Fig. 3(a). Let $\mu_2(x, y) = \mu_2(\overline{xy})$. For any point $x \in \mathcal{C}$ and a real number $r \in [0, 1]$, there is a unique point $y \in \mathcal{C}$ such that $\mu_2(x, y) = r$. Let $p_1(x)$ be the point y such that $\mu_2(x, y) = \alpha_1$. Let $p_2(x)$ be the point z such that $\mu_2(x, z) = \alpha_1 + \alpha_2$, see Fig. 3(b). Let $A(x, y)$ denote the part of S cut off by the line xy such that the arc $\overline{xy}$ lies in $A(x, y)$. Let $\mu_1(x, y)$ denote the weight of mass μ_1 in the area $A(x, y)$, i.e. $\mu_1(x, y) = \mu_1(A(x, y))$. We define the map $\beta : \mathcal{C} \to \mathbb{R}^3$ by $\beta(x) = (\beta_1, \beta_2, \beta_3)$ where

$$\beta_1(x) = \mu_1(x, p_1(x)), \beta_2(x) = \mu_1(p_1(x), p_2(x)), \beta_3(x) = \mu_1(p_2(x), x),$$

see Fig. 3(c). Clearly,

$$0 \leq \beta_1(x) + \beta_2(x) + \beta_3(x) \leq 1. \tag{1}$$

Lemma 2. *Let x be a point of $\mathcal{C}$ such that $\beta_1(x) \leq \alpha_1$, $\beta_2(x) \leq \alpha_2$, $\beta_3(x) \leq \alpha_3$. Then there exists a point p in the triangle $xp_1(x)p_2(x)$ such that the 3-cutting with the center at p and rays $px, pp_1(x)$ and $pp_2(x)$ is a perfect α-partition of S.*

Proof. For every point q in the triangle $xp_1(x)p_2(x)$ we construct 3-cutting using rays $qx, qp_1(x)$ and $qp_2(x)$. If q is an interior point of the triangle $xp_1(x)p_2(x)$ then 3-cutting is convex. Let $\delta_1(q)$ be μ_1-weight of the part of S cut off by the wedge $xqp_1(x)$, see Fig. 4 Similarly, we define two weights $\delta_2(q)$ and $\delta_3(q)$ for

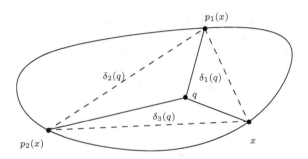

Fig. 4. Perfect α-partition.

other wedges. Let r be a point of the segment $p_1(x)p_2(x)$. When q moves from x to r the weight $\delta_2(q)$ changes from $\delta_2(q) = 1 - \beta_1(x) - \beta_3(x) \geq 1 - \alpha_1 - \alpha_3 = \alpha_2$ to $\delta_2(r) = \beta_2(x) \leq \alpha_2$. There is an interval $[r_1, r_2]$ which is a subsegment of $[x, r]$ such that, for every point $r' \in [r_1, r_2]$, $\beta_2(r') = \alpha_2$.

As we move r from $p_1(x)$ to $p_2(x)$ the intervals $[r_1, r_2]$ form a connected region R. There are two points $p_1' \in [x, p_1(x)]$ and $p_2' \in [x, p_2(x)]$ that lie in the region R. The points p_1' and p_2' are connected in R by a Jordan curve π. Note that $\delta_1(p_1') = \beta_1(x) \leq \alpha_1$ and $\delta_1(p_2') \geq \alpha_1$. According to the intermediate value theorem, there is a point q in the curve π such that $\delta_1(q) = \alpha_1$. Then $\delta_3(q) = 1 - \delta_1(q) - \delta_2(q) = 1 - \alpha_1 - \alpha_2 = \alpha_3$. The lemma follows.

A path $\pi_1 \subseteq \mathcal{C}$ *covers* a path $\pi_2 \subseteq \mathcal{C}$ if $\pi_2 \subseteq \pi_1$. Let $P_i = \{\overline{x_i y_i}\}, i = 1, 2, 3$ be the set of paths $\overline{x_i y_i} \subseteq \mathcal{C}$ such that the line $x_i y_i$ is a perfect $(\alpha_i, 1 - \alpha_i)$-partition of S and $\mu_1(x_i, y_i) = \alpha_1$.

Lemma 3. *Suppose that two masses $\mu_1 : S \to \mathbb{R}$ and $\mu_2 : \delta(S) \to \mathbb{R}$ have three $(\gamma, 1 - \gamma)$-perfect partitions for every $\gamma \in \{\alpha_1, \alpha_2, \alpha_3\}$. Then at least one of the following conditions holds*

(i) *there is a convex α-partition of S, or*
(ii) *there is a path π and an index $i \in \{1, 2, 3\}$ such that every path $\pi_i \in P_i$ intersects π and every path $\pi_j \in P_j, j \neq i$ covers π.*

Proof. Without loss of generality we assume that $\alpha_1 \leq \alpha_2 \leq \alpha_3$. We also assume that P_1, P_2 and P_3 satisfy some properties.

Intersection Property. Any two paths $\overline{x_i y_i} \in P_i$ and $\overline{x_j y_j} \in P_j, i \neq j$ intersect properly (their interiors intersect), otherwise the line segments $x_i y_i$ and $x_j y_j$ form the convex α-partition of S.

Cover Property. Any three paths $\pi_1 \in P_1, \pi_2 \in P_2$ and $\pi_3 \in P_3$ do not cover $\mathcal{C}$, i.e. $\mathcal{C} \not\subseteq \pi_1 \cup \pi_2 \cup \pi_3$. Suppose to the contrary that three paths π_1, π_2 and π_3 cover $\mathcal{C}$. Note that any pair π_i and $\pi_j, i \neq j$ do not cover $\mathcal{C}$, otherwise $\mu_2(\mathcal{C}) \leq \mu_2(\pi_i) + \mu_2(p_j) < 1$. Let $\pi_i = \overline{x_i y_i}$, see Fig. 5 a). By symmetry we can assume that the points x_1, x_2 and x_3 are in clockwise order. The point x_1 satisfies the conditions of Lemma 2 (we can assume that $p_1(x_1) = x_2$ and $p_2(x_1) = x_3$) and S has a perfect α-partition.

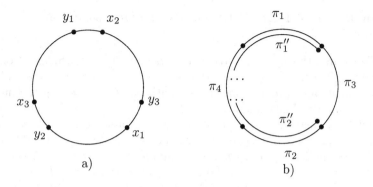

Fig. 5. a) Cover property. b) Paths $\pi_1, \ldots, \pi_4$.

The set P_1 (and P_2) satisfies a property that the intersection of two paths of P_1 is either empty set or a path (since $\alpha_1 < 1/2$).

Suppose that there are two disjoint paths $\pi_1, \pi_2 \in P_1$, i.e. $\pi_1 \cap \pi_2 = \emptyset$. The set $\mathcal{C} \setminus (\pi_1 \cup \pi_2)$ consists of two paths π_3 and π_4 (without endpoints), see Fig. 5 b). Every path of $P_2 \cup P_3$ intersects π_1 and π_2. Therefore it covers either π_3 or π_4. We show that all the paths of $P_2 \cup P_3$ cover the same path, either π_3 or π_4. It suffices to show that any two paths $\pi_5 \in P_2$ and $\pi_6 \in P_3$ cover the same path. Suppose to the contrary that there are two paths π_5 and π_6 that cover π_3 and π_4 respectively. Since $\pi_5 \cap \pi_6 \neq \emptyset$, the paths π_5, π_6 and π_1 cover $\mathcal{C}$ or the paths π_5, π_6 and π_2 cover $\mathcal{C}$. Contradiction. This implies that the intersection $\cap_{\pi_a \in P_2 \cup P_3} \pi_a$ is a non-empty path, say π. We show that every path of P_1 intersects π. The proof is similar to the argument above. If we assume that $\pi \cap \pi_7 = \emptyset$ for some $\pi_7 \in P_1$, then the paths of $P_2 \cup P_3$ cover one of the paths of $\mathcal{C} \setminus (\pi \cup \pi_7)$ and this contradicts the definition of π. The lemma follows for $i = 1$.

It remains to consider the case where $\cap_{\pi_a \in P_1} \pi_a \neq \emptyset$ (the case where three paths from P_1 cover $\mathcal{C}$ is impossible since $\alpha_1 \leq 1/3$). Similar argument allows us to assume that $\cap_{\pi_a \in P_2} \pi_a \neq \emptyset$ (the lemma follows for $i = 2$. The case where three paths π_a, π_b, π_c from P_2 cover $\mathcal{C}$ is impossible since any path π_d from P_1 (i) intersects each path π_a, π_b and π_c, and thus (ii) π_d and two paths from $\{\pi_a, \pi_b, \pi_c\}$ cover $\mathcal{C}$ but $\alpha_1 + 2\alpha_2 \leq 1$. The equality happens when $\alpha_1 = \alpha_2 = \alpha_3 = 1/3$ and $P_1 = P_2$.) Let $\pi_1 = \cap_{\pi_a \in P_1} \pi_a$ and $\pi_2 = \cap_{\pi_a \in P_2} \pi_a$. Suppose that π_1 and π_2 are disjoint. Let π_3 and π_4 be two paths of $\mathcal{C} \setminus (\pi_1 \cup \pi_2)$. Let π_1' be a path of P_1 and let π_2' be a path of P_2. Without loss of generality $\pi_1' \cup \pi_2'$ covers π_3. Every path $\pi \in P_3$ covers π_3 (if $\pi \in P_3$ covers π_4 then π, π_1', and π_2' cover $\mathcal{C}$). Therefore every pair of paths from P_1 and P_2 cover π_3. Let $\pi_1'' \in P_1$ and $\pi_2'' \in P_2$ be two paths whose endpoints are endpoints of π_3, see Fig. 5 b). They cover π_3 and π_4. Contradiction (π_1'' and π_2 cover $\mathcal{C}$). Therefore π_1 and π_2 intersect.

Let $\pi = \pi_1 \cap \pi_2$. The path π is the intersection of two arcs $\pi_a, \pi_b \in P_1 \cup P_2$ since $\pi = \cap_{\pi_c \in P_1 \cup P_2} \pi_c$. Let π_3 be a path of P_3. We have $\pi_3 \cap \pi_a \neq \emptyset$ and $\pi_3 \cap \pi_b \neq \emptyset$. The paths π and π_3 intersect since π, π_a and π_b cover $\mathcal{C}$ otherwise. The lemma follows for $i = 3$.

Theorem 3. *If $\alpha_k \leq 1/2, k \in \{1,2,3\}$ then there is a point x on $\mathcal{C}$ such that $\beta_k(x) \leq \alpha_k$.*

Proof. We assume that $k = 1$. If $\alpha_1 = 1/2$ then the theorem follows from the Ham Sandwich Theorem. We assume that $\alpha_1 < 1/2$. We prove a discrete version of the statement by approximating masses with points and letting the number of points to infinity. Each mass $\mu_i, i = 1,2$ is represented as a set of points and is uniformly distributed among its points. We assume that μ_2 contains n points $p_0, \ldots, p_{n-1}$ on $\mathcal{C}$ and μ_1 contains N points $q_0, \ldots, q_{N-1}$ inside $\mathcal{C}$. We also assume that the indices of points p_i are modulo n. Let $m = \lceil \alpha_1 n \rceil$. We choose n sufficiently large so that $m < n/2$. This is possible since $\alpha_1 < 1/2$.

Let $\bar{\rho} = \sum_{0 \leq i < n} \rho_1(i)$ where $\rho_1(i)$ is the number of μ_1-points cut off by $p_i p_{i+m}$, $\rho_1(i) = N\mu_1(p_i, p_{i+m})$. For $i = 0, \ldots, n-1$ and $j = 0, \ldots, N-1$ we define

$$\chi(i,j) = \begin{cases} 1 & \text{if } q_j \in A(p_i, p_{i+m}) \\ 0 & \text{otherwise.} \end{cases}$$

Then $\rho_1(i) = \sum_{j=0}^{N-1} \chi(i,j)$ and

$$\bar{\rho} = \sum_{i=0}^{n-1} \rho_1(i) = \sum_{i=0}^{n-1} \sum_{j=0}^{N-1} \chi(i,j) = \sum_{j=0}^{N-1} \sum_{i=0}^{n-1} \chi(i,j).$$

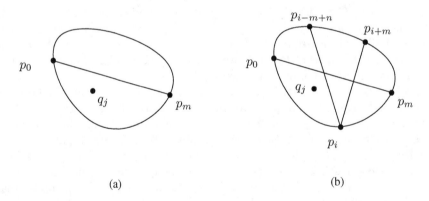

(a)   (b)

Fig. 6. a) $\chi(0,j) = 1$. b) Two lunes $A(p_i, p_{i+m})$ and $A(p_{n-m+i}, p_i)$.

Let $\rho_2(j) = \sum_{0 \leq i < n} \chi(i,j), 0 \leq j < N$. We want to find an upper bound for all $\rho_2(j)$. Let us fix j. We can assume that $\rho_2(j) > 0$; otherwise $\rho_2(j)$ has any positive number as an upper bound. We also assume that $\chi(0,j) = 1$, see Fig 6 (a). We count the indices i such that $\chi(i,j) = 1$. If $i \in \{m, m+1, \ldots, n-m\}$ (clearly $m < n-m$ since $m < n/2$) then $q_j \notin A(p_i, p_{i+m})$ and therefore $\chi(i,j) = 0$. The remaining indices can be broken into two groups $G_1 = \{0, 1, \ldots, m-1\}$ and $G_2 = \{n-m+1, n-m+2, \ldots, n-1\}$. These indices except $i = 0$ can be paired $(1, n-m+1), (2, n-m+2), \ldots, (m-1, n-1)$. Each pair can be written as $(i, n-m+i), i = 1, \ldots, m-1$. We show that $\chi(i, j) + \chi(n-m+i, j) \le 1$. Indeed two lunes $A(p_i, p_{i+m})$ and $A(p_{n-m+i}, p_i)$ intersect by the point p_i only, see Fig 6 (b). Therefore one of the lunes does not contain q_j.

The total number of pairs is $m-1$. Then $\rho_2(j) \le m$ since $\chi(0, j) = 1$. Therefore $\bar{\rho} = \sum_{j=0}^{N-1} \rho_2(j) \le Nm$. There exists an index i such that $\rho_1(i) \le Nm/n$ and $\mu_1(p_i, p_{i+m}) = \rho_1(i)/N \le m/n \le \alpha_1$. Let n and N go to the infinity. Among all the paths $\overline{p_i p_{i+m}}$ there is a sequence that tends to a path $\pi \in \mathcal{C}$ since $\mathcal{C}$ is compact. Then $\mu_1(\pi) \le \alpha_1$ and $\mu_2(\pi) = \alpha_1$ since $m/n = \lceil \alpha_1 n \rceil / n \to \infty$. The theorem follows.

Theorem 4. *If there are three $(\gamma, 1-\gamma)$-perfect partitions of S for every $\gamma \in \{\alpha_1, \alpha_2, \alpha_3\}$ then there is a convex α-partition of S.*

Proof. Assume that there is no convex α-partition of S. By Lemma 3 there is a path π and an index i such that every path $\pi_i \in P_i$ intersects π and every path $\pi_j \in P_j, j \ne i$ covers π. We assume without loss of generality that $i = 3$. Let z_1 and z_2 be the endpoints of π and $\pi = \overline{z_1 z_2}$. Let I_1 be the interval $\overline{z_1 z_1'} \setminus \{z_1, z_1'\}$ where $\mu_1(z_1', z_2) = \alpha_1$, see Fig. 7. We claim that either $\beta_1(x) < \alpha_1$ for all $x \in I_1$ or $\beta_1(x) > \alpha_1$ for all $x \in I_1$. Suppose to the contrary that there are two points $x, y \in I_1$ such that $\beta_1(x) > \alpha_1$ and $\beta_1(y) < \alpha_1$. Then there exists a point $t \in \overline{xy}$ such that $\beta_1(t) = \alpha_1$. Since $x \ne y$ and t is in between, it should be in I_1. On the other hand, $tp_1(t)$, being a perfect $(\alpha_1, 1-\alpha_1)$-partition of S, should contain whole π. This is impossible.

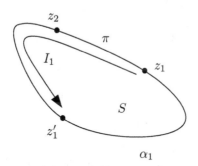

Fig. 7. Paths π and I_1.

We define I_2 similarly to I_1. We define I_3 as the interval $\overline{z_2 z_2'} \setminus \{z_2, z_2'\}$ where $\mu_1(z_2', z_2) = \alpha_3$. For each $i \in \{1, 2, 3\}$, there is a sign $\nabla_i \in \{<, >\}$ such that $\beta_i(t) \nabla_i \alpha_i$ for all $t \in I_i$. Clearly, there are two equal signs among three signs ∇_1, ∇_2 and ∇_3.

Suppose that $\nabla_2 = \nabla_3$. Let $\overline{xy}$ be a path of P_1. We define the triangle Δ with vertices $x, p_1(x)$ and $p_2(x)$. Then $p_1(x) \in I_2$ and $p_2(x) \in I_3$. The sign ∇_2 is $" <"$ otherwise

$$\mu_1(\Delta) = \mu_1(S) - \beta_1(x) - \beta_2(x) - \beta_3(x) =$$

$$1 - \alpha_1 - \beta_2(x) - \beta_3(x) < 1 - \alpha_1 - \alpha_2 - \alpha_3 = 0.$$

By Lemma 2 there is a perfect α-partition of S.

By symmetry it remains to consider the case $\nabla_1 = \nabla_2$. Let $\overline{xy}$ be a path of P_3. We define the triangle Δ with vertices $y, p_1(y)$ and $p_2(y)$. Then $y \in I_1$ and $p_1(y) \in I_2$. The sign ∇_1 is " $<$" by an argument as above. By Lemma 2 there is a perfect α-partition of S.

Lemmas 2, 3 and Theorems 3, 4 establish properties of α-partitions for any α. We use them to prove Theorem 2.

Proof. (Theorem 2). Note that $\alpha_3 = 1 - 2\alpha_1 \le 1/2$ since $\alpha_1 \ge 1/4$. Suppose to the contrary that there is a cake S which has no convex α-partition. By Theorem 4 S has no $(\gamma, 1 - \gamma)$-partition for $\gamma = \alpha_1$ or $\gamma = \alpha_3$. If S does not admit both partitions then by Theorem 3, for any $x \in \mathcal{C}$, $\beta_1(x) < \alpha_1, \beta_2(x) < \alpha_2$ and $\beta_3(x) < \alpha_3$. Then there is a perfect α-partition by Lemma 2. This implies that there are two cases.

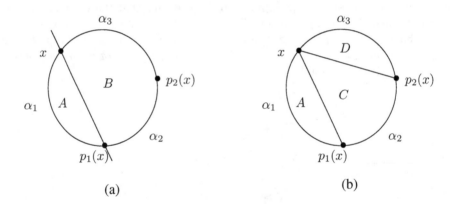

Fig. 8. Case 1.

Case 1. S has a $(\alpha_1, 1 - \alpha_1)$-partition but has no $(\alpha_3, 1 - \alpha_3)$-partition. Let $xp_1(x)$ be a $(\alpha_1, 1 - \alpha_1)$-partition of S into A and B such that $\mu_1(A) = \alpha_1$ and $\mu_1(B) = \alpha_2 + \alpha_3$, see Fig. 8 (a). Let C be the region bounded by the segments $xp_1(x)$, $xp_2(x)$ and the curve $\overline{p_1(x)p_2(x)}$, i.e. $C = B \setminus A(p_2(x)x)$, see Fig. 8 (b). Suppose that $\mu_1(C) > \alpha_2$. Consider the region B and two masses on B, μ'_1 and μ'_2, induced by μ_1 and μ_2 respectively, i.e. $\mu'_i(X) = \mu_i(X)/\mu_i(B)$ for a region $X \subseteq B$ and $i = 1, 2$. By Theorem 3 there is a partition of B by a line into two sets X and $B \setminus X$ such that $\mu'_2(X) = \alpha_2/\mu_2(B) = \alpha_2/(\alpha_2 + \alpha_3)$ and $\mu'_1(X) \le \alpha_2/(\alpha_2 + \alpha_3)$ since $\alpha_2/(\alpha_2 + \alpha_3) \le 1/2$. By the intermediate value theorem (since $\mu'_2(C) = \alpha_2/(\alpha_2 + \alpha_3)$ and $\mu'_1(C) > \alpha_2/(\alpha_2 + \alpha_3)$) the set B has a perfect $(\alpha_2/(\alpha_2 + \alpha_3), \alpha_3/(\alpha_2 + \alpha_3))$-partition. Let B_1 and B_2 be the parts of such a partition. Then the subdivision of S into A, B_1 and B_2 is a convex α-partition.

It remains to consider the case $\mu_1(C) \leq \alpha_2$ for any $x \in C$. Let $D = B \setminus C$, see Fig. 8 (b). Then $\mu_1(D) = 1 - \mu_1(A) - \mu_1(C) \geq 1 - \alpha_1 - \alpha_2 = \alpha_3$. By Theorem 3 there is a point $x \in C$ such that $\beta_3(x) \leq \alpha_3$. By the intermediate value theorem the cake has a $(\alpha_3, 1 - \alpha_3)$-partition. Contradiction.

Case 2. S has a $(\alpha_3, 1 - \alpha_3)$-partition. Let $S = A \cup B$ be a $(\alpha_3, 1 - \alpha_3)$-partition such that $\mu_1(A) = \alpha_3$ and $\mu_1(B) = 1 - \alpha_3$. We apply the ham sandwich cut to B and obtain a convex α-partition of S.

Remark. The constraint $\alpha_1 \geq 1/4$ in Theorem 2 is essential. If $\alpha_1 < 1/4$ then $\alpha_3 > 1/2$ and the example of masses shown in Fig. 2 a) does not admit a convex α-partition.

4   Negative Result for Convex Partitions

We show that there is a cake that does not have a convex α-partitions for some triples $\alpha \in \mathbb{R}^3$.

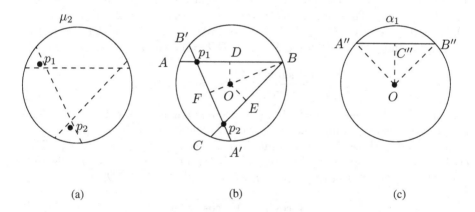

(a)   (b)   (c)

Fig. 9. a) Construction. b) Notation. c) $(\alpha_1, 1 - \alpha_1)$-partition.

In this section we consider the case where $\alpha_1 < \alpha_2 = \alpha_3$. Clearly, $\alpha_2 = \alpha_3 = 1/2 - \alpha_1/2$ and $\alpha_1 \in (0, 1/3)$. The idea of the construction is the following. Let μ_2 be the uniform mass on the unit circle. The other mass μ_1 has two points p_1 and p_2 inside the circle such that the weight of each point is $1/2$, see Fig. 9 (a). The points p_1 and p_2 are restricted to inside a triangle formed by three segments each of which cuts off α_2-portion of the unit circle. We assume that the points p_i are very close to the vertices of the triangle. For analysis purposes we even assume that the points p_i are two vertices of the triangle and the third vertex lies on the circle, see Fig. 9 (b).

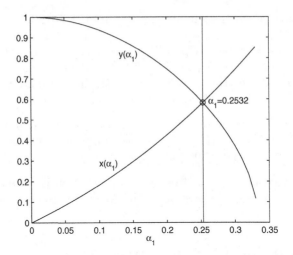

Fig. 10. Functions $x()$ and $y()$ and the value α_1^*.

Let D be the midpoint of AB and let x be the distance between D and p_1. The angles $\angle AOB, \angle BOC$ and $\angle A'OB'$ are equal to $2\pi\alpha_2$. The angle $\angle DOB = \angle AOB/2 = \pi\alpha_2$. The angle $\angle OBD = \pi/2 - \angle DOB = \pi(1/2 - \alpha_2) = \pi\alpha_1/2$. Then $|BD| = \cos(\pi\alpha_1/2)$ and $|DO| = \sin(\pi\alpha_1/2)$. Then $x + |BD| = |BF|/\cos(\pi\alpha_1/2)$ and $|BF| = |BO| + |OF| = 1 + |DO| = 1 + \cos(\pi\alpha_2) = 1 + \sin(\pi\alpha_1/2)$ and
$$x = \frac{1 + \sin(\pi\alpha_1/2)}{\cos(\pi\alpha_1/2)} - \cos(\pi\alpha_1/2).$$

Let $A''B''$ be the segment of a $(\alpha_1, 1 - \alpha_1)$-partition, see Fig. 9 (c). The midpoint C'' is at distance $\cos(\pi\alpha_1)$ from O. The locus of the midpoints of all segments of length $|A''B''|$ in the unit circle is the circle of radius $\cos(\pi\alpha_1)$. This circle crosses the segment AB in Fig. 9 (b) at two points. Let y be the distance from D to any of these two points. Clearly $y = \sqrt{|OC''|^2 - |OD|^2}$ and
$$y = \sqrt{\cos^2(\pi\alpha_1) - \cos^2(\pi\alpha_2)}. \tag{2}$$

x and y as functions of α_1 are monotone, see Fig. 10, and the equation $x(\alpha_1) = y(\alpha_1)$ has a unique root in the interval $(0, 1/3)$. Let α_1^* denote the root. We wrote a program for finding the root that gives $\alpha_1^* \approx 0.2532$.

Theorem 5. *For any $\alpha_1 \in (0, \alpha_1^*)$ there is a cake that does not admit a convex α-partition where $\alpha_2 = \alpha_3 = 1/2 - \alpha_1/2$.*

Proof. Let α_1 be a real number in $(0, \alpha_1^*)$. By the construction every half plane that has α_1-portion of the unit circle does not contain both points p_1 and p_2. For a point $p \in \mathcal{C}$, let $a_i(p)$ be the point on $\mathcal{C}$ such that $\mu_2(p, a_i(p)) = \alpha_i$. Let $A_i(p)$ denote the part of S cut off by the line $pa_i(p)$ such that the arc $\overline{pa_i(p)}$ lies in $A_i(p)$.

Let i be an index in $\{2,3\}$. For any point p in the arc $\overline{CB}$, $\mu_1(A_i(p)) < \alpha_i$. For any point p in the arc $\overline{BB'}$, $A_i(p)$ contains p_1 and $\mu_1(A_i(p)) \geq 1/2 > \alpha_i$. For any point p in the arc $\overline{B'C}$, $A_i(p)$ contains p_2 and $\mu_1(A_i(p)) \geq 1/2 > \alpha_i$. This implies that, for any point $p \in \mathcal{C}$, either $\beta_2(p) > \alpha_2$ or $\beta_3(p) > \alpha_3$. Therefore there is no perfect α-partition of S.

It remains to consider the case where $\mathcal{C}$ is partitioned into four arcs. There are two arcs $\overline{pa_i(p)}$ and $\overline{qa_j(q)}$, $p, q \in \mathcal{C}$ and $1 \leq i < j \leq 3$, that make $(\alpha_i, 1-\alpha_i)$-partition and $(\alpha_j, 1 - \alpha_j)$-partition of S. This is impossible from the assumed configuration of the points p_j and the theorem follows.

5   Conclusion

We proved several results for α-partitioning a generalized cakes. An interesting open question is to determine all α such that any cake admits a convex α-partition.

Acknowledgments

I would like to thank the anonymous referees for their valuable comments.

References

1. J. Akiyama, A. Kaneko, M. Kano, G. Nakamura, E. Rivera-Campo, S. Tokunaga, and J. Urrutia. Radial perfect partitions of convex sets in the plane. In *Proc. Japan Conf. Discrete Comput. Geom.'98*, volume 1763 of *Lecture Notes Comput. Sci.*, pp. 1–13. Springer-Verlag, 2000.
2. J. Akiyama, G. Nakamura, E. Rivera-Campo, and J. Urrutia. Perfect divisions of a cake. In *Tenth Canadian Conference on Computational Geometry*, 1998.
3. I. Bárány and J. Matoušek. Simultaneous partitions of measures by k-fans. *Discrete Comput. Geom.*, 25(3):317–334, 2001.
4. S. Bespamyatnikh, D. Kirkpatrick, and J. Snoeyink. Generalizing ham sandwich cuts to equitable subdivisions. *Discrete Comput. Geom.*, 24(4):605–622, 2000.
5. H. Ito, H. Uehara, and M. Yokoyama. 2-dimension ham sandwich theorem for partitioning into three convex pieces. In *Proc. Japan Conf. Discrete Comput. Geom.'98*, volume 1763 of *Lecture Notes Comput. Sci.*, pp. 129–157. Springer-Verlag, 2000.
6. A. Kaneko and M. Kano. Balanced partitions of two sets of points in the plane. *Comput. Geom. Theory Appl.*, 13:253–261, 1999.
7. A. Kaneko and M. Kano. Generalized balanced partitions of two sets of points in the plane. In *Proc. Japan Conf. Discrete Comput. Geom.'00*, volume 2098 of *Lecture Notes Comput. Sci.*, pp. 176–186. Springer-Verlag, 2001.
8. A. Kaneko and M. Kano. Perfect partitions of convex sets in the plane. *Discrete Comput. Geom.*, 28:211–222, 2002.
9. T. Sakai. Balanced convex partitions of measures in $\mathbb{R}^2$. *Graphs and Combinatorics*, 18(1):169–192, 2002.

Constrained Equitable 3-Cuttings

Sergei Bespamyatnikh[1] and David Kirkpatrick[2]

[1] Department of Computer Science, University of Texas at Dallas
Box 830688, Richardson, TX 75083, USA
besp@utdallas.edu
[2] Department of Computer Science, University of British Columbia
201-2366 Main Mall, Vancouver, B.C., V6T 1Z4, Canada
kirk@cs.ubc.ca

Abstract. We investigate equitable 3-cuttings of two mass distributions in the plane (partitions of the plane into 3 sectors with a common apex such that each sector contains $1/3$ of each mass). We prove the existence of a continuum of equitable 3-cuttings that satisfy some closure property. This permits us to generalize earlier results on both convex and non-convex equitable 3-cuttings with additional constraints.

1 Introduction

This paper continues recent results [2,4,8,13,9] generalizing the Ham Sandwich Theorem for the plane. The planar case of the discrete Ham Sandwich Theorem [16] states that, for finite sets of red and blue points in the plane, there exists a line dividing both red and blue points into sets of equal size. The Ham Sandwich problem is well studied from an algorithmic point of view [1,5,6,7,10,11,12,14,17]. An optimal algorithm of Lo et al. [11] finds a Ham Sandwich cut in linear time. Kaneko and Kano [9] considered balanced partitions of two sets in the plane. They conjectured that, for any gn red and gm blue points ($g, n, m > 0$ are integers) in the plane in general position, there are g disjoint convex polygons with n red and m blue points in each of them. The conjecture of Kaneko and Kano has been independently proven by Ito et al. [8] (the case $g = 3$), Sakai [13] and Bespamyatnikh et al. [4]. Only [4] considered algorithmic issues related to this conjecture. In particular they showed that a balanced partition can be constructed in $O(N^{4/3} \log^3 N \log g)$ time, where $N = n + m$.

For the case $g = 3$, it is shown in [4] that it suffices to consider 3-cuttings, partitions of the plane into three wedges with a common apex. More specifically, either the points of R and B admit a *T-shaped* equitable 3-cutting, formed by two successive 2-cuttings, or there is a *Y-shaped* convex equitable 3-cutting constrained to have one (reference) ray directed vertically downward (see Figure 1). (From an algorithmic perspective the second case presents the main challenge; the first reduces to the standard Ham Sandwich construction.) Bárány and Matoušek [2] establish a number of results concerning the existence (and nonexistence) of (not necessarily convex) 3-cuttings with specific triples that define (not necessarily equal) portions of the points in the wedges. In this paper we

J. Akiyama and M. Kano (Eds.): JCDCG 2002, LNCS 2866, pp. 72–83, 2003.
© Springer-Verlag Berlin Heidelberg 2003

restrict our attention to the case $g = 3$ (with the equal portions $(\frac{1}{3}, \frac{1}{3}, \frac{1}{3})$) but admit arbitrary directions of the reference ray. A central objective of our investigation is to provide a characterization all such 3-cuttings that partition one color (red) equitably. We show that these can be represented using 3-colored arrangements in the plane. Assuming that the masses do not admit a T-shaped equitable 3-cutting, the space of possible convex 3-cuttings with fixed direction of the reference ray is neatly captured by what we call a *triangle diagram* that is of interest in its own right. We prove the existence of a special 3-cutting in a perfect triangle diagram that we call a *centroid*.

We establish an invariant of the triangle diagram under rotation of the reference direction. We use it to prove the Centroid Theorem which allows a direct proof of the existence of an equitable 3-cutting with a variety of reasonable constraints (in place of a fixed direction of the reference ray).

Theorem 1 (Constrained convex 3-cutting). *For any two Borel probability measures either there exists a T-shaped convex equitable 3-cutting or there is an equitable 3-cutting satisfying any one of the following constraints:*

(a) *One of the rays passes through any specific point, not necessarily in the plane. (The special case of the point $(0, -\infty)$ corresponds to the the constraint of the starting ray going directly down).*
(b) *The angle of one of the wedges is $2\pi/3$.*
(c) *One of the wedges has weight $1/3$ of a third mass.*
(d) *Two rays are symmetric with respect to the line through the third ray.*
(e) *The angles of the wedges form an arithmetic progression.*

A T-shaped 3-cutting or a 3-cutting satisfying the condition 3 of Theorem 1 generates $(1/3, 2/3)$ 2-fan partition of three masses which was first shown in [2]. For general (not necessarily convex) 3-cuttings we prove the existence of equitable 3-cutting without a condition of T-shaped cutting.

Theorem 2 (Constrained 3-cutting). *For any two Borel probability measures there exists an equitable 3-cutting satisfying any one of the constraints (a)-(e) from Theorem 1.*

Theorem 2 is a strengthening of a result by Bárány and Matoušek [2] that states the existence of unconstrained equitable 3-cutting.

We also mention a recent paper by Vrećica and Živaljević [15] on conical equipartitions and a paper by Bárány and Matoušek [3] on equipartitioning two masses by 4-fans.

2   Preliminaries

Let P_R be a set of n red points in the plane and let P_B be a set of m blue points in the plane such that all the points are in general position (the points are distinct and no three points are collinear). A *3-cutting* is a partition of the plane into 3 wedges W_1, W_2 and W_3 by 3 rays with a common point that is called the *apex* of the 3-cutting. A 3-cutting is *convex* if its wedges are convex. A 3-cutting is *equitable* if each wedge (as open set) contains at most $n/3$ of red points and at most $m/3$ blue points. One of the rays is defined as the *reference ray*. We call the wedges adjacent to the reference ray the *left* and *right* wedges. The remaining wedge is the *upper* wedge . The rays different from the reference ray are called *left, right*. (Note that at most one of the wedges can be non-convex.)

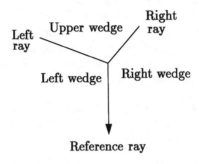

Fig. 1. The notation of a 3-cutting.

Canonical 3-cutting. Let μ be a mass in the plane. For a point p and a direction d, we define a *canonical 3-cutting* as follows. The reference ray emanates from p in the direction d. The left ray maximizes the angle of the left wedge W_1 such that the weight of the interior of W_1 is at most $1/3$. The right ray maximizes the angle of the right wedge W_3 such that the weight of the interior of W_3 is at most $1/3$. If μ' is another mass in the plane we define a *sign* $s_i(\mu, \mu'), i = 1, 2, 3$ to be the μ'-weight of the wedge W_i minus $1/3$.

Pseudo-discrete masses. A *mass* μ in the plane is defined by a Borel probability measure $\mu : \mathbb{R}^2 \to \mathbb{R}$. A *discrete mass* $\mu(P)$ induced by a set of points P is defined so that a weight of a region R is the number of points in $P \cap R$ divided by the number of points in P. It suffices to prove Theorems 1 and 2 for discrete masses, using a reduction from continuous to discrete mass distributions described in [4]. We would like the masses to behave like continuous ones in the sense that every canonical 3-cutting changes continuously when its apex moves on the plane. We introduce *pseudo-discrete masses* that are generated by finite sets P_R and P_B of red and blue points, and whose canonical 3-cuttings are continuous. We put a disk $D(p)$ of sufficiently small radius at every red and blue point $p \in P_B \cup P_R$. We draw two concentric circles C_R and C_B of a large radius such that all disks $D(p)$ are inside both C_R and C_B. We assign a small portion of red mass to C_R and a small portion of blue mass to C_B. We assume that these masses are distributed uniformly although one can use non-uniform distributions. The remaining red and blue masses are distributed among disks so that each red (resp. blue) disk is assigned the same portion of the remaining red (resp. blue) mass. We assume that the distributions of masses in the disks are defined by centrally symmetric polynomial functions. The pseudo-discrete masses capture discrete masses (if the weights of the circles C_R and C_B and the disk radii tend to 0) and allow us to achieve properties useful for proving Theorems 1 and 2.

Hereafter we assume that masses are pseudo-discrete. Let $s_i(p), i = 1, 2, 3$ be the sign of the canonical 3-cutting with apex at p with respect to the pseudo-discrete masses. We partition the plane according to the signs $s_i(p), i = 1, 2, 3$. There 27 possible combinations of signs (each sign $s_i, i = 1, 2, 3$ can be either < 0 or $= 0$ or > 0). We call this partition a *sign partition* of the plane.

Sign Components. Red and blue point sets P_R and P_B admit pseudo-discrete masses such that there are finitely many connected components in the sign partition and each component has dimension corresponding to constraints[1]. For example, a component constrained as $s_1(p) = s_2(p) = s_3(p) = 0$ is a point (dimension 0), a component with constraints $s_1(p) = 0, s_2(p) < 0, s_3(p) > 0$ is a curve (dimension 1) and a component with constrains $s_1(p) < 0, s_2(p) > 0, s_3(p) > 0$ is a two-dimensional region bounded by a curve.

3   Triangle Diagram

In this section we introduce a notion of *triangle diagram* that allows us to characterize all equitable 3-cuttings with vertical reference ray. A study of triangle diagrams may be of independent interest.

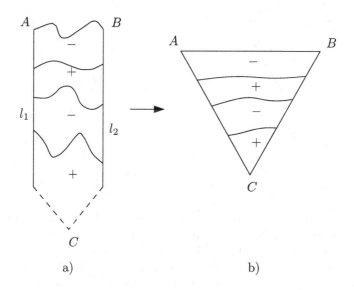

a)   b)

Fig. 2. The levels of s_2.

---

[1] We can show this property using polynomial distributions of constant degree.

We consider only convex 3-cuttings in this section. We assume that the masses satisfy a *sign property*: any halfplane with red weight $1/3$ has blue weight less than $1/3$. The sign property eliminates the case of T-shaped 3-cuttings in Theorem 1.

There are two vertical lines l_1, l_2 cutting the red mass into 3 pieces of equal weight $1/3$. One can show that the locus of apexes of convex red 3-cuttings with one ray going down is a region R sandwiched by these lines from left and right side. The upper boundary of R is defined by a x-monotone curve connecting the left and right sides, say points A and B, see Fig. 2 a). Two parallel lines l_1 and l_2 "meet" at point C with $y = -\infty$. We represent R as a regular triangle, see Fig. 2 b).

The upper wedge of red 3-cuttings with apexes on the curve AB has angle 2π. The sign property implies that the blue weight of the upper wedge is less than $1/3$. In other words $s_2(p) < 0$ for all points p on the curve AB. Note that $s_2(p)$ becomes positive if p goes far enough down in the region R because the left and right rays of red 3-cutting go to the left and right sides of R. Let $k_2(p)$ be the smallest number of times the sign $s_2()$ changes along a path from a point p to AB. Let Ξ be the set of paths from C to AB with exactly $k_2(C)$ changes of the sign $s_2()$ along the path. We define i-th s_2-level, $1 \le i \le k_2(C)$ as the locus of points p from the paths $\xi \in \Xi$ such that $k_2(p) = i - 1$ and $s_2(p) = 0$.

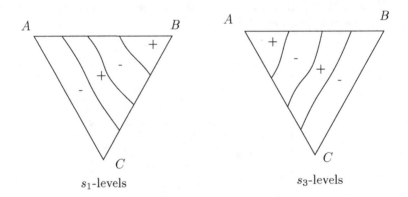

s_1-levels s_3-levels

Fig. 3. The levels of s_1 and s_3.

The sign $s_1()$ is negative for the points in the side AC and is positive at B. We define s_1-levels similarly to s_2-levels. s_1-levels connect the sides AB and BC, see Fig. 3 a). Similarly we define s_3-levels using signs $s_3()$, see Fig. 3 b). The arrangement of s_1-, s_2- and s_3-levels is a *triangle diagram*. We are interested in topological properties of triangle diagrams so the mapping from the plane to the triangle diagram does not need to be uniquely defined.

Lemma 1. *For every $i = 1, 2, 3$, the s_i-levels are disjoint curves.*

We color s_1-, s_2- and s_3-levels in red, blue and green colors respectively producing a diagram. The levels satisfy the following properties.

Oddness Property. The number of levels of each color is odd.

Intersection Property. If two levels of different colors intersect at point p then a level of the third color intersects p.

The intersection property follows from the fact that, for every point p

$$s_1(p) + s_2(p) + s_3(p) = 0 \tag{1}$$

(if, say $s_1(p) = s_2(p) = 0$, then $s_3(p) = 0$). Do the above properties capture sufficient conditions of a triangle diagram? The answer is "no" and Fig. 4 illustrates an unrealizable diagram. It is unrealizable diagram because the middle triangle has all positive signs contradicting with Equation 1. A new condition can be expressed in the following way.

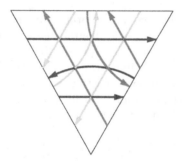

Fig. 4. Impossible diagram.

We orient levels such that the left side of an oriented level has negative sign s_i corresponding to the level. It is easy to show that the levels intersecting in the same point have *alternating directions* in-out-in-out-in-out, see Fig. 5. Note that the vertices of the middle triangle in Fig. 4 do not have alternating directions of the levels.

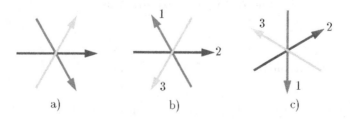

Fig. 5. Orientations of intersecting levels. a) Infeasible orientations. b), c) Feasible orientations.

We define a *triangle diagram* as a digram satisfying the oddness property, the intersection property and the alternating directions. Simple examples of triangle diagrams are shown in Fig. 6[2].

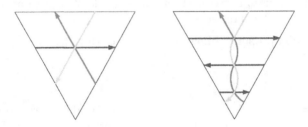

Fig. 6. Simple triangle diagrams.

The triangle diagrams satisfy a chessboard property, i.e. the faces can be colored in two colors, black and white, such that any two faces sharing an edge have different colors.

The triangle diagram changes under rotation of the initial direction of the reference ray. We say that a triangle diagram is *perfect* if it is topologically invariant under rotation, see Fig. 7. The perfect diagrams in Fig. 7 have unique central vertices that play key role in rotating 3-cuttings. Each 3-cutting appears 3 times under rotation. An equitable 3-cutting is a *centroid* if its corresponding 3-cuttings are connected under rotation. Our main result in the triangle diagrams is the existence of a centroid in a perfect diagram[3].

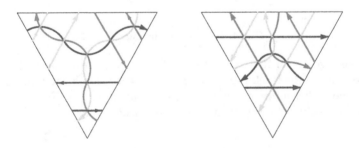

Fig. 7. Perfect triangle diagrams.

Theorem 3 (Centroid). *Each perfect triangle diagram has a unique centroid vertex.*

---

[2] Colored pictures of triangle diagrams are available in the website
   http://www.utd.edu/~besp/3diagram.
[3] Due to lack of space some proofs will be in full paper.

4   General (Not Necessarily Convex) 3-Cuttings

In this section we consider only 3-cuttings with the reference ray directed down. For every point in the plane there is an unique red 3-cutting with reference ray directed down. This red 3-cutting is not necessary convex but, of course, at most one of its wedges can be non-convex. We also allow 3-cuttings whose apexes are at infinity, see Fig. 8.

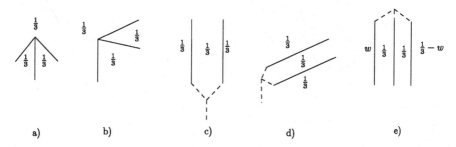

a)   b)   c)   d)   e)

Fig. 8. a), b) non-convex 3-cuttings. c), d) e) converging 3-cuttings.

Consider the arrangement $\mathcal{A}$ defined by the levels on the entire plane. Let l_1 and l_2 be the vertical lines partitioning the red mass into 3 parts of equal weight $1/3$. Let a, b and c be the blue weights of corresponding parts. The boundary of the plane can be viewed as a "shield" $ABCDEA$, see Fig. 9 a), where B and C are top points of the lines l_1 and l_2; E is a bottom point of l_1 and l_2; A and D represent the top line points going to the left and to the right respectively. Note that A and D represent the same 3-cuttings. The 3-cuttings with apexes at A, B, C, D andf E have blue weights a, b and c as depicted in Fig. 9 a). Let S (shield) denote the space of all 3-cuttings with glued points A and D.

Internal vertices. Each internal vertex has degree 6 and is formed by 3 levels. There are two types of the internal vertices of the arrangement $\mathcal{A}$, see Fig. 5 b) and c). We label a vertex as "+" if the outcoming levels form the sequence 123 in clockwise order, see Fig. 5 c). Otherwise it has label "-".

Edges. The internal vertices split level curves into *internal* and *external* edges. An internal edge is a part of level curve between two internal vertices. External edges remain when internal edges are removed from levels. By the alternating directions property the endpoints of an internal edge are internal vertices with opposite labels.

Faces. The faces are labeled by labels that are shown in Fig. 9 b). The face labeling satisfies the transition property that two faces separated by an edge have two consecutive labels on the label diagram.

Let v^+, v^- denote the number of the internal vertices labeled by "+" and "-" respectively. Let b^+, b^- denote the number of the external edges adjacent to internal vertices with corresponding labels.

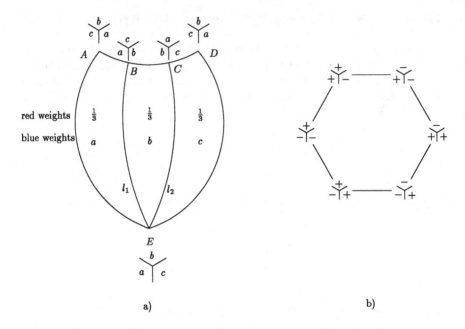

a)   b)

Fig. 9. a) The boundary of the plane. b) The label diagram.

Theorem 4. *Let t be the number of turns of the closed path on label diagram 9 b) when one walks on the boundary of $\mathcal{A}$ in clockwise order. Then*

$$v^+ - v^- = \frac{b^+ - b^-}{6} = t \tag{2}$$

and $t \equiv -1 \bmod 3$.

Proof. Let e be the number of edges in the arrangement $\mathcal{A}$ that are incident to an internal vertex. These edges can be counted using positive endpoints. Each internal vertex labeled "+" participates in six edges and remaining edges are external. Similar counting can be done for negative vertices. Hence

$$e = 6v^+ + b^- = 6v^- + b^+$$

and the first equality of (2) follows.

To prove the second equality of the equation (2) we consider the clockwise order closed path on the boundary $\mathcal{A}$. Suppose we cross the external edge e incident to the internal vertex p labeled "+". The face labels around p are shown in Fig. 10 a). There are six faces and their labels have the same order as the one in the label diagram, see Fig. 9 b). Hence the label transition is the rotation on clockwise angle $\pi/3$ in label diagram.

Suppose we cross the external edge e incident to the internal vertex p labeled "−". The face labels around p are shown in Fig. 10 b). The order of labels

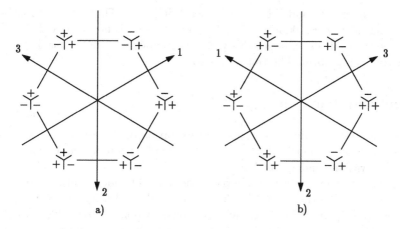

Fig. 10. The vertex labels a) "+" b) "-".

is opposite to the one in the label diagram. Hence the label transition is the rotation on counterclockwise angle $\pi/3$ in label diagram. It follows that each edge counted in b^+ participates as $1/6$ of clockwise turn in the label diagram. The contribution of an edge counted in b^- is $-1/6$. It implies $(b^+ - b^-)/6 = t$. Note that edges whose endpoints are on the boundary contribute 0 into t.

There is a bijection between the boundaries EA and DE. Each 3-cutting corresponding to a point in EA is defined by 2 parallel non-vertical lines equipartitioning the red mass, see Fig. 8 d). Consider such a pair of lines. Let $a + 1/3$, $b + 1/3$ and $c + 1/3$ be the blue weights of the upper halfplane, the middle slab and the lower halfplane respectively. The triple (s_1, s_2, s_3) of signs for the point in EA is (a, b, c). These lines define 3-cutting for a point in the left boundary ED and the corresponding sign triple is (c, b, a).

Consider the motion through an external edge in EA. Let $(a, b, c) - (a', b', c')$ be the label transition. The corresponding label transition in DE is $(c, b, a) - (c', b', a')$. These transitions are opposite in the label diagrams and they are cancelled in total rotation of angle 0 along DEA.

Consider the motion along the boundary $ABCD$. Each 3-cutting corresponding to a point in AB is defined by three parallel vertical lines equipartitioning the red mass, see Fig. 8 e). Each such three lines define three 3-cuttings with one of the lines specifying the reference ray. The labels of these 3-cuttings form regular triangle in the label diagram. Hence the rotation on the label diagram defined by walk along AB is the same as ones along BC and CD. The labels in A, B and C also form a regular triangle in the label diagram. Suppose r is the label rotation along AB in terms of clockwise turns (the clockwise turn is 1). Hence $r = t' - 1/3$ where t' is integer. It follows $t = 3(t' - 1/3) \equiv -1 \bmod 3$ and we are done.

Theorem 5 (Existence of general 3-cutting). *For any red and blue masses in the plane and for any direction d there is an equitable (not necessarily convex) 3-cutting with one ray in the direction d.*

Proof. We can assume that d is the downward direction. Theorem 4 establishes property that $v^+ - v^- \not\equiv 0 \bmod 3$. It implies that $v^+ - v^-$ is non-zero and there is at least one internal vertex in $\mathcal{A}$. This vertex is an apex of an equitable 3-cutting with one ray going down. The theorem follows.

5   Centroid Theorems

In this section we consider 3-cuttings under rotation of the reference ray. Theorem 4 establishes property of $v^+ - v^-$. We extend Theorem 4 and prove that $v^+ - v^-$ is invariant under rotation.

Using Theorem 4 we prove a property of the centroid under rotation. Consider an arrangement $\mathcal{A}'$ in the space $S' = S \times S^1$ induced by $\mathcal{A}$ where S^1 is the space of rotations (the space of reference ray directions). A point (p, α) in the space S' determines the red 3-cutting $c(p, \alpha)$ in the plane with apex p and the reference ray with slope α. Let $\alpha_1(p, \alpha)$ and $\alpha_2(p, \alpha)$ denote the slopes of the left and right rays of the 3-cutting $c(p, \alpha)$. There is the map $\gamma : S' \to S'$ defined by $\gamma(p, \alpha) = (p, \alpha_1(p, \alpha))$. In other words each 3-cutting appears 3 times under rotation. So we call the map γ a *rotation*.

We prove[4] the existence of a centroid for pseudo-discrete masses. It can be extended to Borel probability measures, see discussion in [2,4].

Theorem 6 (Centroid of general 3-cuttings). *For any two Borel probability measures in $\mathbb{R}^2$ there is a centroid.*

It is straightforward to verify using Theorem 6 that each constraint of Theorem 2 can satisfied by a 3-cutting. Theorem 6 also can be used to prove the following generalization of Theorem 3. We call a centroid *convex* if all its 3-cuttings are convex.

Theorem 7 (Convex 3-cuttings). *For two Borel probability measures with the sign property there is a unique convex centroid.*

Theorem 1 follows from Theorem 7.

References

1. I. Bárány. Geometric and combinatorial applications of Borsuk's theorem. In J. Pach, editor, *New Trends in Discrete and Computational Geometry*, volume 10 of *Algorithms and Combinatorics*, pp. 235–249. Springer-Verlag, 1993.
2. I. Bárány and J. Matoušek. Simultaneous partitions of measures by k-fans. *Discrete Comput. Geom.*, 25(3):317–334, 2001.
3. I. Bárány and J. Matoušek. Equipartition of two measures by a 4-fan. *Discrete Comput. Geom.*, 27(3):293–301, 2002.
4. S. Bespamyatnikh, D. Kirkpatrick, and J. Snoeyink. Generalizing ham sandwich cuts to equitable subdivisions. *Discrete Comput. Geom.*, 24(4):605–622, 2000.

---

[4] in full version of the paper.

5. M. Díaz and J. O'Rourke. Ham-sandwich sectioning of polygons. In *Proc. 2nd Canad. Conf. Comput. Geom.*, pp. 282–286, 1990.
6. D. P. Dobkin and H. Edelsbrunner. Ham-sandwich theorems applied to intersection problems. In *Proc. 10th Internat. Workshop Graph-Theoret. Concepts Comput. Sci.*, pp. 88–99, 1984.
7. H. Edelsbrunner and R. Waupotitsch. Computing a ham-sandwich cut in two dimensions. *J. Symbolic Comput.*, 2:171–178, 1986.
8. H. Ito, H. Uehara, and M. Yokoyama. 2-dimension ham sandwich theorem for partitioning into three convex pieces. In *Proc. Japan Conf. Discrete Comput. Geom.'98*, volume 1763 of *Lecture Notes Comput. Sci.*, pp. 129–157. Springer-Verlag, 2000.
9. A. Kaneko and M. Kano. Balanced partitions of two sets of points in the plane. *Comput. Geom. Theory Appl.*, 13:253–261, 1999.
10. C.-Y. Lo, J. Matoušek, and W. Steiger. Ham-sandwich cuts in $\Re^d$. In *Proc. 24th Annu. ACM Sympos. Theory Comput.*, pp. 539–545, 1992.
11. C.-Y. Lo, J. Matoušek, and W. L. Steiger. Algorithms for ham-sandwich cuts. *Discrete Comput. Geom.*, 11:433–452, 1994.
12. C.-Y. Lo and W. Steiger. An optimal-time algorithm for ham-sandwich cuts in the plane. In *Proc. 2nd Canad. Conf. Comput. Geom.*, pp. 5–9, 1990.
13. T. Sakai. Balanced convex partitions of measures in $\mathbb{R}^2$. *Graphs and Combinatorics*, 18(1):169–192, 2002.
14. W. Steiger. Algorithms for ham sandwich cuts. In *Proc. 5th Canad. Conf. Comput. Geom.*, p. 48, 1993.
15. S. T. Vrećica and R. T. Živaljević. Conical equipartitions of mass distributions. *Discrete Comput. Geom.*, 25(3):335–350, 2001.
16. R. T. Živaljević. Topological methods. In J. E. Goodman and J. O'Rourke, editors, *Handbook of Discrete and Computational Geometry*, chapter 11, pp. 209–224. CRC Press LLC, Boca Raton, FL, 1997.
17. R. T. Živaljević and S. T. Vrećica. An extension of the ham sandwich theorem. *Bull. London Math. Soc.*, 22:183–186, 1990.

On the Minimum Perimeter Triangle
Enclosing a Convex Polygon

Binay Bhattacharya[1,*] and Asish Mukhopadhyay[2,**]

[1] School of Computing Science
Simon Fraser University
Burnaby, Canada
binay@cs.sfu.ca
[2] Department of Computer Science
University of Windsor
Windsor, Canada
asishm@davinci.newcs.uwindsor.ca

Abstract. We consider the problem of computing a minimum perimeter triangle enclosing a convex polygon. This problem defied a linear-time solution due to the absence of a property called the *interspersing property*. This property was crucial in the linear-time solution for the minimum area triangle enclosing a convex polygon. We have discovered a non-trivial interspersing property for the minimum perimeter problem. This resulted in an optimal solution to the minimum perimeter triangle problem.

1 Introduction

Geometric optimization is an active area of research in the fast-growing field of Computational Geometry [7], [6], [2]. The motivation for this study comes from a variety of practical problems such as stock-cutting, certain packing and optimal layout problems. An interesting problem in this area that has attracted a lot of attention is that of minimizing various measures (area, perimeter etc.) associated with a convex k-gon that encloses a convex n-gon. In particular, researchers have studied the problem of finding an enclosing triangle of minimum area or perimeter. Klee et al [9] proposed an $O(n \log^2 n)$ algorithm for the minimum area triangle; subsequently, O'Rourke et al [10] proposed an optimal linear-time algorithm for this problem. The question of computing a minimum perimeter triangle turned out to be more difficult and very few results are known. Such computations are interesting from the point of view of approximation of convex objects [8].

De Pano [4] proposed an $O(n^3)$ algorithm for this problem. Aggarwal and Park [1] used poweful matrix searching technique in higher dimension to reduce the complexity of the problem to $O(n \log n)$. The existence of a linear-time algorithm has been a matter of conjecture for a very long time.

* Research supported by NSERC and MITACS
** Research supported by NSERC

J. Akiyama and M. Kano (Eds.): JCDCG 2002, LNCS 2866, pp. 84–96, 2003.
© Springer-Verlag Berlin Heidelberg 2003

In this paper, we propose a linear-time algorithm to settle the conjecture. The paper is organized into five sections. In the next two sections, we discuss two subsidiary problems. The first of these is to compute the minimum perimeter triangle determined by a wedge and a point contained in it. The second is to compute a triangle with minimum perimeter determined by a line and two points lying on one side of it. We will then solve the problem of finding a minimum perimeter enclosure for a convex polygon by repeatedly applying the above two subsidiary problems. These are described in the third section. The last section contains the summary and a discussion on some open problems.

2   The First Subsidiary Problem

A wedge is an interval of directions denoted by $W(s_1, s_2)$, where s_1 and s_2 are two line segments determining the two arms of the wedge. The angle of the wedge is the counterclockwise angle from s_1 to s_2.

Let p be a point inside a wedge $W(BA, BC)$. We define a segment AC through p to be *balanced* if the excircle of triangle $\triangle ABC$ contained in W is tangent to AC at p (Fig. 1).

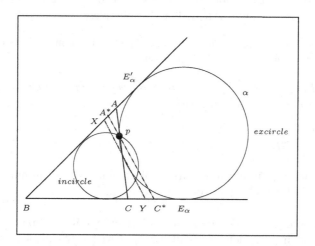

Fig. 1. *The excircle of triangle $\triangle ABC$ touches AC at p.*

The problem of determining a minimum perimeter triangle, whose two sides are bounded by the arms of the wedge, and the third side is incident on p is of fundamental importance to us. The following lemma characterizes such a minimum perimeter triangle and suggests a construction.

Lemma 1. *The triangle $\triangle ABC$ whose side AC includes p is of minimum perimeter only if AC is balanced.*

Proof. Suppose that the side AC of triangle $\triangle ABC$ touches the excircle α and it is not the minimum perimeter triangle. Let E_α and E'_α be the touching points of the excircle α with the wedge. The perimeter of the triangle $\triangle ABC$ is $|AB| + |BC| + |AE'_\alpha| + |CE_\alpha|$ which is $2|BE_\alpha|$. Suppose if possible, the triangle $\triangle A^*BC^*$ other than $\triangle ABC$ is the minimum perimeter triangle. Therefore, A^*C^* through p intersects the excircle. Consider the triangle $\triangle XBY$ where XY, parallel to A^*B^*, touches the excircle (Figure 1). Since XY touches the excircle, the perimeter of triangle $\triangle XBY$ is also $2|BE_\alpha|$. Thus $Perimeter(\triangle ABC) = Perimeter(\triangle XBY) < Perimeter(\triangle A^*BC^*)$, which contradicts the fact that $\triangle A^*BC^*$ is the minimum perimeter triangle. Therefore, triangle $\triangle ABC$ is the minimum perimeter triangle. $\square$

It is possible to construct the minimum perimeter triangle in $O(1)$ time, once $W(BC, CA)$ and p are known.

3   The Second Subsidiary Problem

We now consider the following *variant* of the first subsidiary problem.

Given two points p and q at heights h_p and h_q, $h_p \geq h_q$, respectively above a line L, p to the left of q, find a triangle $\triangle ABC$ of minimum perimeter such that the side BC is incident on L, while the points p and q are interior to the sides AB and AC respectively.

How do we characterize such a minimum perimeter triangle? The following lemma provides a set of *necessary* conditions.

Lemma 2. *If triangle $\triangle ABC$ is a minimum perimeter triangle, satisfying the conditions of the above problem, then:*

(a) $|pB| = |qC|$
(b) $|Ap| + |Aq| = |BC|$.

Proof. This follows from two applications of Lemma 1 on the minimum perimeter triangle, once for each wedge determined by L. $\square$

We label a wedge with the apex on L as left-wedge if it is open to the right. If an wedge is open to the left, it is called a right-wedge. We always assume that L is horizontal.

It is important to note that if a triangle is of minimum perimeter, it is minimum for both the left-wedge and the right-wedge whose apexes are on L; however, the other way round is not true in general. This is shown in Figure 2.

3.1   Triangle Configuration with Minimum Perimeter

As shown in Figure 2, more than one triangle configuration satisfy Lemma 2. In the following we will answer the following questions:

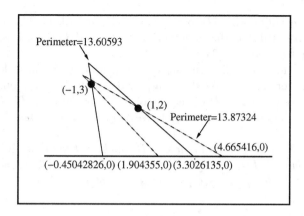

Fig. 2. *Two configurations satisfying Lemma 2.*

(a) How many triangle configurations satisfying Lemma 2 are possible?
(b) Which one realizes the minimum perimeter?

We now prove a few useful properties of the triangles generated by the excircles. Suppose the side AC in triangle $\triangle ABC$ is balanced (see Figure 3). Here the wedge $W(BC, BA)$ is a left-wedge. Suppose the side $A'B'$ in triangle $\triangle A'B'C$ is balanced where $W(CA, CB)$ is a right-wedge. Then

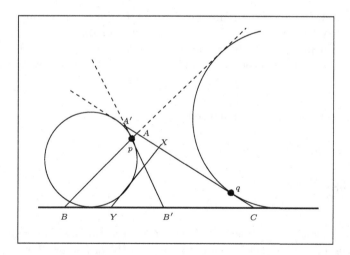

Fig. 3. *The perimeter decreases as we change from wedge $W(BC, BA)$ to wedge $W(CA, CB)$.*

Lemma 3. *The perimeter of $\triangle A'B'C$ is smaller than that of $\triangle ABC$.*

Proof. The excircle of $\triangle A'B'C$ touches $A'B'$ at p, while AB through p intersects the excircle (see Figure 3). Consider the tangent to the excircle touching

$A'B'$, obtained by sliding the AB, parallel to itself, towards C. The new triangle $\triangle XYC$, thus obtained, has the same perimeter as $\triangle A'B'C$ which is, therefore, smaller than that of $\triangle ABC$. $\square$

The following lemma, called the *separation lemma*, is useful in locating the minimum perimeter triangle. We define a pair of points B and C on L *separate* another pair B' and C' on L, B to the left of B', if the interval $[B, C]$ intersects the interval $[B', C']$ without containing it. We call (B, C) *separates* (B', C') on L.

Lemma 4. *Let B be a point to the left of B' on L. Let the edge AC of $\triangle ABC$ through q be balanced. Let the edge $A'C'$ of triangle $\triangle A'B'C'$ through q be balanced. Then (on L) the pair (B, C) separates the pair (B', C').*

Proof. Omitted $\square$

Let B_{left} and C_{limit} be two points on L such that $|pB_{left}| = |qC_{limit}|$ as shown in Figure 4. Here B_{left} is the left most such point on L, and we assume that $|qC_{limit}| \geq h_p$, if not, there is no triangle satisfying the conditions of Lemma 2. We are now ready to show that

Theorem 1. *There can exist at most two triangle configurations satisfying Lemma 2 and the triangle configuration $\triangle ABC$ where B is closest to B_{left} is the minimum perimeter triangle.*

Proof. Let $\triangle A^*B^*C^*$ be the minimum perimeter triangle determined by L, p and q. It is easy to see that B^* must lie to the right of B_{left}.

Let B_x be a point on L, to the right of B_{left} at a distance x from B_{left}. Given x, we consider two types of triangles that are determined by L, p and q:

1. $\triangle A_x^f B_x C_x^f$ where $|pB_x| = |qC_x^f|$
2. $\triangle A_x^g B_x C_x^g$ where $A_x^g C_x^g$ is balanced.

We now define the following two functions

1. $f(x) = 2|qC_x^f| + 2|B_x C_x^f|$
2. $g(x) = 2|qC_x^g| + 2|B_x C_x^g|$

Here $g(x)$ represents the perimeter of $\triangle A_x^g B_x C_x^g$. Here $f(x)$ and $g(x)$ will capture the properties of the necessary conditions (a) and (b) of Lemma 2, respectively. Our idea is to separate out the necessary conditions, study them individually, and then study when they are jointly satisfied by intersecting $f(x)$ and $g(x)$.

We first show that

Lemma 5. *$x = \alpha$ is an extreme point of $g(x)$ iff $g(\alpha) = f(\alpha)$*

Proof. The *only if* part:

Suppose $g(x)$ has an extreme point at $x = \alpha$. Therefore, $\triangle A_x^g B_x C_x^g$ satisfies the conditions of Lemma 2. Since $|pB_x| = |qC_x^g|$, $\triangle A_x^f B_x C_x^f = \triangle A_x^g B_x C_x^g$. Therefore $f(\alpha) = g(\alpha)$.

The *if* part:

If $\triangle A_x^f B_x C_x^f \neq \triangle A_x^g B_x C_x^g$, either C_x^f lies to the left of C_x^g or to the right of C_x^g. It is easy to see that in the former case, $f(\alpha) < g(\alpha)$ and in the latter case $f(\alpha) > g(\alpha)$. This follows from the fact that $\angle C_x^f$ is always acute. Since $f(\alpha) = g(\alpha)$, $\triangle A_x^f B_x C_x^f = \triangle A_x^g B_x C_x^g$. $\square$

Corollary 1. *The intersection points of $f(x)$ and $g(x)$ are the only extreme points of $g(x)$.*

Let $x = l$ be an extreme point of $g(x)$ for the smallest x in $[0, |B_{left}B_{right}|]$. We have the following characterization of this extreme point.

Lemma 6. $g(x)$ *has a local minimum at $x = l$.*

Proof. Here we represent the point B_{left} by B_0, i.e. the point at a distance $x = 0$. We first notice that $f(0) > g(0)$. This follows from the fact that $f(0) = 2|qC_{limit}| + 2|B_{left}C_{limit}|$ is greater than the perimeter of the triangle $\triangle A_0^f B_0 C_0^f$ and $g(0)$ is smaller than the perimeter of $\triangle A_0^f B_0 C_0^f$. Let $A'B'$ be balanced in $\triangle A'B'C_0^g$ (Figure 4). According to Lemma 3, the perimeter of triangle $\triangle A'B'C_0^g$ is less than that of $\triangle A_0^g B_0 C_0^g$, since $A_0^g C_0^g$ of $\triangle A_0^g B_0 C_0^g$ is balanced. The proof of the lemma is complete once we show that $g(x)$ has no extreme point for any x in $(0, |B_{left}B'|)$.

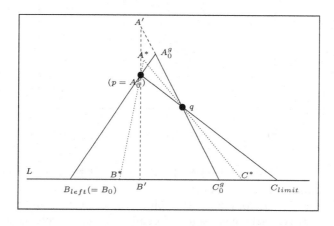

Fig. 4. *The B^* can not lie between B_{left} and B'.*

If possible, suppose $\triangle A^* B^* C^*$ satisfies the conditions of Lemma 2 and B^* lies between B_{left} and B'. C^* must lie to the right of C_0^g, since (B_{left}, c_0^g) separates (B^*, C^*). In this case (B', C_0^g) can not separate (B^*, C^*). Therefore, no such triangle $\triangle A^* B^* C^*$ can exist. Therefore, the first extreme point of $g(x)$ is a minimum. $\square$

This leads to the following claim

Lemma 7. $g(x)$ can intersect $f(x)$ at most twice.

Proof. It is easy to show that $f(x)$ is convex. However, we don't know much about $g(x)$.

Let $x_1, x_2, x_3, \ldots, x_k$ be the intersection points of $f(x)$ and $g(x)$, where $x_1 < x_2 < x_3 < \ldots < x_k$. Since $g(x_1)$ is a local minimum, $g(x_2) > g(x_1)$. Since $x_2 > x_1$ and both $(x_1, g(x_1))$ and $(x_2, g(x_2))$ lie on $f(x)$, we have the situation shown in Fig 5.

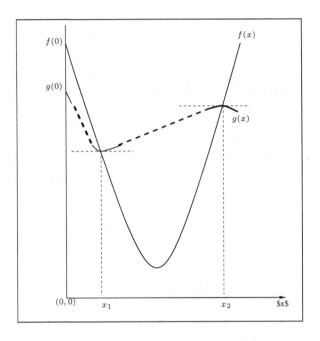

Fig. 5. *Intersection of $g(x)$ and $f(x)$.*

After x_2, $g(x)$ cannot intersect $f(x)$ again. If it does, at the next extreme point x_3, a local minimum, $g(x_3) = f(x_3)$ and $g(x_3) < g(x_2)$. This is not possible since $x_2 < x_3$ and $f(x)$ is convex. This implies that there could be at most two configurations satisfying the conditions of Lemma 2 and the left most extreme point of $g(x)$ is the global minimum. $\square$

We have thus established Theorem 1. $\square$

We will also refer these two triangle configurations satisfying Lemma 2 as the extreme triangle configurations.

Computing the Optimal Solution. In the following we show an explicit form of the function $g(x)$. We assume that

- L is the x-axis.
- (a, h_p) and (b, h_q) are the respective coordinates of p and q, and $h_p \geq h_q$.
- The coordinate of B_x is $(x, 0)$.

The following expression for $g(x)$ can be obtained by first computing the intersections of the angular bisector of the wedge $W(B_x A_x^g, B_x C_x^g)$ and the parabola equidistant to q and L. The intersection point with the largest x-coordinate is the x-coordinate of the center of the excircle determined by the wedge $W(B_x A_x^g, B_x C_x^g)$ and the point q. The expression for $g(x)$ is

$$g(x) = -2x + \frac{2h_p h_q}{A(x) + a - x} + 2b+$$

$$2\sqrt{(\frac{h_p h_q}{A(x) + a - x} + b)^2 - h_q^2 - b^2 - \frac{2h_p h_q x}{A(x) + a - x}}$$

where $A(x) = \sqrt{h_p^2 + (a - x)^2}$.

The results of theorem 1 imply that the minimum perimeter triangle can be determined by computing the left most extremum point of $g(x)$. The theorem also imples that the minimum perimeter triangle can be determined by computing the left most intersection point of $f(x)$ and $g(x)$. In our case $f(x)$ is a convex quadratic function of x. Thus we conclude that the minimum perimeter triangle can be computed from $f(x)$ and $g(x)$ in constant time.

4 Main Problem

We now return to our main problem: computing a minimum perimeter triangle circumscribing an n-sided convex polygon, P. Finding the minimum perimeter triangle for a pair of points, as described in the previous section, could be construed as a degenerate case of the problem of this section.

A minimum perimeter triangle has the following property proved in De Pano [4]:

Theorem 2. *There exists a minimum perimeter triangle whose one side is flush with an edge of the polygon.*

The main idea of our construction is to find a minimum perimeter triangle of the n-gon by considering each edge in anticlockwise order. We repeat this by going around the polygon in clockwise order. Below, we give the details of the anticlockwise search. This discussion will make clear the need for a clockwise search.

Let $< p_0, e_1, p_1, e_2, \ldots, p_i, e_{i+1}, p_{i+1}, \ldots, p_{n-1}, e_n, p_n (= p_0) >$ be the vertices and edges of the polygon P in anticlockwise order; here, e_i denotes an edge with endpoints p_{i-1} and p_i.

Let the line L be flush with the edge e_1 of P. Consider any left-wedge $W(BC, BA)$ containing P where B and C lie on L. We always assume that the arms of the wedge touch P. We are interested in finding AC which supports P at q such that the perimeter of $\triangle ABC$ is the smallest for the given wedge.

We find this point q by exhaustively searching the boundary of the polygon. We find an excircle for the triangle, two of whose sides, BC and BA, are the arms of the wedge $W(BC, BA)$ and the third side AC is incident on e_2. If the point of tangency belongs to e_2 we stop. Else, we find a circle through p_2 inscribed in $W(BC, CA)$. If the tangent to the circle at this point is also tangent to P we stop, else we move to the edge e_3 and find an excircle for the triangle whose edges are the arms of the wedge $W(BC, BA)$ and the edge incident on e_3. The process terminates when the excircle also touches P. The time to compute the touching point q is proportional to the size of the traversed polygonal chain from e_2 to q. It should be noted that q can also be determined by traversing the polygonal chain in clockwise order. Using the arguments similar to Lemma 1, we can show that

Lemma 8. *If the excircle touches P at q and AC touches both P and the circle at q, the perimeter of $\triangle ABC$ is the smallest for the wedge $W(BC, CA)$.*

Using the arguments similar to the one used in Lemma 2 we can show that

Lemma 9. *If p and q are the points where the excircles of the $\triangle A^*B^*C^*$ touch the sides A^*B^* and A^*C^* internally, then the points p and q lie on P such that*

(a) $|pB^*| = |qC^*|$.
(b) $|B^*C^*| = |A^*p| + |A^*q|$

We call the side A^*C^* (A^*B^*) of $\triangle A^*B^*C^*$ as balanced for the polygon P and the left-wedge $W(B^*C^*, B^*A^*)$ (right-edge $W(C^*A^*, C^*B^*)$). Sometimes we will just say that A^*B^* and A^*C^* of $\triangle A^*B^*C^*$ are balanced. Again these conditions are necessary but not sufficient.

We now state below some results, similar to the ones established for the pair of points described earlier. The proofs are also similar, and therefore omitted.

Lemma 10. *If AC in $\triangle ABC$ is balanced and $A'B'$ in $\triangle A'B'C$ is balanced (see Lemma 3 for the point case), $Perimeter(\triangle A'B'C) < Perimeter(\triangle ABC)$.*

Lemma 11. (Separation Lemma) *Let B be a point to the left of B' on L (see Lemma 4 for the point case). Let edge AC of $\triangle ABC$ be balanced. Let edge $A'C'$ of triangle $\triangle A'B'C'$, containing P, be balanced. Then on L, the pair (B, C) separates the pair (B', C').*

Let us consider the following process which computes a triangle $\triangle ABC$ containing P with BC on L and $\triangle ABC$ satisfies the necessary conditions of Lemma 9. We consider e_i as the flush edge. We first consider a wedge whose arms are L and L', where L' is parallel to L and supports P at a_i. We denote this left-wedge by $W(BC, BA)$ where B is assumed to be on both L and L'. Thus BC is on L and BA is on L'. We find the balanced edge AC that touches P at q. According to Lemma: 8, $\triangle ABC$ is the minimum perimeter of P for the given wedge $W(BC, BA)$.

In the following we always denote the current triangle by $\triangle ABC$. We denote by p (q) the left (right) touching point of $\triangle ABC$. We write h_p (h_q) to indicate the height of p (q) from the base line L.

We now switch to the right-wedge $W(CA, CB)$ and find the excircle that fits into this wedge and touches P. This determines the balanced edge AB that touches P at a point, say p.

The above search is looking for the left most triangle configuration that satisfies the necessary conditions of Lemma 9. The processes of circle-fitting and wedge-switiching terminates when $pB = qC$, where p is the point on P, where the excircle for the right-wedge $W(CA, CB)$ touches AB and q is the point on P where the excircle for the left-wedge $W(BA, BC)$ touches BC and p, or when $h_p < h_q$.

We give a formal description of this algorithm below where the horizontal line L is flush with the edge e_i. The procedure **FindRightBalanced** (**FindLeftBalanced**) searches for the right (left) touching point of P by the excircle by examining the polygonal chain in anticlockwise order. Initially, the search in **FindRightBalanced** starts from p_i and we use the pointer *rightStartingPoint* to indicate the starting point of the search. For the subsequent search, the search process starts from where it stopped last. The part of the polygonal chain searched by **FindRightBalanced** is called the *right searched chain* for the flush edge e_i. Similarly, we use the pointer *leftStartingPoint* to indicate the starting point of the search for the procedure **FindLeftBalanced**. The part of the chain searched by **FindLeftBalanced** is called the *left searched chain* for the flush edge e_i. Initially, leftStartingPoint has the value a_i, the antipodal point of the flush edge e_i. Both the procedures return the balanced sides along with the touching points. When either one procedure fails to return the touching point, it indicates that there does not exist any triangle $\triangle ABC$ satisfying Lemma 9 where $h_p \geq h_q$.

Algorithm Minimum Perimeter(e_i, P)

Input A convex polygon P and an edge e_i of this polygon
Output A triangle $\triangle ABC$, with e_i lying on side BC.

- Initialize $\triangle ABC$;
- Set $u \leftarrow \infty$; $h_p \leftarrow \infty$; $h_q \leftarrow 0$;
- Set $leftStartingPoint \leftarrow a_i$; $rightStartingPoint \leftarrow p_i$;

1: $(AC, q) \leftarrow$ **FindRightBalanced**$(W(BC, BA), rightStartingPoint, h_p)$;

- if q is not defined, then STOP;
  { There is no $\triangle ABC$ satisfying Lemma 9 with $h_q \leq h_p$. }
- $v \leftarrow |qC|$;
- if $(u = v)$ go to **2**;
- $(AB, p) \leftarrow$ **FindLeftBalanced**$(W(CA, CB), leftStartingPoint, h_q)$;
- if p is not defined, then STOP;
  { There is no $\triangle ABC$ satisfying Lemma 9 with $h_p \geq h_q$. }
- $u \leftarrow |pB|$;
- if $(u \neq v)$ go to **1**;

2: return $\triangle ABC$;

We can show that

Lemma 12. *The triangle $\triangle ABC$ returned by* **Minimum Perimeter** *algorithm is the minimum perimeter configuration for the line that is flush with the edge e_i and the points p and q on P at which the excircles touch AB and AC respectively.*

Proof. Omitted.

Now we prove what may be appropriately called a *left-interspersing* lemma.

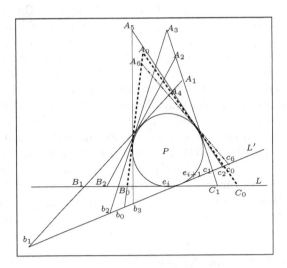

Fig. 6. *Checking polygon edges in anticlockwise order.*

Lemma 13. *The search for a minimum perimeter circumscribing triangle never backtracks as we traverse the polygon P in counter-clockwise order.*

Proof. Let triangle $\triangle A_0 B_0 C_0$ be of minimum perimeter where $B_0 C_0$ contains the edge e_i of P; let e_{i+1} be the edge anticlockwise to e_i. Let L and L' be the support lines of e_i and e_{i+1} respectively (see Figure 6).

Let A_0B_0 produced meet L' at b_0. If there is no such point, the left searched chain for the flush edge e_i is not going to be searched again by **FindLeftBalanced** for the flush edge e_{i+1}. Thus there is no backtracking in this case. We now consider the case when b_0 exists. Let L' be the line through e_{i+1}. We show that we cannot have a minimum perimeter triangle $\triangle a'b'c'$, with the apex b' of the left-wedge on L', such that b' is to the left of b_0. Effectively, this means that we can start the **FindRightBalanced** procedure for edge e_{i+1} with the left-wedge having apex at b_0 on L'. To obtain a contradiction, let b_1 be an arbitrary point on L' to the left of b_0. The thrust of the proof is to show that the circle-fitting and wedge-switching process, starting with the left-wedge with the apex at b_1, will force the apex of the left-wedge of the minimum perimeter triangle to lie to the right of b_1 on L'. We first draw a supporting line to P from b_1. Let this supporting line intersect L at B_1.

Let A_1C_1 be balanced for the left-wedge at B_1 that contains P. Let c_1 be its intersection with L'. From Lemma 11 we know that (on L) the pair (B_1, C_1) separates the pair (B_0, C_0).

Let A_2B_2 be balanced for the right wedge at C_1 that contains P. The point B_2 lies between B_1 and B_0 - this follows from Lemma 11 and the fact that triangle $\triangle A_0B_0C_0$ is of minimum perimeter for the edge e.

Let A_3b_2 be balanced for the right-wedge at c_1 that contains P. Since the pair (A_2, C_1) separates the pair (A_3, c_1), the point B_2 must lie above the supporting line of A_3b_2. We therefore conclude that the point b_2 is to the right of b_1 on L'.

Now let A_4c_2 be balanced for the left-wedge at b_1 that contains P. On the supporting line of b_1B_1 the pair (A_4, b_1) must separate the pair (A_1, B_1). Hence, on L', the point c_2 will lie to the right of c_1.

Next, consider the balanced edge A_5b_3 for the right-wedge at c_2 that contains P. Since the pair (b_2, c_1) separates the pair (b_3, c_2) on L', the point b_3 is to the right of b_1.

Thus the minimum configuration $\triangle a'b'c'$ for the edge e_{i+1} will have b' lying to the right of the point b_1 on L'. Since b_1 was an arbitrary point to the left of b_0 on L, b' can not lie to the left of b_0 on L'. Hence, we can start searching from b_0 for the flush edge e_{i+1} and thus during the execution of **FindLeftBalanced** procedure no backtracking takes place.

Consider the balanced edge A_6c_3 determined by the left-wedge at b_0 and P. By the separation lemma, (b_0, A_6) separates (B_0, A_0). Therefore, c_6 must lie to the right of c_0. Therefore, we do not backtrack during the execution of **FindRightBalanced** procedure also as we switch flush edge from e_i to e_{i+1}. $\square$

In an exactly analogous manner we can prove a *right-interspersing* lemma. This would justify that when we repeat the above search by moving clockwise around the polygon, we will not have to backtrack. And as we have noticed a clockwise search is necessary because the left-search aborts if the height of the left contact point p becomes smaller than that of the right contact point q, when we are looking for a minimum perimeter triangle anchored at any fixed edge.

4.1   Complexity of the Algorithm

It is obvious from the discussion of the previous section that as we move around the polygon P, the task of finding a minimum perimeter polygon, anchored at a given edge, can be carried over from where we left off at the previous edge. Since the cost of finding a minimum perimeter triangle, anchored at each edge, is proportional to the number of vertices and edges scanned, the linearity of the algorithm is obvious.

Theorem 3. *The smallest perimeter triangle of a convex polygon can be computed in linear time.*

5   Conclusions

We have described a simple but intricate algorithm for finding a minimum perimeter triangle that circumscribes a convex polygon of n sides. Here we have of course assumed that solutions to the subsidiary problems discussed at the beginning of the paper require constant time. The challenge is to find a nice solution than this.

The most interesting open question is that of designing an algorithm for finding a minimum perimeter k-gon that circumscribes a convex n-gon along the lines of the above algorithm to improve on the result in Aggarwal et al [1].

References

1. A. Aggarwal and J. K. Park, *Notes on searching in multi-dimensional monotone arrays*, FOCS, pp. 497-512, 1988.
2. P.K. Agarwal and M. Sharir, *Efficient Algorithms for geometric optimization*, ACM Computing Surveys, Vol. 30, pp 412-458, 1998.
3. J.E. Boyce, D.P. Dobkin, R.L. Drysdale and Leo Guibas, *Finding extremal polygons*, SIAM J. Comput., Vol. 14, pp 134-147, 1985.
4. N. A. A. De Pano, *Polygon Approximation with optimized polygonal enclosures: applications and algorithms*, Ph. D thesis, 1987.
5. D. P. Dobkin and L. Snyder, *On a general method for maximizing and minimizing among certain geometric problems*, Proc. IEEE Symp. FOCS, 1979, pp. 9-17.
6. D. Dori and M. Ben-Bassat, *Circumscribing a convex polygon by a polygon of fewer sides with minimal area addition*, Computer Vision Graphics and Image Processing, Vol. 24, pp 131-159, 1983.
7. H. Freeman and R. Shapira, *Determining the minimum area encasing rectangle for an arbitrary closed curve*, CACM, Vol. 18, pp 409-413, 1975.
8. P.M. Gruber, *Approximation of convex bodies*, In Convexity and its Applications, Editor P.M. Gruber, Birkhauser, 1983.
9. V. Klee and M. C. Laskowski, *Finding the smallest triangles containing a given convex polygon*, J. Algorithms, Vol. 6, pp 359-375, 1985.
10. J, O'Rourke, A. Aggarwal, S. Madilla, and M. Baldwin, *An optimal algorithm for finding minimal enclosing triangles*, Journal of Algorithms, Vol. 7, pp 258-269, 1986.
11. A. W. Roberts and D. E. Varberg, *Convex Functions*, Academic Press, 1973.
12. G.T. Toussaint, *Solving Geometric Problems with the "Rotating Calipers"*, Proceedings IEEE MELECON '83, Athens, Greece.

Succinct Data Structures for Approximating Convex Functions with Applications*

Prosenjit Bose[1], Luc Devroye[2], and Pat Morin[1]

[1] School of Computer Science, Carleton University, Ottawa, Canada, K1S 5B6,
{jit,morin}@cs.carleton.ca
[2] School of Computer Science, McGill University, Montréal, Canada, H3A 2K6,
luc@cs.mcgill.ca

Abstract. We study data structures for providing ϵ-approximations of convex functions whose slopes are bounded. Since the queries are efficient in these structures requiring only $O(\log(1/\varepsilon) + \log \log n)$ time, we explore different applications of such data structures to efficiently solve problems in clustering and facility location. Our data structures are succinct using only $O((1/\varepsilon) \log^2(n))$ bits of storage. We show that this is optimal by providing a matching lower bound showing that any data structure providing such an ϵ-approximation requires at least $\Omega((1/\varepsilon) \log^2(n))$ bits of storage.

1 Introduction

We consider the problem of approximating convex functions of one variable whose slopes are bounded. We say that a non-negative number y is an ε-approximation to a non-negative number x if $(1 - \varepsilon)x \leq y \leq x$ [1]. We say that a function g is an ε-approximation to a function f if $g(x)$ is an ε-approximation to $f(x)$ for all x in the domain of f.

Let $f : \mathbb{R} \to \mathbb{R}^+$ be a convex function that is non-negative everywhere. In this paper we show that, if the absolute value of the slope of f is bounded above by n, then there exists a piecewise-linear function g that ε approximates f at all points x except where the slope of f is small (less than 1) and that consists of $O(\log_E n)$ pieces, where $E = 1/(1 - \varepsilon)$. The function g can be computed in $O(K \log_E n)$ time, where K is the time it takes to evaluate expressions of the form $\sup\{x : f'(x) \leq t\}$ and f' is the first derivative of f. Once we have computed the function g, we can store the pieces of g in an array sorted by x values so that we can evaluate $g(x)$ for any query value x in $O(\log \log_E n)$ time. Since we are interested in the joint complexity as a function of $\varepsilon < 1/2$ and $n \geq 10$, it is worth noting that $log_E n = \Theta((1/\varepsilon) \log n)$ and thus that $\log \log_E n = \Theta(\log(1/\varepsilon) + \log \log n)$.

As an application of these results, we consider functions defined by sums of Euclidean distances in d dimensions and show that that they can be approximated using the above results. To achieve this, we use a random rotation

---

* This research was partly supported by NSERC.
[1] This definition is a bit more one-sided that the usual definition, which allows any y such that $|x - y| \leq \varepsilon x$.

J. Akiyama and M. Kano (Eds.): JCDCG 2002, LNCS 2866, pp. 97–107, 2003.
© Springer-Verlag Berlin Heidelberg 2003

technique similar to the method of random projections [6]. We show that the sum of Euclidean distances from a point to a set of n points can be closely approximated by many sums of Manhattan distances from the point to the set. This technique is very simple and of independent interest.

The remainder of the paper is organized as follows. Section 2 presents our result on approximating convex functions using few linear pieces. Section 3 discusses how these results can be interpreted in terms of data structures for approximating convex functions. Section 4 gives lower bounds on the space complexity of approximating convex function. Section 5 describes applications of this work to facility location and clustering problems. Finally, Section 6 summarizes and concludes the paper.

2   Approximating Convex Functions

Let $h(x) = c + |nx|$, for some $c, n \geq 0$. Then, it is clear that the function g such that $g(x) = c + (1 - \varepsilon)|nx|$ is an ε-approximation of h. Furthermore, g is an ε-approximation for any function h_2 such that $g(x) \leq h_2(x) \leq h(x)$ for all $x \in \mathbb{R}$. (see Fig. 1). This trivial observation is the basis of our data structure for approximating convex functions.

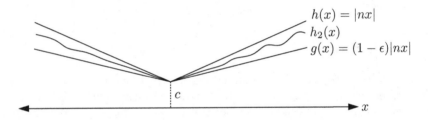

$h(x) = |nx|$
$h_2(x)$
$g(x) = (1 - \epsilon)|nx|$

c

x

Fig. 1. The function g is an approximation of h and of h_2.

Let f be a non-negative convex function and let f' be the first derivative of f. Assume that $f'(x)$ is defined for all but a finite number of values of x and that $|f'(x)| \leq n$ for all x in the domain of f'. For convenience, we define the *right derivative* $f^*(x)$ as follows: If $f'(x)$ is defined, then $f^*(x) = f'(x)$. Otherwise, $f^*(x) = \lim_{\delta \to 0+} f'(x + \delta)$.

Let a be the largest value at which the slope of f is at most $-(1 - \varepsilon)n$, i.e.,

$$a = \max\{x : f^*(x) \leq -(1 - \varepsilon)n\} .$$

(Here, and throughout, we use the convention that $\max \emptyset = -\infty$ and $\min \emptyset = \infty$.) Similarly, let $b = \min\{x : f^*(x) \geq (1 - \varepsilon)n\}$. Then, from the above discussion, it is clear that the function

$$g(x) = \begin{cases} f(a) + (1 - \varepsilon)(x - a)n \text{ if } x \leq a \\ f(b) + (1 - \varepsilon)(b - x)n \text{ if } x \geq b \\ f(x) \qquad\qquad\qquad \text{otherwise} \end{cases} \qquad (1)$$

is an ε-approximation of f (see Fig. 2).

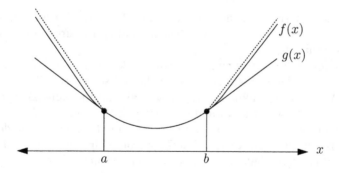

Fig. 2. The function g is a $(1 - \varepsilon)$ approximation of f.

Equation (1) tells us that we can approximate f by using two linear pieces and then recursively approximating f in the range (a, b). However, in the range (a, b), f^* is in the range $(-(1 - \varepsilon)n, (1 - \varepsilon)n)$. Therefore, if we recurse $\lceil \log_E n \rceil$ times, we obtain a function g with $O(\log_E n) = O((1/\varepsilon) \log n)$ linear pieces that approximates f at all points except possibly where f^* is less than one.

Theorem 1 *Let f and f^* be defined as above. Then there exists a piecewise-linear function g with $O(\log_E n)$ pieces that is an ε-approximation to f at all values except where $|f^*(x)| \leq 1$.*

3   Data Structures

In this section, we consider the consequences of Theorem 1 in terms of data structures for approximating convex functions. By storing the pieces of g in an array sorted by x values, we obtain the following.

Theorem 2 *Let f and f^* be defined as in Section 2. Then there exists a data structure of size $O((1/\varepsilon) \log n)$ that can compute an ε-approximation to $f(x)$ in $O(\log(1/\varepsilon) + \log \log n)$ time for any query value x where $|f^*(x)| \geq 1$.*

Next, we consider a more dynamic model, in which the function f is updated over time. In particular, we consider the following operations that are applied to the initial function $f(x) = 0$, for all $x \in \mathbb{R}$.

1. QUERY(x): Return an ε-approximation to $f(x)$.
2. INSERT(a): Increase the slope of f by 1 in the range (a, ∞), i.e., set $f(x) \leftarrow f(x) + x - a$ for all $x \in [a, \infty)$.
3. DELETE(x): Decrease the slope of f by 1 in the range (x, ∞). In order to maintain convexity, the number of calls to DELETE(x) may not exceed the number of calls to INSERT(x) for any value of x.

Note that a sequence of INSERT and DELETE operations can only produce a monotonically increasing function f whose slopes are all integers. This is done to simplify the exposition of the data structure. If an application requires that f be allowed to decrease and increase then two data structures can be used and their results summed.

The function f has some number m of breakpoints, where the slope of f changes. We store these breakpoints in a balanced search tree T, sorted by x-coordinate. With each breakpoint x, we also maintain the value $\Delta(x)$ by which the slope of f increases at x. In addition, we link the nodes of T in a doubly-linked list, so that the immediate successor and predecessor of a node can be found in constant time. It is clear that T can be maintained in $O(\log n)$ time per operation using any balanced search tree data structure.

In addition to the search tree T, we also maintain an array A of size $O((1/\varepsilon) \log n)$ that contains the piecewise linear approximation of f. The ith element in this array contains the value x_i such that $x_i = \min\{x : f^*(x) \geq E^i\}$, a pointer to the node in T that contains x_i, and the values of $f(x_i)$ and $f^*(x_i)$, i.e., the value of f at x_i and slope of f at x_i. To update this array during an INSERT or DELETE operation, we first update the values of $f(x_i)$ and $f^*(x_i)$ for each i. Since there are only $O((1/\varepsilon) \log n)$ array entries, this can be done in $O((1/\varepsilon) \log n)$ time.

Next, we go through the array again and check which values of x_i need to be changed (recall that $x_i = \min\{x : f^*(x) \geq E^i\}$). Note that, since INSERT or DELETE can only change the value of $f^*(x)$ by 1, if the value of x_i changes then it changes only to its successor or predecessor in T. Since the nodes of T are linked in a doubly-linked list, and we store the values of $f(x_i)$ and $f^*(x_i)$ we can detect this and update the value of x_i, $f(x_i)$ and $f^*(x_i)$ in constant time. Therefore, over all array entries, this takes $O((1/\varepsilon) \log n)$ time.

To evaluate an approximation to $f(x)$, we do a binary search on A to find the index i such that $[x_i, x_{i+1})$ contains x and then output $f(x_i) + (x - x_i)f^*(x_i)$. By the choice of x_i, this is a ε-approximation to $f(x)$. We have just proven the following:

Theorem 3 *There exists a data structure of size $O(n)$ that supports the operations* INSERT, DELETE *in* $O((1/\varepsilon) \log n)$ *time and* QUERY *in* $O(\log(1/\varepsilon) + \log \log n)$ *time, where n is the maximum slope of the function f being maintained.*

4 A Lower Bound on Storage

In this section we prove an $\Omega((1/\varepsilon) \log^2 n)$ lower bound on the number of bits required by any data structure that provides an ε-approximation for convex functions. The idea behind our proof is to make $m = \Theta((1/\varepsilon) \log n)$ choices from a set of n elements. We then encode these choices in the form of a convex function f whose slopes are in $[0, n]$. We then show that given a function g that is an ε-approximation to f we can recover the $m = \Theta((1/\varepsilon) \log n)$ choices. Therefore, any data structure that can store an ε-approximation to convex functions whose slopes lie in $[0, n]$ must be able to encode $\binom{n}{m}$ different possibilities and must therefore store $\Omega((1/\varepsilon) \log^2 n)$ bits in the worst case.

Let $x_1, \ldots, x_n$ be an increasing sequence where $x_1 = 0$ and each x_i, $2 \leq i \leq n$ satisfies

$$x_{i-1} + (1 - \varepsilon) \left(\frac{x_i - x_{i-1}}{1 - 2\varepsilon} \right) > x_{i-1} + \frac{x_i - x_{i-1}}{1 - 2\varepsilon} \; , \tag{2}$$

which is always possible since $(1 - \epsilon)/(1 - 2\varepsilon) > 1$.

Let $p_1, \ldots, p_m$ be any increasing sequence of $m = \lfloor \log_{E'} n \rfloor$ integers in the range $[1, n]$, where $E' = 1/(1 - 2\varepsilon)$. We construct the function f as follows:

1. For $x \in [-\infty, 0)$, $f(x) = 0$.
2. For $x \in (x_{p_i}, x_{p_{i+1}})$, $f(x)$ has slope $1/(1 - 2\varepsilon)^i$.
3. For $x > p_m$, $f(x)$ has slope n.

The following lemma, illustrated in Fig. 3 allows us to decode the values of $p_1, \ldots, p_m$ given an ε-approximation to f.

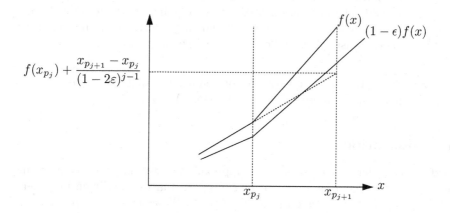

Fig. 3. An illustration of Lemma 1.

Lemma 1 *For the function f defined above and for any i such that $i = p_j$ for some $1 \leq j \leq m$,*

$$(1 - \varepsilon)f(x_{p_{j+1}}) > f(x_{p_j}) + \frac{x_{p_{j+1}} - x_{p_j}}{(1 - 2\varepsilon)^{j-1}} \; .$$

Proof. The lemma follows (with some algebra) from (2).

Suppose that g is an ε-approximation to f, i.e, for all $x \in \mathbb{R}$, $g(x)$ satisfies $(1 - \varepsilon)f(x) \leq g(x) \leq f(x)$. Then Lemma 1 can be used to recover the values of $p_1, \ldots, p_m$ from g. Suppose, that we have already recovered $p_1, \ldots, p_j$ and that we now wish to recover p_{j+1}. Note that, since we have recovered $p_1, \ldots, p_j$ we can compute the exact value of $f(x_{p_j})$. We then evaluate $g(x_{p_j} + 1)$, $g(x_{p_j} + 2)$, and so on until encountering a value k such that

$$g(x_{p_j} + k) > f(x_{p_j}) + \frac{x_{p_j} + k - x_{p_j}}{(1 - 2\varepsilon)^j}$$

Lemma 1 then guarantees that $p_{j+1} = p_j + k - 1$. In this way, we can reconstruct the entire function f and recover the values of $p_1, \ldots, p_m$.

Although in the above discussion the slopes used in the construction of f are not always integral it is clear that carefully rounding values appropriately would yield the same results using only integer valued slopes. Since we can encode $\binom{n}{m}$ different choices of $p_1, \ldots, p_m$ in this manner and $\log \binom{n}{m} = \Omega((1/\varepsilon) \log^2 n)$, we conclude the following:

Theorem 4 *Any data structure that can represent an ε-approximation to any convex function whose slopes are integers in the range $[0, n]$ must use $\Omega((1/\varepsilon) \log^2 n)$ bits of storage in the worst case.*

Remark 1 Some readers may complain that the function used in our lower bound construction uses linear pieces whose lengths are exponential in n. However, one should take into account that the endpoints of these pieces have x-coordinates that are integral powers of 2 and they can therefore be encoded in $O(\log n)$ bits each using, e.g., a floating point representation.

Remark 2 Another easy consequence of Lemma 1 is that any piecewise linear function that ε-approximates f has $\Omega((1/\varepsilon) \log n)$ pieces.

5   Applications

Next, we consider applications of our approximation technique for convex functions to the problem of approximating sums of distances in d dimensions. Let S be a set of n points in d dimensions. The *Fermat-Weber weight* of a point $q \in \mathbb{R}^d$ is

$$\mathrm{FW}(p) = \sum_{p \in S} \|pq\| \ ,$$

where $\|pq\|$ denotes the distance between points p and q. Of course, different definitions of distance (e.g., Euclidean distance, Manhattan distance) yield different Fermat-Weber weights.

5.1   The 1-Dimensional Case

One setting in which distance is certainly well defined is in one dimension. In this case,

$$\|pq\| = |p - q| \ ,$$

so the Fermat-Weber weight of x is given by

$$\mathrm{FW}(x) = f(x) = \sum_{y \in S} |x - y| \ .$$

Note that the function f is convex (it is the sum of n convex functions) and has slopes bounded below by $-n$ and above by n, so it can be approximated using the techniques Section 3. Furthermore, adding or removing a point p to/from S decreases the slope of f by 1 in the range $(-\infty, p)$ and increases the slope of f by 1 in the range (p, ∞), so the dynamic data structure of the previous section can be used to maintain an ε-approximation of f in $O(\log_E n) = O((1/\varepsilon) \log n)$ time per update.

Given the set S, constructing the ε-approximation for f can be done in $O(n/\varepsilon)$ time by a fairly straightforward algorithm. Using a linear-time selection algorithm, one finds the elements of S with ranks $\lfloor \varepsilon n/2 \rfloor$ and $\lceil (1-\varepsilon/2)n \rceil$. These are the values a and b in (1). Once this is done, the remaining problem has size $(1 - \varepsilon)n$ and is solved recursively. Although some care is required to compute the values $f(a)$ and $f(b)$ at each stage, the deatils are not difficult and are left to the interested reader.

Remark 3 A general result of Agarwal and Har-Peled [1] implies that the Fermat-Weber weight of points in one dimension can actually be ε-approximated by a piecewise-linear function with $O(1/\varepsilon)$ pieces. However, it is not clear how easily this approach can be made dynamic to handle insertion and deletions of points.

5.2   The Manhattan Case

The Manhattan distance between two points p and q in $\mathbb{R}^d$ is

$$\|pq\|_1 = \sum_{i=1}^{d} |p_i - q_i| \; ,$$

where p_i denotes the ith coordinate of point p. We simply observe that Manhattan distance is the sum of d one-dimensional distances, so the Fermat-Weber weight under the Manhattan distance can be approximated using d one-dimensional data structures.

5.3   The Euclidean Case

The Euclidean distance between two points p and q in $\mathbb{R}^d$ is

$$\|pq\|_2 = \left(\sum_{i=1}^{d} (p_i - q_i)^2 \right)^{1/2} .$$

A general technique used to approximate Euclidean distance is to use a polyhedral distance function, in which the unit sphere is replaced with a polyhedron that closely resembles a sphere. For example, the Manhattan distance function is a polyhedral distance function in which the unit sphere is replaced with a unit hypercube. Although this technique works well when d is small, such metrics generally require a polyhedron with a number of vertices that is exponential in d, making them poorly suited for high dimensional applications.

Another technique, that works well when d is very large (greater than $\log n$), and for most distance functions, is that of random projections [6]. Here, a random $O(\log n)$-flat is chosen and the points of S are projected orthogonally onto this flat. With high probability, all interpoint distances are faithfully preserved after the projection, so the problem is reduced to one in which the dimension of the point set is $O(\log n)$. The difficulty with this approach, when using Euclidean distances, is that sums of Euclidean distances are difficult to deal with even when $d = 2$ [2], thus the reduction in dimension does not help significantly.

Here we present a third approach that combines both of these techniques and adds two new twists: (1) we obtain a polyhedral metric as the sum of several Manhattan metrics and (2) our polyhedron is random. The first twist allows us to apply approximate data structures for one-dimensional convex functions while the second allows us to achieve approximation guarantees using an a number of vertices that increases only linearly with d.

Let $f(p)$ denote the Fermat-Weber weight of p under the Euclidean distance function. Choose k independent random orientations of the coordinate axes. Let $f_i(p)$ denote the Fermat-Weber weight of p under the Manhattan distance function after rotating the axes according to the ith random orientation. Then $f_i(p)$ may take on any value in the range $[f(p), \sqrt{d}f(p)]$. In particular, $f_i(p)$ has an expected value

$$\mathbf{E}[f_i(p)] = c_d f(p) \ ,$$

where c_d is a constant, dependent only on d, whose value is derived in Appendix A. Consider the function

$$g(p) = \frac{1}{kc_d} \times \sum_{i=1}^{k} f_i(p)$$

that approximates the Fermat-Weber weight under Euclidean distance.

Lemma 2 $\Pr\{|g(p) - f(p)| \geq \varepsilon f(p)\} = \exp(-\Omega(\varepsilon^2 k))$

Proof. The value of $g(p)$ is a random variable whose expected value is $f(p)$ and it is the sum of k independent random variables, all of which are in the range $[f(p), \sqrt{d}f(p)]$. Applying Hoeffding's inequality [4] immediately yields the desired result.

In summary, $g(p)$ is an ε-approximation of $f(p)$ with probability $1 - e^{-\Omega(\varepsilon^2 k)}$. Furthermore, $g(p)$ is the sum of k Fermat-Weber weights of points under the Manhattan distance function. Each of these Manhattan distance functions is itself a sum of d Fermat-Weber weights in 1 dimension. These 1-dimensional Fermat-Weber weights can be approximated using the results of Section 3 or the results of Agarwal and Har-Peled [1].

5.4   Clustering and Facility Location

Bose *et al* [3] describe data structures for approximating sums of distances. They show how to build a data structure in $O(n \log n)$ time that can $(1 - \varepsilon)$-approximate the Fermat-Weber weight of any point in $O(\log n)$ time. However, the constants in their algorithms depend exponentially on the dimension d.

The same authors also give applications of this data structure to a number of facility-location and clustering problems, including evaluation of the Medoid and AverageDistance clustering measures, the Fermat-Weber problem, the constrained Fermat-Weber problem, and the constrained obnoxious facility-location problem. All of these applications also work with the data structure of Section 3, many with improved running times. A summary of these results is given in Table 1.

Table 1. Applications of the data structure for evaluating the Fermat-Weber weights of points under the Euclidean distance function.

Problem	Exact solution	Previous ε-approx.		Ref.	New ε-approx.
Average distance	$O(n^2)$	$O(n)^a$	$O(n \log n)$	[5,3]	$O(n \log \log n)$
Medoid (1-Median)	$O(n^2)$	$O(n)^a$	$O(n \log n)$	[5,3]	$O(n \log \log n)$
Discrete Fermat-Weber	$O(n^2)$	$O(n)^a$	$O(n \log n)$	[5,3]	$O(n \log \log n)$
Fermat-Weber	–	$O(n)^b$	$O(n \log n)$	[5,3]	$O(n)$
Constrained Fermat-Weber	$O(n^2)$	$O(n)^b$	$O(n \log n)$	[5,3]	$O(n)$
Constrained OFL	$O(n^2)$	$O(n)^a$	$O(n \log n)$	[5,7,3]	$O(n \log \log n)$

[a] Refers to a randomized algorithm that outputs a $(1-\epsilon)$-approximation with constant probability.

[b] Refers to a randomized algorithm that outputs a $(1 - \epsilon)$-approximation with high probability, i.e., with probability $1 - n^{-c}$, for some $c > 0$.

6   Summary and Conclusions

We have given static and dynamic data structures for approximating convex functions of one variable whose slopes are bounded. These data structures have applications to problems involving sums of distances in d dimensions under both the Manhattan and Euclidean distance functions. In developing these applications we have arrived at a technique of independent interest, namely that of approximating Euclidean distance as the sum of several Manhattan distances under several different orientations of the coordinate system.

References

1. P. K. Agarwal and S. Har-Peled. Maintaining the approximate extent measures of moving points. In *Proceedings of the 12th Annual ACM-SIAM Symposium on Discrete Algorithms*, pages 148–157, 2001.
2. C. Bajaj. The algebraic degree of geometric optimization problems. *Discrete & Computational Geometry*, 3:177–191, 1988.
3. P. Bose, A. Maheshwari, and P. Morin. Fast approximations for sums of distances, clustering and the Fermat-Weber problem. *Computational Geometry: Theory and Apllications*, 24:135–146, 2002.
4. W. Hoeffding. Probability inequalities for sums of bounded random variables. *Journal of the American Statistical Association*, 58:13–30, 1963.

5. P. Indyk. Sublinear time algorithms for metric space problems. In *Proceedings of the 31st ACM Symposium on Theory of Computing (STOC'99)*, 1999.
6. P. Indyk. Algorithmic applications of low-distortion geometric embeddings. In *Proceedings of the 42nd IEEE Symposium on Foundations of Computer Science*, pages 10–33, 2001.
7. J. Kleinberg. Two algorithms for nearest neighbour search in high dimensions. In *Proceedings of the 29th ACM Symposium on Theory of Computing (STOC'97)*, pages 599–608, 1997.
8. D. S. Mitrinovic. *Analytic Inequalities*. Springer, New York, 1970.

A The Value of c_d

The value of c_d is given by

$$c_d = \mathbf{E}\left[\sum_{i=1}^{d} |X_i|\right] \ ,$$

where $(X_1, \ldots, X_d)$ is a point taken from the uniform distribution on the surface of the unit ball in $\mathbb{R}^d$. We observe that $(X_1^2, \ldots, X_d^2)$ is distributed as

$$\left(\frac{N_1^2}{N^2}, \ldots, \frac{N_d^2}{N^2}\right) \ ,$$

where $N^2 = \sum_{i=1}^{d} N_i^2$ and $(N_1, \ldots, N_d)$ are i.i.d. normal$(0,1)$. Clearly,

$$\frac{N_1^2}{N^2} = \frac{N_1^2}{N_1^2 + \sum_{i=2}^{d} N_i^2} \overset{\mathcal{L}}{=} \frac{G(\frac{1}{2})}{G(\frac{1}{2}) + G(\frac{d-1}{2})}$$

where $G(\frac{1}{2})$, and $G(\frac{d-1}{2})$ are independent gamma$(\frac{1}{2})$ and gamma$(\frac{d-1}{2})$ random variables, respectively. Thus, N_1^2/N is distributed as a beta$(\frac{1}{2}, \frac{d-1}{2})$ random variable, $\beta(\frac{1}{2}, \frac{d-1}{2})$. We have:

$$\mathbf{E}\left[\sum_{i=1}^{d} |X_i|\right] = d\,\mathbf{E}\left[\sqrt{\beta\left(\frac{1}{2}, \frac{d-1}{2}\right)}\right]$$

$$= d \int_0^1 \frac{x^{\frac{1}{2}-1}(1-x)^{\frac{d-1}{2}-1}}{B(\frac{1}{2}, \frac{d-1}{2})} \cdot \sqrt{x} \ \ dx$$

$$= d \cdot \frac{B(1, \frac{d-1}{2})}{B(\frac{1}{2}, \frac{d-1}{2})}$$

$$= \frac{2}{B(\frac{1}{2}, \frac{d+1}{2})} \ ,$$

where $B(a, b)$ is the beta function.

From Mitrinovic [8, p. 286], we note:

$$\frac{2}{B(\frac{1}{2}, \frac{d+1}{2})} \geq \sqrt{\frac{d}{2} + \frac{1}{4} + \frac{1}{16d + 32}} \cdot \frac{1}{\Gamma(\frac{1}{2})} \tag{3}$$

$$= 2\sqrt{\frac{d}{2} + \frac{1}{4} + \frac{1}{16d + 32}} \cdot \frac{1}{\sqrt{\pi}} \tag{4}$$

$$\geq \sqrt{\frac{2d + 1}{\pi}} \ . \tag{5}$$

Furthermore,

$$\mathbf{E}\left[\sum_{i=1}^{d} |X_i|\right] = d\,\mathbf{E}[|X_1|]$$

$$\leq \frac{d + 1}{\sqrt{\pi} \cdot \sqrt{\frac{d}{2} + \frac{3}{4} + \frac{1}{16d + 48}}}$$

$$\leq \frac{2(d + 1)}{\sqrt{\pi} \cdot \sqrt{2d + 3}}$$

$$\leq \sqrt{\frac{2(d + 1)}{\pi}} \ .$$

In summary,

$$\sqrt{\frac{2d + 1}{\pi}} \leq c_d = \frac{2\Gamma(\frac{d}{2} + 1)}{\sqrt{\pi} \cdot \Gamma(\frac{d+1}{2})} \leq \sqrt{\frac{2(d + 1)}{\pi}} \ .$$

Efficient Algorithms for Constructing a Pyramid from a Terrain

Jinhee Chun[1], Kunihiko Sadakane[2], and Takeshi Tokuyama[3]

[1] Graduate School of Information Sciences, Tohoku University, Sendai, Japan
{jinhee,tokuyama}@dais.is.tohoku.ac.jp
[2] Graduate School of Information Science and Electrical Engineering
Kyushu University, Fukuoka, Japan
sada@csce.kyushu-u.ac.jp

Abstract. In [4], the following *pyramid construction problem* was proposed: Given nonnegative valued functions ρ and μ in d variables, we consider the optimal pyramid maximizing the total parametric gain of ρ against μ. The pyramid can be considered as the optimal unimodal approximation of ρ relative to μ, and can be applied to hierarchical data segmentation. In this paper, we give efficient algorithms for a couple of two-dimensional pyramid construction problems.

1 Introduction

Let ρ and μ be two nonnegative-valued functions defined on a cube $(0, n]^d$ in the d-dimensional Euclidean space $\mathcal{R}^d$. Naturally, we can consider them as distributions or measure functions on the cube. In particular, it is often convenient to regard them as measure functions to get intuition. Consider a family $\mathcal{F}$ of regions in the cube. For a region $R \in \mathcal{F}$, $g(R; \rho, \mu) = \rho(R) - \mu(R)$ is called the *gain* of ρ against μ in R where $\rho(R) = \int_{x \in R} \rho(x) dx$. More generally, introducing a nonnegative parameter t, $g_t(R, \rho, \mu) = \rho(R) - t \cdot \mu(R)$ is called the parametric gain of ρ against μ within the region R; this can be considered as the gain value in which we replace μ by $t \cdot \mu$.

The problem of finding the region $R \in \mathcal{F}$ maximizing $g(R; \rho, \mu)$ is a fundamental problem in data segmentation[1]. In particular, the parametric gain is useful for solving segmentation problems in several applications including image processing and data mining; indeed, in those applications we seek for the segmentation maximizing a given concave objective function such as entropy or intercluster valiance, and the optimal segmentation also maximizes parametric gain for a suitable parameter value; this enables to design efficient algorithms applying standard methods of parametric optimization [1]. In data mining applications, we select d numerical attributes to correspond a data record to a point in the cube ($d = 2$ in the literature [6,8]). Given a large size of sample data set, we define μ to give the data distribution in the cube. Among the data set, we

---

[1] $\max_{R \in \mathcal{F}} |g(R; \rho, \mu)|$ is called the *discrepancy* between ρ and μ with respect to the family $\mathcal{F}$.

J. Akiyama and M. Kano (Eds.): JCDCG 2002, LNCS 2866, pp. 108–117, 2003.

consider the subset of data that satisfy certain condition defined by a *target attribute*, and the subset gives another distribution function ρ. An *association rule* (named *d-dimensional rule* or *region rule*) is defined by using a region in R, so that μ and ρ determine *support* $\rho(R)$ and *confidence* $\rho(R)/\mu(R)$ of the rule. For obtaining a region yielding a rule with both large support and high confidence, we apply the above data segmentation.

A *pyramid* $\mathcal{P}$ is a series of regions $\{P_t\}_{t\geq 0}$ in $\mathcal{F}$ satisfying that $P_t \subseteq P_{t'}$ if $t > t'$. Our aim is to compute the *optimal* pyramid $\mathcal{P}$ maximizing the total parametric gain $V(\mathcal{P}) = \int_{t=0}^{\infty} g_t(P_t; \rho, \mu)dt$.

The problem is more intuitive if μ is the unit function $\mu \equiv 1$. In this case, the optimal pyramid can be considered as a unimodal reformation of ρ minimizing loss of positional potential. This is a basic problem in computational geometry and geography (especially for $d = 2$). In general, the pyramid can be considered as a unimodal approximation of the measure ρ relative to μ. Figures 1 and 2 give examples of pyramids (where $\mu \equiv 1$) for $d = 1$ and $d = 2$, respectively. In Figure 2, function values are given by using density of pixels.

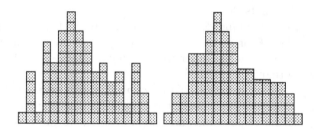

Fig. 1. Input one variable function ρ and its optimal pyramid.

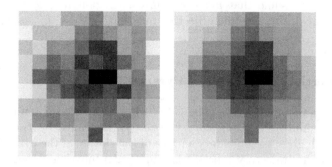

Fig. 2. Optimal pyramid (right) for an input function ρ (left) in two variables.

Construction of an optimal pyramid is a natural extension of region segmentation, and will be useful in several applications such as statistics, geomorphology, computer vision [2], and data mining [3]; our main motivation is to develop a tool for constructing an optimized probabilistic decision tree in our project to extend the SONAR data mining system [5,6,7,8].

In [4], properties of optimal pyramids are studied, and some algorithms for computing pyramids are given. The computational complexity of the problem depends depends on the dimension d, the number of pixels $N = n^d$, the problem precision Γ (i.e. the values of ρ and μ are desicribed by using $O(\log \Gamma)$ bits), and the region family $\mathcal{F}$. In this paper, we consider some two-dimensional pyramid construction problems: We give an $O(N \log^2 n)$ time algorithm for the family of connected lower half regions. Also, we give an $O(N^2 \log(N\Gamma))$ time algorithm and an $O(\epsilon^{-1} N^{1.5} \log \Gamma)$ time approximate algorithm for the family of point-stabbed unions regions of rectangles, where ϵ is the factor of the error of the approximation.

2   Preliminaries

We consider functions on the cube $C = (0, n]^d$ that is decomposed into $N = n^d$ unit cubes (translated copies of $(0, 1]^d$) called *cells* (also often called *voxels*). Let $\mathcal{G}$ be the set of grid points in C.

The definition of a pyramid depends on the family $\mathcal{F}$ of regions, and we are interested in families $\mathcal{F}$ of regions in $C = (0, n]^d$ such that each region is a connected union of cells; in the other words, they are regions in the pixel grid $\mathcal{G}$. We also assume that $\mathcal{F}$ contains $\emptyset$.

A function f on C is called a *step function* if it is a constant function in each cell, that is, $f(x) = f(x')$ holds for any pair x and x' in the same cell. Since we only consider regions that are union of cells, we can assume without loss of generality that ρ and μ are step functions. Thus, ρ and μ are considered as d-dimensional nonnegative-valued arrays of size n^d, and we assume that each entry of the arrays is given as rational numbers using $O(\log \Gamma)$ bits.

We need to assume that $\mu(R) \neq 0$ in each region $R \in \mathcal{F}$ satisfying that $\rho(R) \neq 0$, since otherwise the objective function $V(\mathcal{P}) = \int_{t=0}^{\infty} g_t(P_t; \rho, \mu) dt$ is not bounded.

2.1   Properties of the Optimal Pyramid

We consider a pyramid on the cube $C = (0, n]^d$ with respect to a region family $\mathcal{F}$. For a pyramid $\mathcal{P}$, the *trajectory function* $f_\mathcal{P}$ is defined by $f_\mathcal{P}(x) = \sup\{t : x \in P_t\}$, which gives the boundary surface of the closure of the pyramid. We can observe that the trajectory function is a unimodal function if each region in $\mathcal{F}$ is connected.

The horizontal section P_t of height t of a pyramid $\mathcal{P}$ can be represented by $\{x : f_\mathcal{P}(x) \geq t\}$. The *exposed part* of a section of height t is $\{x : f_\mathcal{P}(x) = t\}$. If the exposed part of a section is non-empty, we call it a *flat* of $\mathcal{P}$. The flat with the maximum height is called the *top flat*. By definition, a flat must be a set difference of two members of $\mathcal{F}$.

A *balancing region* for t is $R \setminus R'$ of a pair $R \supset R'$ in $\mathcal{F}$ satisfying that $\int_{x \in R \setminus R}(\rho(x) - t \cdot \mu(x))dx = 0$. A flat of $\mathcal{P}$ is called a *balancing flat* if it is a balancing region for the height t of the flat.

Lemma 1. *For the optimal pyramid $\mathcal{P}$, the function $f_{\mathcal{P}}$ satisfies the following three conditions. Moreover, a function satisfying them gives the boundary surface of an optimal pyramid.*
(1) $f_{\mathcal{P}}$ is boundary surface of a pyramid.
(2) Each flat F of the above pyramid is a balancing flat;
Consequently, $\int_{x \in F} f_{\mathcal{P}}(x)\mu(x)dx = \int_{x \in F} \rho(x)dx$.
(3) Among all functions satisfying (1) and (2), $f_{\mathcal{P}}$ maximizes the potential $U(\mathcal{P}) = \int_{x \in C}(f_{\mathcal{P}}(x))^2\mu(x)dx$.

Proof. The first condition is trivial, and we will show $f_{\mathcal{P}}$ satisfies (2) and (3). In each flat of height t of the optimal pyramid, the two regions $P = P_{t-\epsilon}$ and $P' = P_{t+\epsilon}$ for an infinitesimally small ϵ should have the same gain at t, and the flat must be $P' \setminus P$; hence the flat must be balanced, and (2) holds.

At each flat F with height t_F, the potential $U(\mathcal{P}, F) = \int_{x \in F}(f_{\mathcal{P}}(x)^2\mu(x))dx$ within F is $(t_F)^2\mu(F)$. On the other hand, since F is balancing, $\rho(F) = t \cdot \mu(F)$; hence, $V(\mathcal{P}, F) = \int_{t < t_F}(\rho(F) - t \cdot \mu(F))dt = t_F \cdot \rho(F) - \frac{t_F^2}{2} \cdot \mu(F) = \frac{t_F^2}{2} \cdot \mu(F)$. Therefore, $U(\mathcal{P}, F) = 2V(\mathcal{P}, F)$. This gives (3), since the sum of $V(\mathcal{P}, F)$ over all flats equals $V(\mathcal{P})$, while that of $U(\mathcal{P}, F)$ gives the potential $U(\mathcal{P})$ defined in (3).

From the above proof, we can also see that a function satisfying the three conditions give the boundary surface of an optimal pyramid, since otherwise there exists a pyramid with a larger value of $V(\mathcal{P})$, and thus a larger potential; a contradiction.

Proposition 1. *Let $f_{\mathcal{P}}$ be a surface function of a pyramid. Then, $\mathcal{P}$ is the optimal pyramid if and only if $E(\mathcal{P}) = \int_{x \in C}(f_{\mathcal{P}}(x) - \rho(x)/\mu(x))^2\mu(x)dx$ is minimized.*

Proof. It is easy to see that each flat of the pyramid must be balanced in order to minimize $E = E(\mathcal{P})$. It suffices to show that $E = \int_{x \in C}\rho^2(x)/\mu(x)dx - U(\mathcal{P})$. $1/2\{E - \int_{x \in C}\rho^2(x)/\mu(x)dx + U(\mathcal{P})\} = \int_{x \in C}f_{\mathcal{P}}(x)^2\mu(x) - f_{\mathcal{P}}(x)\rho(x)dx = \sum_{F:flats}\int_{x \in F}t_F(t_F\mu(x) - \rho(x))dx = 0$ because of the balancing condition.

Consider a special case where $\mu \equiv 1$. Then, (2) of Lemma 1 says that we obtain a flat of $f_{\mathcal{P}}$ by leveling ρ locally, and (3) means that loss of potential is minimized. Indeed, if $d = 2$, we can consider ρ and $f_{\mathcal{P}}$ as surfaces of a terrain and its corresponding pyramid, and their potential (after normalization) corresponds to the positional potential of the terrain and the pyramid assuming that the gravity is a constant. Thus, if $\mu \equiv 1$, regarding ρ as a terrain, we have the optimal pyramid by moving earth from higher cells to lower cells with the minimum loss of positional potential to form a unimodal terrain such that each horizontal section belongs to $\mathcal{F}$.

Also, Proposition 1 says that the surface function of the optimal pyramid is the one approximating $\rho(x)/\mu(x)$ with the minimum L_2 error.

2.2 Closed Family

Let R_t^{opt} be the region in $\mathcal{F}$ maximizing $g_t(R; \rho, \mu)$. Intuitively, if t increases, R_t^{opt} is shrunk. If $\{R_t^{opt}\}$ forms a pyramid , it is obviously the optimal pyramid. Unfortunately, it is not always true that it is a pyramid; however, it turns out to be true if $\mathcal{F}$ satisfies certain conditions.

A (discrete) family $\mathcal{F}$ of regions in $\mathcal{R}^d$ is called a *closed family* if it is closed under intersection and union operations, that is, $R \cap R' \in \mathcal{F}$ and $R \cup R' \in \mathcal{F}$ for every pair R, R' of regions in $\mathcal{F}$.

Proposition 2. *If $\mathcal{F}$ is a closed family, and let P_t^{opt} be the region in $\mathcal{F}$ maximizing $g_t(R, \rho, \mu)$. If there are more than one regions maximizing $g_t(R, \rho, \mu)$, we take any one which is minimal under inclusion. Then, $\{P_t^{opt}\}_{t \geq 0}$ gives the optimal pyramid $\mathcal{P}$.*

Proof. It suffices to show that $A = P_t^{opt} \subseteq B = P_{t'}^{opt}$ if $t > t'$. Note that the function $g_t(R, \rho, \mu) - g_t(R', \rho, \mu)$ is a nonincreasing function in t if $R' \subset R$. If $A \backslash B$ is not empty, $0 \geq g_{t'}(A \cup B, \rho, \mu) - g_{t'}(B, \rho, \mu) = g_{t'}(A, \rho, \mu) - g_{t'}(A \cap B, \rho, \mu) \geq g_t(A, \rho, \mu) - g_t(A \cap B, \rho, \mu) \geq 0$. Since $A = P_t^{opt}$, $g_t(A, \rho, \mu) \leq g_t(A \cap B, \rho, \mu)$; thus, $g_t(A, \rho, \mu) = g_t(A \cap B, \rho, \mu)$ and because of minimality under inclusion, $A = A \cap B$. Thus, $A \subseteq B$.

Thus, if a family $\mathcal{F}$ is constructed by using closed families as its building blocks, we have hope to design an efficient algorithm for computing the optimal pyramid for $\mathcal{F}$.

3 Algorithms for Computing the Optimal Pyramid

In [4], $O(n \log n)$ time algorithm is given for the one-dimensional problem where $\mathcal{F}$ is the set of all integral intervals in $[0, n]$. Moreover, it is shown that in any dimensional case, the optimal pyramid can be computed $O(M^2 N)$ time if the region family $\mathcal{F}$ has M regions. However, the algorithm is not efficient even for some typical problems in two dimensions.

3.1 Connected Lower Half Regions

A subregion R of the $n \times n$ pixel plane is called a rectilinear lower half region (lower half region, in short) if there is a function $f_R(x)$ such that R is the union of pixels satisfying $y \leq f_R(x)$ holds in each of them. In general, a lower half region need not be connected The family of connected lower half regions (called based-monotone regions in [6,8]) is one of region families adopted in SONAR data mining system [6,8] in order to represent two-dimensional association rules The family of connected lower half regions is not a closed family, since the union of two nonintersecting connected lower half regions is not connected. However, given any index i, the family of connected lower half regions having $(i, 0)$ on their boundaries is a closed family, which we call connected i-lower half regions. Each column of a horizontal section of an optimal pyramid at t is an optimal nonempty prefix that maximizes the parametric gain (it may be negative). There are $O(N)$ critical values of t at which the optimal prefix changes. We can compute the sorted set S of these critical values and preprocess the columns in $O(N \log N)$ time to have a prefix query data structure such that the nonempty optimal prefix of each column at t can be queried in $O(\log n)$ time.

Lemma 2. *The optimal pyramid for the family of connected i-lower half regions for a fixed i can be computed in $O(N)$ time, assuming we have the sorted set S.*

Proof. At each t, the optimal region is a union of optimal nonempty prefixes of columns over an interval I_t of column indices, and I_t is monotonic with respect to t. By using our prefix query data structure, the optimal nonempty prefixes at t of all columns can be computed in $O(n \log n) = O(\sqrt{N} \log N)$ time. Now, the problem becomes one-dimensional, and we can compute the horizontal section P_t in $O(n) = O(\sqrt{N})$ time. We now compute a pair $t_0 < t_1$ of parameter values (called *breaking values*) satisfying that the leftmost column index of P_t is greater than $i/2$ if $t < t_0$ and less than $i/2$ if $t \geq t_1$. We first run binary searching on S, and then, once all the column prefixes are determined, we can easily compute t_0 and t_1.

Next, we branch the process, so that for $t < t_0$, we only consider the first $i/2 - 1$ indices, and for $t > t_1$, we consider the indices in $(i/2, i]$. Continuing this branching of process, we can compute all the breaking values in $O(\sqrt{N} \log^3 N)$ time. We analogously consider for the rightmost column index. Now, we can compute the optimal pyramid easily in $O(N)$ additional time.

Since there are $n = \sqrt{N}$ candidates of i, we can compute the optimal pyramid for the family of all connected lower half regions in $O(N^{3/2})$ time. Furthermore, we can improve the time complexity as follows:

Theorem 1. *The optimal pyramid for the family of all connected lower half regions can be computed in $O(N \log^2 N)$ time.*

Proof. First, we improve the query data structure for the nonempty optimal prefixes of columns at t so that the prefixes of k consecutive columns can be queried in $O(k + \log n)$ time. This can be done by applying fractional cascading technique [9], and the preprocessing time is $O(N \log N)$. Thus, we can compute all the breaking values of the locally optimal pyramid $\mathcal{P}(i)$ for the family of connected i-lower half in $O(n \log^2 n) = O(\sqrt{N} \log^2 N)$ time, improving the time complexity given in Lemma 2 by a factor of $\log N$. Now, consider the set $S = S_0 \cup S_1$, where S_0 consists of all the breaking values of n locally optimal pyramids $\mathcal{P}(1), \mathcal{P}(2), \ldots, \mathcal{P}(n)$ and S_2 consisits of all the values of t at which nonempty optimal prefixes of columns change. There are at most $3N$ elements in S, and S is constructed in $O(N \log^2 N)$ time. We consider an interval tree on $[1, n]$ corresponding to the column indices. At each $t \in S$, we have at most $2n$ column indices indicating left and right endpoints of the intervals corresponding to the horizontal sections at t of the local optimal pyramids. We apply sweep method in which we decrease the parameter value t. We design a data structure such that we can compute the sum of nonempty optimal prefixes of columns in any query interval I in $O(\log n)$ time at the current value of t. This is easy by using the interval tree, since we need to maintain the sum within each principal interval as a function of t. The number of update is $O(N)$, and each update of the data structure needs $O(\log n)$ time. Now, by using this data structure, we construct $\mathcal{P}(i)$ for $i = 1, 2, \ldots, n$ by using the sweep algorithm. We can construct a new flat of each pyramid in $O(\log n)$ time, and hence we can compute the pyramids in $O(n^2 \log n) = O(N \log N)$ time. Hence, the total time complexity is $O(N \log^2 N)$, which is dominated by the construction time of S.

3.2   Downstep Regions

In many applications, we want to consider a family of regions whose boundary curves have nice shapes.

A subregion R of $\mathcal{G}$ is called a downstep region if there is a nonincreasing function $f_R(x)$ such that R is the union of pixels satisfying $y \leq f_R(x)$ in each of them. There are exactly 2^n members in the family.

Lemma 3. *The optimal region P_t^{opt} for the family of downstep regions can be computed in $O(N)$ time.*

Proof. We first compute $g(i,j) = \sum_{s=1}^{i} \rho((s,j)) - t\mu((s,j))$, which is the gain in the first i pixels in the j-th column. This can be done in $O(N) = O(n^2)$ total time for all (i,j). We sweep the pixel grid $\mathcal{G}$ from left to right, and compute $f(i,j)$ which is the maximum gain of a downstep region up to the j-th column such that the region contains (i,j)-th pixel as the top pixel at the j-th column. Then it is easy to see that $f(i,j) = \max\{f(i,j-1) + g(i,j), f(i+1,j) - g(i+1,j) + g(i,j)\}$. Thus, we can compute all of $f(i,j)$ in $O(N)$ time by dynamic programming. The gain of the optimal region is $max_{i=0}^n f(i,n)$, and the region can be computed by backtracking the dynamic programming process.

Lemma 4. *Suppose we have the optimal region $P_{t(1)}^{opt}$ and $P_{t(2)}^{opt}$ for a parameter value $t(1) < t(2)$, and $P_{t(1)}^{opt} \setminus P_{t(2)}^{opt}$ has Y pixels. Then, for any t satisfying $t(1) < t < t(2)$, P_t^{opt} can be computed in $O(Y)$ time. Here, we output the boundary curve of P_t^{opt} by using that of $P_{t(2)}^{opt}$ and indicating how to update it.*

Proof. It suffices to consider each connected component of $P_{t(1)}^{opt} \setminus P_{t(2)}^{opt}$ in the pixel grid $\mathcal{G}$ separately. Let C be a component. We can apply a simple dynamic programming algorithm sweeping the component C from left to right.

Theorem 2. *Suppose that the values of ρ and μ are quotient numbers of integers less than Γ. The optimal pyramid for the family of downstep regions can be computed in*
$O(N \log(N\Gamma))$ *time.*

Proof. We consider a binary decomposition process of the interval $I = [a, b]$ containing possible values of t, where $a = \rho(G)/\mu(G)$ and $b = \max_{c \in G} \rho(c)/\mu(c)$. In the k-th level of the binary decomposition process, the interval is decomposed into $s = 2^k$ subintervals $I_1, I_2, \ldots, I_s$. Thus, we have $O(\log N\Gamma)$ levels. The time complexity for computing all optimal pyramids with respect to the center values t_i of the intervals I_i $(i = 1, 2, \ldots, I_s)$ is $O(\sum_{i=1}^{s} Y_i) = O(N)$, where Y_i is the difference of the number of pixels in the optimal regions at the left and right ends of the interval I_i. Thus, we have the time complexity.

Remark: We may apply parametric searching paradigm to have an $O(N^{3/2} \log^2 n)$ time algorithm. We omit it in this paper, since the $O(N \log(n\Gamma))$ time complexity is almost always better in practice.

3.3   Point-Stabbed Union at a Grid Point

In many applications, we want to consider a family of regions whose boundary curves have nice shapes. We fix a grid point $p = (x_p, y_p)$ of $\mathcal{G}$. A region R is called *point-stabbed union* (of rectangles) at p if it is a union of rectangles each of which contains p in its closure. Indeed, it is the "closure under union operations" of the set of rectangles containing p, and hence the smallest closed family containing all rectangles containing p.

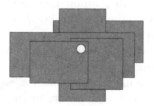

Fig. 3. A stabbed union of rectangles.

Fig. 4. A pyramid using stabbed unions.

A discretized rhombus (whose diagonals are axis-parallel) is a point-stabbed union, thus a horizontal section of a real pyramid (rotated by 45 degrees) is a point-stabbed union. Moreover, discretization of an axis parallel elliptic region is a point-stabbed union. Let $\mathcal{P}(\mathcal{S}(p))$ be the optimal pyramid with respect to the point-stabbed unions at p, and we call the one among $\{\mathcal{P}(\mathcal{S}(p))|p \in \mathcal{G}\}$ maximizing the potential the optimal pyramid with respect to point-stabbed unions without fixing the peak; an example is shown in Figure 2.

It is indeed a restricted class of rectilinear convex regions, such that if we cut a point-stabbed union with lines $x = x_p$ and $y = y_p$, we have four (reflected) downstep regions. Thus, we have the following as a corollary of Theorem 2.

Theorem 3. *The optimal pyramid for the family of point-stabbed unions at p can be computed in $O(N \log(N\Gamma))$ time.*

Thus, we can compute the optimal pyramid of point-stabbed unions without fixing the peak in $O(N^2 \log(N\Gamma))$ time. We also give a faster approximation algorithm.

Lemma 5. *The optimal regions $P_t^{opt}(p)$ for the family of point-stabbed unions at p for all $p \in \mathcal{G}$ can be computed in $O(N^{1.5})$ time for each t.*

Proof. For a point $(i, j) \in \mathcal{G}$, consider the north-left quarter region $NW(P_t^{opt}(i, j))$ (that is an upstep region obtained by reflecting a downstep region at p) of $P_t^{opt}((i, j))$. Fixing i, we design a dynamic programming algorithm that computes the upstep regions for the left of j-the column for all $j = 1, 2, ..., n$ simultaneously. We first compute $\gamma(i, j) = \sum_{s=1}^{i} \rho((s, j)) - t \cdot \mu((s, j))$, which is the gain in the first i pixels in the j-th column. This can be done in $O(N) = O(n^2)$ total time for all (i, j). We sweep the pixel grid $\mathcal{G}$ from left to right, and compute $\alpha(i, j)$ which is the maximum gain of a upstep region up to the j-th column such that the region contains (i, j)-th pixel as the top pixel at the j-th column. Then it is easy to see that $\alpha(i, j) = \max\{\alpha(i, j-1)+\gamma(i, j), \alpha(i-1, j)-\gamma(i-1, j)+\gamma(i, j)\}$. Since this computation is done in constant time for each (i, j), we can compute all of $\alpha(i, j)$ in $O(N)$ time by dynamic programming. The gain of the optimal region is $max_{i=0}^{n}\alpha(i, n)$, and the region can be computed by backtracking the dynamic programming process.

Thus, $NW(P_t^{opt}(i, j)$ for $j = 1, 2, ..., n$ can be computed in $O(N)$ time. Thus, $NW(P_t^{opt}(i, j))$ for all $(i, j) \in \mathcal{G}$ can be computed in $O(N\sqrt{N})$ time. The other quarters of the optimal regions can be analogously computed.

A pyramid $\mathcal{P}$ for a family $\mathcal{F}$ is an ϵ-approximation of the optimal pyramid if $U(\mathcal{P}) > (1 - \epsilon)U(\mathcal{P}_{opt})$, where $\mathcal{P}_{opt}$ is the optimal pyramid.

Theorem 4. *For all $p \in \mathcal{G}$, ϵ-approximations of pyramids of the point-stabbed union at p can be computed in $O(\epsilon^{-1}N^{1.5} \log N)$ time.*

Proof. We simply compute optimal regions for $t = t_1 < t_2 < \ldots < t_l$ where t_l is the maximum value in the input pixels and $t_1 = T/N$, where T is the sum of values of all pixels, such that $t_i = (1 - \epsilon)^{-1}t_{i-1}$ for $2 \leq i \leq i - 1$ and $t_l \leq (1 - \epsilon)^{-1}t_{l-1}$. It is easy to see that $l = O(\epsilon^{-1} \log N)$. Once we compute the optimal regions at these values, we can compute the heights of the regions so that the each flat becomes balanced. This gives the $(1 - \epsilon)$ approximations for all p.

This algorithm can be used as a subroutine for computing the (exact) optimal pyramid with respect to the point-stabbed unions without given the peak p. By running the above approximation algorithm, we can prune away most of grid points from the candidate list of the peaks of the optimal pyramid. Suppose that there remains L candidates. Then, we can run the $O(N \log(N\Gamma))$ time algorithm for compute the optimal pyramid at each of the candidate peaks, and thus the total time complexity becomes $O(NL \log N\Gamma + \epsilon^{-1}N^{1.5} \log n)$. This is usually much more efficient than $O(N^2 \log(N\Gamma))$ time algorithm that simply runs the $O(N \log(N\Gamma))$ time algorithm for each of N grid points.

4   Concluding Remarks

Our approximation algorithm uses $U(\mathcal{P})$ for the objective function. It will be nice if we have an algorithm with a good approximation ratio using $E(\mathcal{P})$ as the objective function. Higher dimensional extension of the results in this paper is considered in [3].

If we remove the stabbing point p from a point-stabbed region, it may become disconnected, thus, the interior of a point-stabbed region need not be connected. Instead of point-stabbed regions, we can consider cell-stabbed regions, which are unions of rectangles containing a cell c. The interior of a cell-stabbed region is always connected. Algorithms for computing pyramids for cell-stabbed regions are given in [3].

References

1. T. Asano, D. Chen, N. Katoh, and T. Tokuyama, Efficient Algorithms for Optimization-Based Image Segmentation, *International Journal of Computational Geometry and Applications* **11** (2001) 145-166.
2. I. Bloch, Spatial Relationship between Objects and Fuzzy Objects using Mathematical Morphology, in *Geometry, Morphology and Computational Imaging*, 11th Dagsthul Workshop on Theoretical Foundations of Computer Vision, April 2002.
3. D. Chen, Z. Chun, N. Katoh, T. Tokuyama, Layered Data Segmentation for Numeric Data Mining, in preparation.
4. Z. Chun, N. Katoh, T. Tokuyama, How to reform a terrain into a pyramid, preprint.
5. T. Fukuda, Y. Morimoto, S. Morishita, and T. Tokuyama, Mining Optimized Association Rules for Numeric Attributes, *Journal of Computer and System Sciences* **58** (1999) 1-12.
6. T. Fukuda, Y. Morimoto, S. Morishita, and T. Tokuyama, Data Mining with Optimized Two-Dimensional Association Rules, *ACM Transaction of Database Systems* **26** (2001) 179-213.
7. Y. Morimoto, H. Ishii and S. Morishita, Construction of Regression Trees with Range and Region Splitting, *The 23rd VLDB Conference* (1997) 166-175.
8. Y. Morimoto, T. Fukuda, S. Morishita, and T. Tokuyama, Implementation and Evaluation of Decision Trees with Range and Region Splitting, *Constraints* (1997) 402-427.
9. F. P. Preparata and M. I. Shamos, *Computational Geometry – An Introduction*, Springer-Verlag, 1988 (2nd ed.).

On the Face Lattice of the Metric Polytope

Antoine Deza[1], Komei Fukuda[2], Tomohiko Mizutani[3], and Cong Vo[3]

[1] McMaster University, Department of Computing and Software, Hamilton, Canada
deza@mcmaster.ca
[2] McGill University, School of Computer Science, Montréal, Canada
fukuda@cs.mcgill.ca
[3] Tokyo Institute of Technology, Math. and Comput. Sci. Dept., Tokyo, Japan

Abstract. In this paper we study enumeration problems for polytopes arising from combinatorial optimization problems. While these polytopes turn out to be quickly intractable for enumeration algorithms designed for general polytopes, algorithms using their large symmetry groups can exhibit strong performances. Specifically we consider the metric polytope m_n on n nodes and prove that for $n \geq 9$ the faces of codimension 3 of m_n are partitioned into 15 orbits of its symmetry group. For $n \leq 8$, we describe additional upper layers of the face lattice of m_n. In particular, using the list of orbits of high dimensional faces, we prove that the description of m_8 given in [9] is complete with 1 550 825 000 vertices and that the LAURENT-POLJAK conjecture [16] holds for $n \leq 8$. Computational issues for the orbitwise face and vertex enumeration algorithms are also discussed.

1 Introduction

A full d-dimensional convex (bounded) polytope P can be defined either by the linear inequalities associated to the set $\mathcal{F}(P)$ of its facets or as the convex hull of its vertex set $\mathcal{V}(P)$. The computation of $\mathcal{F}(P)$ from $\mathcal{V}(P)$ is the facet enumeration problem and the computation of $\mathcal{V}(P)$ from $\mathcal{F}(P)$ is the vertex enumeration problem. These two problems are equivalent by the vertex/facet duality. More generally, any proper face f of P can be defined either by the subset $\mathcal{F}(f)$ of facets containing f or as the convex hull of the vertices $\mathcal{V}(f)$ belonging to f. Given the facet set $\mathcal{F}(P)$, the face enumeration problem consists in enumerating all the faces f of P in terms of facet sets $\mathcal{F}(f)$. These computationally difficult problems have been well studied; see [2,3,14] and references there. In this paper, we consider combinatorial polytope, i.e. polytopes arising from combinatorial optimization problems. These polytopes are often trivial for the very first cases and then the so-called combinatorial explosion occurs even for small instances. On one hand, combinatorial polytopes are quickly intractable for enumeration algorithm designed for solving general polytope, but on the other hand, algorithms using their large symmetry groups allow enumerations which were not possible otherwise. For example, large instances of the traveling salesman polytope, the linear ordering polytope, the cut polytope and the metric polytope were computed in [4,9] using the same algorithm called *adjacency decomposition*

J. Akiyama and M. Kano (Eds.): JCDCG 2002, LNCS 2866, pp. 118–128, 2003.

method in [4] and *orbitwise vertex enumeration algorithm* in [9] which, given a vertex, find the adjacent ones, see Section 5.3 for more details. In this paper, pursuing the same approach, we propose an orbitwise face enumeration algorithm for combinatorial polytope. Focusing on the face lattice of the metric polytope m_n, we compute its upper layers for $n \leq 9$. These results allow us to prove that the description of m_8 given in [9] is complete with 1 550 825 000 vertices and that the *dominating set* and *no cut-set* conjectures, see [9,16], hold for m_8. A description of the faces of codimension 3 for any n is given as well as some preliminary results on the vertices of m_9.

2 Face Enumeration for Combinatorial Polytopes

2.1 Combinatorial Polytopes

Many combinatorial polyhedra are associated to optimization problems arising from the complete directed graph D_n or the undirected graph K_n on n nodes. Well studied combinatorial polyhedra include the cut polytope c_n and the metric polytope m_n. While c_n is the convex hull of the incidence vectors of all the cuts of K_n, m_n can be defined as a relaxation of c_n by the triangular inequalities, see Section 3.1 and see [12] for more details. One important feature of most combinatorial polytopes is their very large symmetry group. We recall that the symmetry group $Is(P)$ of a polytope P is the group of isometries preserving P. For $n \geq 5$, $Is(m_n) = Is(c_n)$ is induced by the $n!$ permutations on V_n and the 2^{n-1} *switching reflections*, see Section 3.2, and $|Is(m_n)| = 2^{n-1}n!$, see [11]. For $n = 4$, $m_4 = c_4$ and $|Is(m_4)| = 2(4!)^2$. Clearly all faces are partitioned into orbits of faces equivalent under permutations and switchings. An orbitwise vertex enumeration algorithm was proposed in [4,9] and, in a similar vein, we propose an orbitwise face enumeration algorithm.

2.2 Orbitwise Face Enumeration Algorithm

The input is a full d-dimensional polytope P defined by its (non-redundant) facet set $\mathcal{F}(P) = \{f_1^{d-1}, \ldots, f_m^{d-1}\}$. The algorithm first computes the list $\mathcal{L}^{d-1} = \{\tilde{f}_1^{d-1}, \ldots, \tilde{f}_{I^{d-1}}^{d-1}\}$ of all the canonical representatives of the orbits of facets. Then the algorithm generates the set $L^{d-2} = \{\tilde{f}_s^{d-1} \cap f_r^{d-1} : s = 1, \ldots, I^{d-1}, r = 1, \ldots, m\}$. After computing the dimension of each subface $\tilde{f}_s^{d-1} \cap f_r^{d-1}$ and keeping only the $(d-2)$-faces, the algorithm reduces L^{d-2} to the list of canonical representatives of orbits of $(d-2)$-faces $\mathcal{L}^{d-2} = \{\tilde{f}_1^{d-2}, \ldots, \tilde{f}_{I^{d-2}}^{d-2}\}$. In general, after generating the list $\mathcal{L}^{d-t+1}$, the algorithm computes $\mathcal{L}^{d-t}$ by:

(i) generating the set L^{d-t} by intersecting each canonical representative $\tilde{f}_s^{d-t+1}$ with each facet F_r for $s = 1, \ldots, I^{d-t+1}$ and $r = 1, \ldots, m$,

(ii) computing the set $\mathcal{F}(\tilde{f}^{d-t+1} \cap f^{d-1})$ of all facets containing $\tilde{f}^{d-t+1} \cap f^{d-1}$ and then its dimension $dim(\tilde{f}^{d-t+1} \cap f^{d-1})$

(iii) for $dim(\tilde{f}^{d-t+1} \cap f^{d-1}) = d - t$, computing the canonical representative $\tilde{f}^{d-t}$ of $\tilde{f}^{d-t+1} \cap f^{d-1}$

The algorithm terminates after the list $\mathcal{L}^0$ of canonical representatives of the orbits of vertices is computed. Clearly the algorithm works faster when the symmetry group $Is(P)$ is larger. The main two subroutines are the computation of the canonical representative $\tilde{f}$ of the orbit O_f generated by a face f and the computation of the dimension $dim(f)$. The determination of $\mathcal{F}(\tilde{f}^{d-t+1} \cap f^{d-1})$ amounts to a redundancy check for the remaining facets of $\mathcal{F}(P)\backslash\{\mathcal{F}(\tilde{f}^{d-t+1} \cap f^{d-1})\}$. This operation can be done using $ccclib$ ($redcheck$), see [13], and is polynomially equivalent to linear programming; see [3]. The rank of $\mathcal{F}(\tilde{f}^{d-t+1} \cap f^{d-1})$ directly gives $dim(\tilde{f}^{d-t+1} \cap f^{d-1})$. The computation of the canonical representative $\tilde{f}^{d-t}$ is done using a brute-force approach; that is, by generating all the elements belonging to the orbit $O_{\tilde{f}^{d-t+1} \cap f^{d-1}}$.

Remark 1.

1. With I^{d-t} the number of orbits of $(d-t)$-faces and m the number of facets, the dimension (resp. canonical representative) computation subroutine is called exactly (resp. at most) $m(1 + \sum_{t=1,\ldots,d-1} I^{d-t})$ times.

2. The output; that is, for $t = 1,\ldots,d$ the list $\mathcal{L}^{d-t}$ of canonical representatives $\tilde{f}_s^{d-t} : s = 1,\ldots,I^{d-t}$, is extremely compact. The full list of $(d-t)$-faces can be generated by the action of the symmetry group on each representative face $\tilde{f}_s^{d-t}$. With $|O_{\tilde{f}_s^{d-t}}|$ the size of the orbit generated by $\tilde{f}_s^{d-t}$, the total number of faces is $\sum_{t=1,\ldots,d} \sum_{s=1,\ldots,I^{d-t}} |O_{\tilde{f}_s^{d-t}}|$.

Item 1 of Remark 1 indicates that the algorithm runs smoothly as long as the number I^{d-t} of orbits of $(d-t)$-faces is relatively small. The number of $(d-t)$-faces usually grows extremely large with t getting close to $\lfloor \frac{d}{2} \rfloor$; that is: "Face lattices are very fat". Therefore the computation of the full face lattice of a polytope is generally extremely hard. Besides small dimensional polytopes and specific cases such as the d-cube, we can expect a similar pattern for the values of I^{d-t}; that is: "Orbitwise face lattices are also fat". On the other hand, one can expect the combinatorial explosion to occur at a deeper layer for the orbitwise face lattice than for the ordinary one. Actually, this algorithm is particularly suitable for the computation of the upper τ layers of the orbitwise face lattice for a small given τ. In that case the algorithm stops when $\mathcal{L}^{d-\tau}$ is computed. The computation of the orbitwise upper face lattice can be efficiently combined with classical vertex enumeration. See Section 5.1 for an application to the complete description of the vertices of m_8.

3 Faces of the Metric Polytope

3.1 Cut and Metric Polytopes

The $\binom{n}{2}$-dimensional cut polytope c_n is usually introduced as the convex hull of the incidence vectors of all the cuts of K_n. More precisely, given a subset S of $V_n = \{1,\ldots,n\}$, the *cut* determined by S consists of the pairs (i,j) of elements of V_n such that exactly one of i, j is in S. By $\delta(S)$ we denote both the cut and its incidence vector in $\mathbb{R}^{\binom{n}{2}}$; that is, $\delta(S)_{ij} = 1$ if exactly one of i, j is in S and 0 otherwise for $1 \leq i < j \leq n$. By abuse of notation, we use the term cut for both the cut itself and its incidence vector, so $\delta(S)_{ij}$ are considered as coordinates of a point in $\mathbb{R}^{\binom{n}{2}}$. The cut polytope c_n is the convex hull of all 2^{n-1} cuts, and the *cut cone* C_n is the conic hull of all $2^{n-1} - 1$ nonzero cuts. The cut polytope and one of its relaxation - the metric polytope - can also be defined in terms of a finite metric space in the following way. For all 3-sets $\{i, j, k\} \subset \{1, \ldots, n\}$, we consider the following inequalities:

$$x_{ij} - x_{ik} - x_{jk} \leq 0, \tag{1}$$

$$x_{ij} + x_{ik} + x_{jk} \leq 2. \tag{2}$$

(1) induce the $3\binom{n}{3}$ facets which define the *metric cone* M_n. Then, bounding the latter by the $\binom{n}{3}$ facets induced by (2) we obtain the metric polytope m_n. The facets defined by (1) (resp. by (2)) can be seen as triangle (resp. perimeter) inequalities for distance x_{ij} on $\{1, \ldots, n\}$ and are denoted by $\Delta_{i,j,\bar{k}}$ (resp. by $\Delta_{i,j,k}$). While the cut cone is the conic hull of all, up to a constant multiple, $\{0, 1\}$-valued extreme rays of the metric cone, the cut polytope c_n is the convex hull of all $\{0, 1\}$-valued vertices of the metric polytope. For a detailed study of those polytopes and their applications in combinatorial optimization we refer to DEZA AND LAURENT [12] and POLJAK AND TUZA [17].

3.2   Combinatorial and Geometric Properties

The polytope c_n is a $\binom{n}{2}$-dimensional $\{0, 1\}$-polyhedron with 2^{n-1} vertices and m_n is a polytope of the same dimension with $4\binom{n}{3}$ facets inscribed in the cube $[0, 1]^{\binom{n}{2}}$. We have $c_n \subseteq m_n$ with equality only for $n \leq 4$. Any facet of the metric polytope contains a face of the cut polytope and the vertices of the cut polytope are vertices of the metric polytope. In fact, the cuts are precisely the integral vertices of the metric polytope. The metric polytope m_n wraps the cut polytope c_n very tightly. Indeed, in addition to the vertices, all edges and 2-faces of c_n are also faces of m_n, for 3-faces it is false for $n \geq 4$, see [7]. Any two cuts are adjacent both on c_n and on m_n; in other words m_n is *quasi-integral*; that is, the skeleton of the convex hull of its integral vertices, i.e. the skeleton of c_n, is an induced subgraph of the skeleton of the metric polytope itself. We recall that the skeleton of a polytope is the graph formed by its vertices and edges. While the diameters of the cut polytope and the dual metric polytope satisfy $\delta(c_n) = 1$ and $\delta(m_n^*) = 2$, the diameters of their dual are conjectured to be $\delta(c_n^*) = 4$ and $\delta(m_n) = 3$, see [6,16]. One important feature of the metric and cut polytopes is their very large symmetry group. More precisely, for $n \geq 5$, $Is(m_n) = Is(c_n)$ is induced by the permutations on $V_n = \{1, \ldots, n\}$ and the switching reflections by a cut and $|Is(m_n)| = 2^{n-1} n!$, see [11]. Given a cut $\delta(S)$, the switching reflection $r_{\delta(S)}$ is defined by $y = r_{\delta(S)}(x)$ where $y_{ij} = 1 - x_{ij}$ if $(i, j) \in \delta(S)$ and $y_{ij} = x_{ij}$ otherwise. For $n = 4$, $c_4 = m_4$ and there are some additional symmetries: $|Is(m_4)| = 2(4!)^2$. For the symmetry group of the cones C_n, M_n and some relatives, see [10]. Note that the symmetries preserve the adjacency relations and the linear independency.

3.3   Faces of the Metric Polytope

We recall some results and conjectures on the faces of the metric polytope. The cuts are the only integral vertices of m_n. Consider the following map ϕ_0 : $\mathbb{R}^{\binom{n-1}{2}} \longrightarrow \mathbb{R}^{\binom{n}{2}}$, defined by: $\phi_0(v)_{ij} = v_{ij}$ for $1 \leq i < j \leq n-1$, $\phi_0(v)_{i,n} = v_{1,i}$ for $2 \leq i \leq n-1$ and $\phi_0(v)_{1,n} = 0$. Both $\phi_0(v)$ and its switching by $\delta(\{n\})$ are called *trivial extensions* of v. Note that a trivial extension of a vertex of m_{n-1} is a vertex of m_n. Besides the cuts, all vertices with are not fully fractional are trivial extensions; that is, the *new vertices* of m_n are the fully fractional ones. The $(\frac{1}{3}, \frac{2}{3})$-valued fully fractional vertices are well studied, see [12,15,16], and include the anticut orbit formed by the 2^{n-1} *anticuts* $\bar{\delta}(S) = \frac{2}{3}(1, \ldots, 1) - \frac{1}{3}\delta(S)$. If $G = (V_n, E)$ is a connected graph, we denote by d_G its path metric, where $d_G(i,j)$ is the length of a shortest path from i to j in G for $i \neq j \in V_n$. Then $\tau(d_G) = max(d_G(i,j) + d_G(i,k) + d_G(j,k) : i,j,k \in G)$ is called the *triameter of* G and we set $x_G = \frac{2}{\tau(d_G)} d_G$. Any vertex of m_n of the form x_G for some graph is called a *graphic* vertex, see [12,15,16] and Fig. 1 for the graphs of 2 graphic $(\frac{1}{3}, \frac{2}{3})$-valued vertices of m_8. Note that for any connected graph $G = (V_n, E)$, we have $\tau(d_G) \leq 2(n-1)$ and that any $(\frac{1}{3}, \frac{2}{3})$-valued vertex v of m_n is (up to switching) graphic; that is, there exist a graph G and a cut $\delta(S)$ such that $v = r_{\delta(S)}(x_G)$. Since $m_3 = c_3$ and $m_4 = c_4$, the vertices of m_3 and m_4 are made of 4 and 8 cuts forming 1 orbit. The 32 vertices of m_5 are 16 cuts and 16 anticuts, i.e., form 2 orbits. The metric polytope m_6 has 544 vertices, see [16], partitioned into 3 orbits: cuts, anticuts and 1 orbit of trivial extensions; and m_7 has 275 840 vertices, see [8], partitioned into 13 orbits: cuts, anticuts, 3 orbits of trivial extensions, 3 $(\frac{1}{3}, \frac{2}{3})$-valued orbits and 5 other fully fractional orbits. For m_8, 1 550 825 600 vertices partitioned into 533 orbits (cuts, anticuts, 28 trivial extensions, 37 $(\frac{1}{3}, \frac{2}{3})$-valued and 466 other fully fractional) were found assuming Conjecture 2, see [9]. The description was conjectured to be complete.

Conjecture 1. [16] Any vertex of the metric polytope m_n is adjacent to a cut.

Conjecture 2. [9] For $n \geq 6$, the restriction of the skeleton of the metric polytope m_n to the non-cut vertices is connected.

Conjecture 2 can be seen as complementary to the Conjecture 1 both graphically and computationally: For any pair of vertices, while Conjecture 1 implies that there is a path made of cuts joining them, Conjecture 2 means that there is a path made of non-cuts vertices joining them. In other words, the cut vertices would form a *dominating set* but not a *cut-set* in the skeleton of m_n. On the other hand, while Conjecture 1 means that the enumeration of the extreme rays of the metric cone M_n is enough to obtain the vertices of the metric polytope m_n; Conjecture 2 means that we can obtain the vertices of m_n without enumerating the extreme rays of M_n. Note that for arbitrary graphs these are clearly independent. Conjecture 1 underlines the extreme connectivity of the cuts. Recall that the cuts form a clique in both the cut and metric polytopes. Therefore, if Conjecture 1 holds, the cuts would be a dominant clique in the skeleton of m_n implying that its diameter would satisfy $\delta(m_n) \leq 3$. The orbitwise description of the facets and ridges (faces of codimension 2) of m_n for any n was given in [6] as well as the face $\Delta_{1,2,3} \cap \Delta_{1,2,\bar{3}}$ of codimension $n-1$ and the face $\Delta_{1,2,3} \cap \Delta_{1,\bar{3},4}$ of codimension 3. We have $\mathcal{L}^{d-1}(m_n) = \{\Delta_{1,2,3}\}$ and $\mathcal{L}^{d-2}(m_{n\geq 6}) = \{\Delta_{1,2,3} \cap \Delta_{1,2,4},\ \Delta_{1,2,3} \cap \Delta_{1,4,5},\ \Delta_{1,2,3} \cap \Delta_{4,5,6}\}$, $\mathcal{L}^{d-2}(m_5) = \{\Delta_{1,2,3} \cap \Delta_{1,2,4},\ \Delta_{1,2,3} \cap \Delta_{1,4,5}\}$ and $\mathcal{L}^{d-2}(m_4) = \{\Delta_{1,2,3} \cap \Delta_{1,2,4}\}$. The full orbitwise face lattices of m_4 and m_5 were given in [7]. In Section 4.1 we compute additional orbits of faces of small metric polytopes and in Section 4.2 we characterize $\mathcal{L}^{d-3}(m_n)$ for any n.

4 Generating Faces of the Metric Polytope

4.1 Faces of Small Metric Polytopes

As stated earlier, generating the full face lattice is usually extremely hard. We restricted the computation to the enumeration of the upper τ layers of the orbitwise face lattice of m_n. We choose to set $\tau = 4$ for the partial orbitwise enumeration of m_6 (resp. m_7 and m_8). The first 4 entries of the f-vectors of m_6, m_7 and m_8 are: $f(m_6) = \{1, 3, 10, 34, \ldots\}$, $f(m_7) = \{1, 3, 13, 61, \ldots\}$ and $f(m_8) = \{1, 3, 14, 79, \ldots\}$, Due to space limitation, we refer to [5] for a detailed presentation. The set $\mathcal{L}^{d-3}(m_n)$ is easy to check for reasonable values of n as $I^{d-3}(m_n) \leq 15$, see Theorem 1. Additional properties of m_n can be used to increase the efficiency of the algorithm. In particular, the set L^{d-t} can be generated by considering for each s only the facets which are not equivalent under isometries preserving $\tilde{f}_s^{d-t+1}$. The *support* of $\Delta_{i,j,k}$ (or $\Delta_{i,j,\bar{k}}$) is $\sigma(\Delta_{i,j,k}) = \sigma(\Delta_{i,j,\bar{k}}) = \{i, j, k\}$. Let assume that, as in Section 5.2, we are interested only in the upper $n-1$ layers of the face lattice of m_n. In that case, when generating $\tilde{f}_s^{d-t+1} \cap \Delta_r$ with $t < n$, we can disregard Δ_r if $\sigma(\Delta_r) = \sigma(\Delta)$ for any $\Delta \in \mathcal{F}(\tilde{f}_s^{d-t+1})$ as for such Δ_r we have $codim(\tilde{f}_s^{d-t+1} \cap \Delta_r) \geq n-1$.

4.2 Faces of Codimension 3 of the Metric Polytope

As recalled earlier the first 2 upper layers of m_n are known for any n. We have $I^{d-1}(m_n) = 1$, $I^{d-2}(m_{n\geq 6}) = 3$ and, by Theorem 1, we get $I^{d-3}(m_{n\geq 9}) = 15$.

Theorem 1. *For $n \geq 9$, the faces of codimension 3 of the metric polytope m_n are partitioned into 15 orbits equivalent under permutations and switchings.*

The first 15 (resp. 14, 13, 10 and 6) representatives given in Table 1 generate the 15 (resp. 14, 13, 10 and 6) orbits of faces of codimension 3 of $m_{n\geq 9}$ (resp. m_8, m_7, m_6 and m_5). The first 2 representatives and the last one generate the 3 orbits of faces of codimension 3 of m_4.

Proof. For $n \leq 9$ Theorem 1 can be directly checked using the orbitwise face enumeration algorithm with $\tau = 3$; that is, the algorithm is set to compute only

Table 1. The orbits of faces of codimension 3 of m_n for $n \geq 4$.

Orbit $O_{f_i^3}$	Representative f_i^3	m_n for which f_i^3 is a $(d-3)$-face	$|O_{f_i^3}|$
$O_{f_1^3}$	$\Delta_{1,2,3} \cap \Delta_{1,2,4} \cap \Delta_{1,3,4}$	$m_{n\geq4}$	$32\binom{n}{4}$
$O_{f_2^3}$	$\Delta_{1,2,3} \cap \Delta_{1,2,4} \cap \Delta_{1,\bar{3},4}$	$m_{n\geq4}$	$24\binom{n}{4}$
$O_{f_3^3}$	$\Delta_{1,2,3} \cap \Delta_{1,2,4} \cap \Delta_{1,2,5}$	$m_{n\geq5}$	$160\binom{n}{5}$
$O_{f_4^3}$	$\Delta_{1,2,3} \cap \Delta_{1,2,4} \cap \Delta_{1,3,5}$	$m_{n\geq5}$	$960\binom{n}{5}$
$O_{f_5^3}$	$\Delta_{1,2,3} \cap \Delta_{1,2,4} \cap \Delta_{3,4,5}$	$m_{n\geq5}$	$480\binom{n}{5}$
$O_{f_6^3}$	$\Delta_{1,2,3} \cap \Delta_{1,2,4} \cap \Delta_{\bar{3},4,5}$	$m_{n\geq5}$	$480\binom{n}{5}$
$O_{f_7^3}$	$\Delta_{1,2,3} \cap \Delta_{1,2,4} \cap \Delta_{1,5,6}$	$m_{n\geq6}$	$5\,760\binom{n}{6}$
$O_{f_8^3}$	$\Delta_{1,2,3} \cap \Delta_{1,2,4} \cap \Delta_{3,5,6}$	$m_{n\geq6}$	$5\,760\binom{n}{6}$
$O_{f_9^3}$	$\Delta_{1,2,3} \cap \Delta_{1,4,5} \cap \Delta_{2,4,6}$	$m_{n\geq6}$	$3\,840\binom{n}{6}$
$O_{f_{10}^3}$	$\Delta_{1,2,3} \cap \Delta_{1,4,5} \cap \Delta_{\bar{2},4,6}$	$m_{n\geq6}$	$3\,840\binom{n}{6}$
$O_{f_{11}^3}$	$\Delta_{1,2,3} \cap \Delta_{1,2,4} \cap \Delta_{5,6,7}$	$m_{n\geq7}$	$6\,720\binom{n}{7}$
$O_{f_{12}^3}$	$\Delta_{1,2,3} \cap \Delta_{1,4,5} \cap \Delta_{1,6,7}$	$m_{n\geq7}$	$6\,720\binom{n}{7}$
$O_{f_{13}^3}$	$\Delta_{1,2,3} \cap \Delta_{1,4,5} \cap \Delta_{2,6,7}$	$m_{n\geq7}$	$40\,320\binom{n}{7}$
$O_{f_{14}^3}$	$\Delta_{1,2,3} \cap \Delta_{1,4,5} \cap \Delta_{6,7,8}$	$m_{n\geq8}$	$53\,760\binom{n}{8}$
$O_{f_{15}^3}$	$\Delta_{1,2,3} \cap \Delta_{4,5,6} \cap \Delta_{7,8,9}$	$m_{n\geq9}$	$17\,920\binom{n}{9}$
$O_{f_{16}^3}$	$\Delta_{1,2,3} \cap \Delta_{1,2,4} \cap \Delta_{1,3,4}$	m_4	$2\binom{n}{2}$

the upper 3 layers of the face lattice of m_n. Let assume $n \geq 9$, the faces of codimension 2 of m_n are partitioned into 3 orbits generated by $\Delta_{1,2,3} \cap \Delta_{1,2,4}$, $\Delta_{1,2,3} \cap \Delta_{1,4,5}$ and $\Delta_{1,2,3} \cap \Delta_{4,5,6}$. Any faces of codimension 3 of m_n can therefore be written as the intersection of a facet Δ of m_n with one of these 3 faces $\Delta' \cap \Delta''$ of codimension 2. If the support $\sigma(\Delta) \not\subset \{1, \ldots, 9\}$, by elementary permutations preserving Δ' and Δ'' we can generate $\tilde{\Delta} \in O_\Delta$ with $O_{\Delta' \cap \Delta'' \cap \tilde{\Delta}} = O_{\Delta' \cap \Delta'' \cap \Delta}$ and $\sigma(\tilde{\Delta}) \subset \{1, \ldots, 9\}$. In other words, to generate orbitwise all the subfaces of the canonical faces of codimension 2 it is enough to consider the case $n = 9$. This way one can easily obtain 28 faces f_i of codimension at least 3. Then, as for the orbitwise face enumeration algorithm, we have to compute for $i = 1, \ldots, 28$ and for any n the dimension $dim(f_i)$ and - if $codim(f_i) = 3$ - to compute the canonical representative $\tilde{f}_i$. Therefore we have to first determine the set $\mathcal{F}_n(f_i)$ of facets of m_n containing f_i. Clearly, if an inequality (i) defining a facet of m_n is forced to be satisfied with equality by the inequalities defining Δ', Δ'' and $\tilde{\Delta}$ being satisfied with equality, then the same inequality (i) - now seen as defining a facet of m_{n+1} - will also be forced to be satisfied with equality. In other words the set $\mathcal{F}_n(f_i)$ can only increase with n and $dim(f_i)$ can only decrease with n. Therefore, among the 28 faces f_i, only the 15 first faces of codimension 3 for m_9

given in Table 1 are candidates for being faces of codimension 3 for $m_{n \geq 9}$. A case by case study of the 15 faces, gives $\mathcal{F}_n(f_i)$ and proves that indeed these 15 faces generate 15 orbits of faces of codimension 3 for $n \geq 9$. The idea is simply to notice that the pattern of $\mathcal{F}_n(f_i)$ is essentially given by the value of $\mathcal{F}_{12}(f_i)$. Since all the cases are similar, we only present the computation of $\mathcal{F}_n(f_{15})$ where $f_{15} = \Delta_{1,2,3} \cap \Delta_{4,5,6} \cap \Delta_{7,8,9}$. Using the orbitwise face enumeration algorithm with $\tau = 3$, one can easily check that $\mathcal{F}_{12}(f_{15}) = \{\Delta_{1,2,3}, \Delta_{4,5,6}, \Delta_{7,8,9}\}$. Let $n \geq 12$ and Δ be a facet of m_n with $\sigma(\Delta) \not\subset \{1, \ldots, 12\}$. By elementary permutations preserving $\mathcal{F}_{12}(f_{15})$ we can generate $\tilde{\Delta} \in O_\Delta$ with $\sigma(\tilde{\Delta}) \subset \{1, \ldots, 12\}$. Let now consider $\tilde{\Delta}$ as a facet of m_{12}. Since $\tilde{\Delta} \notin \mathcal{F}_{12}(f_{15})$ at least one vertex v of m_{12} satisfies $v \in f_{15}$ and $v \notin \tilde{\Delta}$. Then, the $(n-12)$-times 0-extension v_{ext} of v is a vertex of m_n satisfying $v_{ext} \in f_{15}$ but $v_{ext} \notin \tilde{\Delta}$ where $\tilde{\Delta}$ is now considered as a facet of m_n. Thus, $\tilde{\Delta} \notin \mathcal{F}_n(f_{15})$ and, by the same elementary permutations, $\Delta \notin \mathcal{F}_n(f_{15})$; that is, $\mathcal{F}_n(f_{15}) = \{\Delta_{1,2,3}, \Delta_{4,5,6}, \Delta_{7,8,9}\}$ and $codim(f_{15}) = 3$ for any $n \geq 9$. In the same way, for $\mathcal{F}_n(f)$ increasing with n, the pattern of $\mathcal{F}_n(f)$ is essentially given by small values of n. Consider for example $\mathcal{F}_n(\Delta_{1,2,3} \cap \Delta_{1,2,\bar{3}})$: We have $\mathcal{F}_n(\Delta_{1,2,3} \cap \Delta_{1,2,\bar{3}}) = \{\Delta_{1,2,i}, \Delta_{1,2,\bar{i}} : i = 3, \ldots, n\}$ and therefore $|\mathcal{F}_n(\Delta_{1,2,3} \cap \Delta_{1,2,\bar{3}})| = 2(n-2)$ and $codim(\Delta_{1,2,3} \cap \Delta_{1,2,\bar{3}}) = n-1$. As for $\mathcal{F}_n(f_{15})$, one can compute $\mathcal{F}_4(\Delta_{1,2,3} \cap \Delta_{1,2,\bar{3}})$ and notice that $\Delta \in \mathcal{F}_{n \geq 5}(\Delta_{1,2,3} \cap \Delta_{1,2,\bar{3}}) \iff \tilde{\Delta} \in \mathcal{F}_4(\Delta_{1,2,3} \cap \Delta_{1,2,\bar{3}})$; that is, $\Delta = \Delta_{1,2,i}$ or $\Delta_{1,2,\bar{i}} : i = 4, \ldots, n$. □

Remark 2. The proof of Theorem 1 indicates that the number $I^{d-k}(m_n)$ of orbits of faces of codimension k of the metric polytope is probably constant for $n \geq 3k$. Another interesting issue is the determination of an upper bound for $I^{d-k}(m_n)$ for any k and n.

5   Generating Vertices of the Metric Polytope

5.1   Combining Orbitwise Face Enumeration with Classical Vertex Enumeration

As emphasized earlier, the face lattice is usually much larger than the number of vertices. Therefore, computing the full face lattice in order to obtain the vertices is extremely costly. On the other hand, the upper layers of the orbitwise face lattice might be relatively small. In that case the orbitwise face enumeration can be efficiently combined with a classical vertex enumeration methods in the following way. First, for an appropriate small τ, compute the upper orbitwise face lattice till the list $\mathcal{L}^{d-\tau}$ of canonical $(d-\tau)$-faces is obtained. Then for $s = 1, \ldots, I^{d-\tau}$, compute by a classical vertex enumeration method the set $\mathcal{V}(\tilde{f}_s^{d-\tau})$ of vertices belonging to $\tilde{f}_s^{d-\tau}$. Finally, compute the canonical representative $\tilde{v}$ for each vertex $v \in \mathcal{V}(\tilde{f}_s^{d-\tau})$. The set of all such vertices $\tilde{v}$ is exactly $\mathcal{L}^0$ as each canonical vertex of $\mathcal{L}^0$ belongs, up to an isometry of $Is(P)$, to at least one of the $(d-\tau)$-faces $\tilde{f}_s^{d-\tau}$. Clearly, the choice of τ is critical. Typically, for the first values of t, by going down one layer from $\mathcal{L}^{d-t+1}$ to $\mathcal{L}^{d-t}$ the number of orbits increases ($I^{d-t} \geq I^{d-t+1}$) and the average sizes of faces decreases ($\mathcal{V}(\tilde{f}_{average}^{d-t}) \leq$

$\mathcal{V}(\tilde{f}^{d-t+1}_{average})$). Therefore, a *good* τ should be such that $I^{d-\tau}$ and $\mathcal{V}(\tilde{f}^{d-\tau}_{average})$ are relatively small: In particular the largest $\tilde{f}^{d-\tau}_s$ should within problems currently solvable by vertex enumeration algorithms. In Section 5.2, assuming that the computation of m_{n-1} is just within current vertex enumeration abilities, we indicate that for m_n a good τ should satisfy $n-1 \leq \tau$ and that $\tau = 7$ is actually enough for the description of m_8. Note that $n-1 = 7 = \lfloor \frac{d}{4} \rfloor$ for m_8.

5.2   Vertices of the Metric Polytope on 8 Nodes

As mentioned earlier, the face $\tilde{f}^{d-n+1}_\mu = \Delta_{1,2,3} \cap \Delta_{1,2,\bar{3}}$ generates one orbit of faces of codimension $n-1$ of m_n which are combinatorially equivalent to m_{n-1}. In other words, the orbitwise face lattice of m_n contains a copy of m_{n-1} in $\mathcal{L}^{d-n+1}$. This implies that some canonical faces of $\mathcal{L}^{d-n+2}$ are quite larger than m_{n-1} and therefore intractable if we assume that m_{n-1} is just within current vertex enumeration methods abilities. For m_8, it means that we should compute at least $\mathcal{L}^{21}$ and it turns out to be enough as $\tilde{f}^{21}_\mu$ (which we do not need to enumerate since $\tilde{f}^{21}_\mu \simeq m_7$) and other elements of $\mathcal{L}^{21}$ are tractable. The whole computation is quite long as $\mathcal{L}^{21}$ is large as well as $\mathcal{V}(\tilde{f}^{21}_{average})$. For the same reasons, skipping $\tilde{f}^{21}_\mu$, the computation of the canonical vertices for each $\mathcal{V}(\tilde{f}^{21}_\mu)$ is also long. Insertion algorithms usually handle high degeneracy better than pivoting algorithms, see [2] for a detailed presentation of the main vertex enumeration methods. The metric polytope m_n is quite degenerate as the cut incidence $Icd_{\delta(S)} = 3\binom{n}{3}$ is much larger than the dimension $d = \binom{n}{2}$. We recall that the incidence $Icd_v = |\mathcal{F}(v)|$. Thus we choose an insertion algorithm for the enumeration of each $\tilde{f}^{21}_s$: the *cddlib* implementation of the double description method [13]. The ordering of the facet is lexicographic with the rule $-1 \prec 1 \prec 0$. The result shows that $\mathcal{L}^0$ is made of the 533 canonical vertices found in [9]. Due to space limitation, we refer to [5] for a detailed presentation. The conjectured description of m_8 being complete, the following is straightforward.

Proposition 1.

1. The metric polytope m_8 has exactly 1 550 825 600 vertices and its diameter is $\delta(m_8) = 3$. The metric cone M_8 has exactly 119 269 588 extreme rays.
2. The LAURENT-POLJAK dominating set Conjecture 1 and the no cut-set Conjecture 2 hold for m_8.

A vertex of a d-dimensional polytope is simple if $|\mathcal{F}(v)| = d$. While most of the vertices of m_8 are almost simple, the only simple vertices of m_8 belong to the orbits $O_{\tilde{v}_{532}}$ and $O_{\tilde{v}_{533}}$ of size $|O_{\tilde{v}_{532}}| = 368\,640$ and $|O_{\tilde{v}_{533}}| = 430\,080$; that is, only 0.05% of the total number of vertices of m_8 are simple. Both canonical representative $\tilde{v}_{532}$ and $\tilde{v}_{533}$ are graphic $(\frac{1}{3}, \frac{2}{3})$-valued vertices, see Fig. 1. The largest denominator among vertices of m_8 is 15 and occurs only for vertices of $O_{\tilde{v}_{451}}$ with $\tilde{v}_{451} = \frac{1}{15}(2,4,4,5,5,7,8,6,6,5,5,5,10,8,5,5,5,4,9,9,3,4,10,10,5,10,5,5)$ and $|O_{\tilde{v}_{451}}| = 2\,580\,480$. All vertices of m_8 are adjacent to at least 2 cuts and the vertices adjacent to exactly 2 cuts belong to $O_{\tilde{v}_{531}}$ with $|O_{\tilde{v}_{531}}| = 1\,290\,240$ and $\tilde{v}_{531} = \frac{1}{9}(2,2,3,3,4,4,5,4,3,3,6,6,3,5,5,2,6,3,6,3,3,6,3,3,6,4,5,3)$.

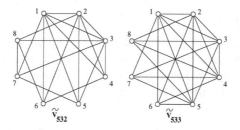

Fig. 1. Graphic canonical vertices of the only two orbits of simple vertices of m_8.

5.3   Vertices of the Metric Polytope on 9 Nodes

The computation of the vertices of m_9 is most probably intractable as we expect this extremely degenerate 36-dimensional polytope to have around 10^{14} vertices partitioned among several hundred thousand orbits. In this section, we present some computational results concerning the vertices of m_9. Given a vertex v, after computing the canonical representative $\tilde{v}$, the orbitwise vertex enumeration algorithm computes the set $N_{\tilde{v}}$ of vertices adjacent to $\tilde{v}$, then identifies all the orbits of vertices contained in $N_{\tilde{v}}$ and picks up the next representative whose neighborhood is not yet computed. The algorithm terminates when all the orbitwise neighborhoods $N_{\tilde{v}}$ are computed. To compute those neighborhoods, the algorithm performs one classic vertex enumerations for each orbit $O_{\tilde{v}}$ of vertices. The complexity of computing $N_{\tilde{v}}$ is closely related to the incidence of $\tilde{v}$. While the computation is easy for vertices having a small incidence, highly degenerated vertices can be intractable. For example, the algorithm failed to computed $N_{\delta(S)}$ the set of vertices adjacent to a cut $\delta(S)$ for m_8; we recall that $Icd_{\delta(S)} = 168$ and that $dim(m_8) = 28$. This remark leads to the following *skipping high degeneracy* heuristic: compute $N_{\tilde{v}}$ only for $Icd_{\tilde{v}} \leq Icd_{max}$ where Icd_{max} is an arbitrarily set in advance upper bound for the considered incidences, see [9]. In particular, assuming Conjecture 2 means that we can enumerate m_n by setting $Icd_{max} = Icd_{\delta(S)} - 1$. Setting $Icd_{max} = 44$, we computed 253 210 orbits of vertices of m_9. We wish to set $Icd_{max} = \frac{1}{2}\binom{n+1}{3} = 60$ (i.e. halfway from the dimension $\binom{n}{2}$ to the anticut incidence $Icd_{\bar{\delta}(S)} = \binom{n}{3}$, see [9]), but significantly raising the value of Icd_{max} is currently beyond our computational capacities. The largest denominator found is 39 and most of the vertices are almost simple but the lowest incidence is 37, i.e. no simple vertex was found so far. Conjecture 2 has been checked for the 253 210 orbits of vertices of m_9 computed so far.

Remark 3. None of the currently known vertices of m_9 is simple. Since m_6 and m_7 have no simple vertex, the only known simple vertices of m_n for $n \geq 6$ belong to the orbits $O_{\tilde{v}_{532}}$ and $O_{\tilde{v}_{533}}$. We believe that, while we can obtain nearly all the vertices of m_9 by setting $Icd_{max} = 44$, the algorithm can not reach all of them unless the value of Icd_{max} is raised significantly. We also believe that m_9 has simple vertices, see [8], which are among the currently unreachable vertices of m_9.

Acknowledgments

We would like to thank an anonymous referee for many helpful suggestions which improved the scientific value and the presentation of the paper. Many thanks to Professor Satoshi Matsuoka, Tokyo Institute of Technology, for letting us use his laboratory advanced PC cluster.

References

1. D. Avis: lrs homepage, School of Computer Science, McGill University, Canada (2001) http://www.cgm.cs.mcgill.ca/~avis/C/lrs.html
2. D. Avis, D. Bremner, R. Seidel: How good are convex hull algorithms? Computational Geometry: Theory and Applications 7 (1997) 265–301
3. D. Avis, K. Fukuda, S. Picozzi: On canonical representations of convex polyhedra. In: A.M. Cohen, X.S. Gao, N. Tokuyama (eds.): Mathematical Software, World Scientific (2002) 351–360
4. T. Christof, G. Reinelt: Decomposition and parallelization techniques for enumerating the facets of combinatorial polytopes. International Journal of Computational Geometry and Applications 11-4 (2001) 423–437
5. A. Deza: Metric polytope homepage, Math. and Comput. Sci. Dept., Tokyo Institute of Technology, Japan (2002)
   http://www.is.titech.ac.jp/~deza/metric.html
6. A. Deza, M. Deza: The ridge graph of the metric polytope and some relatives. In: T. Bisztriczky, P. McMullen, R. Schneider, A.I. Weiss (eds.): Polytopes: Abstract, Convex and Computational (1994) 359–372
7. A. Deza, M. Deza: The combinatorial structure of small cut and metric polytopes. In: T.H. Ku (eds.): Combinatorics and Graph Theory, World Scientific (1995) 70–88
8. A. Deza, M. Deza, K. Fukuda: On skeletons, diameters and volumes of metric polyhedra. In: M. Deza, R. Euler, Y. Manoussakis (eds.): Lecture Notes in Computer Science 1120 Springer-Verlag, Berlin Heidelberg New York (1996) 112–128
9. A. Deza, K. Fukuda, D. Pasechnik, M. Sato: On the skeleton of the metric polytope. In: J. Akiyama, M. Kano, M. Urabe (eds.): Lecture Notes in Computer Science 2098 Springer-Verlag, Berlin Heidelberg New York (2001) 125–136
10. A. Deza, B. Goldengorin, D. Pasechnik: e-print math.MG/0306049 at arXiv.org (2003) http://arXiv.org/abs/math.MG/0306049
11. M. Deza, V. Grishukhin, M. Laurent: The symmetries of the cut polytope and of some relatives. In: P. Gritzmann, B. Sturmfels (eds.): Applied Geometry and Discrete Mathematics, the "Victor Klee Festschrift" DIMACS Series in Discrete Mathematics and Theoretical Computer Science 4 (1991) 205–220
12. M. Deza, M. Laurent: Geometry of cuts and metrics. Algorithms and Combinatorics 15 Springer-Verlag, Berlin Heidelberg New York (1997)
13. K. Fukuda: cddlib reference manual, cddlib Version 0.92, ETHZ, Zürich, Switzerland (2001) http://www.ifor.math.ethz.ch/~fukuda/cdd_home/cdd.html
14. K. Fukuda, V. Rosta: Combinatorial face enumeration in convex polytopes. Computational Geometry: Theory and Applications 4 (1994) 191–198
15. M. Laurent: Graphic vertices of the metric polytope. Discrete Mathematics 151 (1996) 131–153
16. M. Laurent, S. Poljak: The metric polytope. In: E. Balas, G. Cornuejols, R. Kannan (eds.): Integer Programming and Combinatorial Optimization (1992) 247–286
17. S. Poljak, Z. Tuza: Maximum cuts and large bipartite subgraphs. In: W. Cook, L. Lovasz, P. Seymour (eds.): DIMACS Series 20 (1995) 181–244

Partitioning a Planar Point Set into Empty Convex Polygons

Ren Ding[1,*], Kiyoshi Hosono[2,**], Masatsugu Urabe[2,***], and Changqing Xu[1,†]

[1] Hebei Normal University, Shijiazhuang 050016, P.R. China
rending@heinfo.net, chqxu@sina.com
[2] Tokai University, 3-20-1 Shimizu-Orido, Shizuoka, 424-8610 Japan
{hosono,qzg00130}@scc.u-tokai.ac.jp

Abstract. For a planar n point set P in general position, a convex polygon of P is called empty if no point of P lies in its interior. We show that P can be always partitioned into at most $\lceil 9n/34 \rceil$ empty convex polygons and that $\lceil (n+1)/4 \rceil$ empty convex polygons are occasionally necessary.

1   Introduction

Let P be a set of n points in general position in the plane. A convex polygon determined by a subset of P is called an *empty polygon* if no points of P lie in the interior of its convex hull. Notice that we always regard here point sets consisting of at most two elements as empty polygons. More generally, we also call a closed region in the plane *empty* if the interior contains no points of P. In [2] the following definition was proposed: Let $g(P)$ be the minimum number of empty polygons for a given P such that the empty polygons covers all the elements of P, and define $G(n) = \max\{g(P)\}$, over all the sets of n points. We claim that this idea is only a vertex-partition, that is, the convex hull of any empty polygon may allow to intersect one of other empty polygon. See Fig. 1. Then it has been shown that $\lceil (n-1)/4 \rceil \leq G(n) \leq \lceil 3n/11 \rceil$. In this paper, we improve on these bounds.

Theorem 1. $\lceil (n+1)/4 \rceil \leq G(n) \leq \lceil 9n/34 \rceil$ *for* $n \geq 1$.

Theorem 2. $G(n) \leq (5n+1)/19$ *for* $n = 19 \cdot 2^{k-1} - 4$ $(k \geq 1)$.

To prove the upper bound of Theorem 1 we need the following lemma.

Lemma 3. $G(15) \leq 4$.

* This study was supported in part by Hebei Science Foundation 199174.
** This study was supported in part by Research and Study Program of Tokai University Educational System General Research Organization.
*** This research was partially supported by Grant-in-Aid for Scientific Reserch of the Ministry of Education, Science, Culture, Sports, Science and Technology of Japan, 13640137.
† This study was supported in part by Hebei Science Foundation 199174.

J. Akiyama and M. Kano (Eds.): JCDCG 2002, LNCS 2866, pp. 129–134, 2003.
© Springer-Verlag Berlin Heidelberg 2003

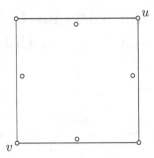

Fig. 1. This configuration P has an empty convex hexagon and a line segment $\overline{uv}$; $g(P) = 2$.

2   Upper Bound

We use the following propositions in [2] to prove Lemma 3. Note that this lemma implies $G(15) = 4$ by $G(n) \geq \lceil (n-1)/4 \rceil$ in [2].

Proposition A. *An 11 point set can be partitioned into at most 3 empty polygons.*

Proposition B. *A 7 point set can be partitioned into at most 2 empty polygons.*

We begin with definitions of terms and notation that will be used in our proof. For a given point set P we denote the subset of P on the boundary of $CH(P)$ by $V(P) = \{v_1, v_2, \ldots, v_t\}$ with the order anti-clockwise, where "CH" stands for the convex hull. We use the notation $\overline{ab}$ to refer to the line segment between a and b, and ab refers to the extended straight line associated with two points a and b. In particular, we denote the edge of $CH(P)$ by $\overline{v_i v_{i+1}}$ for any i. Denote the *convex cone* by $C(a; b, c)$ determined for $\{a, b, c\} \subset P$ such that a is the apex and b and c are on its boundary. Then let $A(a; b, c)$ be defined as the point of P in $C(a; b, c)$ if $C(a; b, A(a; b, c))$ is empty. We call $A(a; b, c)$ the *attack point* from the half-line ab to ac. Note that if $C(a; b, c)$ is empty, $A(a; b, c) = c$.

Let P be a set of 15 points in general position in the plane. If an empty convex 4-gon can be separated by a line from the remaining 11 points, our result follows immediately from Proposition A. Thus consider a configuration of 15 points that do not have this property. We show that this implies the following assumptions:

Assumption 1. *There exists a point p_i of P for every edge $\overline{v_i v_{i+1}}$ of $CH(P)$ such that $C(v_i; p_i, v_{i+1}) \cup C(v_{i+1}; p_i, v_i)$ is empty.*

We call such p_i the *characteristic point* to $\overline{v_i v_{i+1}}$.

We explain the validity of this assumption. If $\triangle v_{i-1} v_i v_{i+1}$ is empty, we can separate the empty quadrilateral $v_{i-1} v_i v_{i+1} A(v_{i+1}; v_{i-1}, v_{i-2})$ from the remaining 11 points. Thus we can partition at most 4 empty polygons of P by Prop.A. Let $p_i = A(v_{i+1}; v_i, v_{i-1})$. If $C(v_i; p_i, v_{i+1})$ is not empty, the empty quadrilateral $p_i v_i v_{i+1} A(p_i; v_{i+1}, v_{i+2})$ can be separated. Therefore $C(v_i; p_i, v_{i+1})$ is also empty. Next we show a point p_i is different from p_{i-1}. If $C(v_{i-1}; p_i, v_i)$ is also empty, the empty quadrilateral $v_{i-1}v_ip_iA(p_i; v_{i-1}, v_{i-2})$ can be also separated. Thus we can assume the existence of $p_{i-1} = A(v_{i-1}; v_i, p_i)$. By the same argument, $C(v_i; p_{i-1}, v_{i-1})$ can be also supposed to be empty. Hence there exists a characteristic point to each edge of $CH(P)$.

By Assumption 1, the quadrilateral $Q_i = p_iv_{i+1}v_{i-1}p_{i-1}$ is convex, where both p_i and p_{i-1} are in $\triangle v_{i-1}v_iv_{i+1}$. Now if Q_i is empty, Q_i itself can be separated since the remaining points are contained in $C(v_i; p_i, p_{i-1})$. Therefore we may assume

Assumption 2. *Every convex quadrilateral* $Q_i = p_iv_{i+1}v_{i-1}p_{i-1}$ *is not empty for any i.*

Consider the triangular domain T_i determined by two lines $v_{i+1}p_i$, $v_{i-1}p_{i-1}$ and $\overline{p_ip_{i-1}}$ for every i. Suppose that T_i is empty. Since Q_i is not empty, find the closest point q in Q_i to $\overline{p_ip_{i-1}}$, then the empty quadrilateral $v_ip_iqp_{i-1}$ can be separated from other points. Therefore we may suppose the following.

Assumption 3. *Every triangular domain* T_i *determined by two lines* $v_{i+1}p_i$, $v_{i-1}p_{i-1}$ *and* $\overline{p_ip_{i-1}}$ *for every i is not empty.*
We call such T_i the *characteristic domain* of v_i.

We are now ready to prove Lemma 3, that is, a set P of 15 points in general position can be partitioned into at most 4 empty polygons. The proof is by classification of $|V(P)|$.

(I) $|V(P)| \geq 5$: Assumptions 1 and 3 together imply that $|P\backslash V(P)| \geq 2|V(P)|$. Thus $|V(P)|$ is at most 5. If $|V(P)| = 5$, $g(P) = 4$ holds since $P\backslash V(P)$ constructs an empty 10-gon, by Assumptions 1, 2 and 3.

(II) $|V(P)| = 4$: By Assumptions 1 and 3 there exists a characteristic domain, say T_1 containing exactly one point q_1. Our first claim is that $C(v_2; p_1, p_4) \cup C(v_4; p_1, p_4)$ contains only q_1. We first show that $C(v_2; p_1, q_1)$ is empty. In fact, if $p \neq q_1$ for $p = A(v_2; p_1, q_1)$, the empty quadrilateral $v_1p_1pA(p; v_1, p_4)$ can be separated from the other points. Suppose that $C(v_2; p_1, q_1)$ is empty and let $q = A(v_2; q_1, p_4)$. If q is in $C(v_1; p_1, q_1)$, we obtain the empty quadrilateral $v_1p_1qq_1$. For the case where q is in $C(v_1; q_1, p_4)$, then if $C(q_1; v_2, q)$ is empty, the empty quadrilateral $q_1p_1v_2q$ can be separated. If not, we are done by Prop.B since two empty polygons $\triangle v_1p_4v_4$ and the pentagon $qq_1p_1v_2A(q; v_2, p_2)$ can be separated from the remaining 7 points. Thus by symmetry we can also assume that $C(v_4; p_4, p_1)$ contains only q_1.

Let $r = A(q_1; v_2, p_2)$. If $C(v_1; p_1, q_1)$ is empty, we can choose the empty quadrilateral $q_1p_1v_2r$. Suppose that $C(v_1; p_1, q_1)$ is not empty. Then if $s = A(r; q_1, p_4)$ exists, we obtain the two empty polygons $\triangle v_1p_4v_4$ and pentagon $q_1p_1v_2rs$. Thus we can also suppose that $C(r; q_1, p_4)$ is empty. Then if r is on the opposite side of v_1 with $\overline{v_2v_4}$, the line segment $\overline{v_1v_3}$ and the empty hexagon $p_4q_1p_1v_2rA(r; p_4, v_4)$ are obtained. Therefore we can assume that r is on the same side of v_1 with $\overline{v_2v_4}$. Then if $t = A(v_2; p_4, r)$ exists, $\triangle v_1p_4v_4$ and the pentagon $tq_1p_1v_2A(t; v_2, r)$ are also obtained. Thus suppose that $C(v_2; p_4, r)$ is also empty.

By symmetry we can also assume the existence of point $r' = A(q_1; v_4, p_3)$ in $C(v_1; q_1, v_4)$ such that both $C(v_4; p_1, r')$ and $C(q_1; v_4, r')$ are empty and r' is on the same side of v_1 with $\overline{v_2v_4}$. Note that both $\{r, r', v_2, v_4\}$ and $\{p_1, p_4, q_1, r, r'\}$ are in convex position. Then if $\triangle q_1rr'$ is not empty, $u = A(r; p_4, r')$ can be supposed to be in $C(v_1; r, q_1)$ by symmetry. Thus $\triangle v_4p_4r'$ and the pentagon $q_1v_1p_1ru$ are obtained. Suppose that $\triangle q_1rr'$ is empty. We can suppose that both $C(p_1; v_2, r)$ and $C(p_4; v_4, r')$ are empty by symmetry since, if $C(p_1; v_2, r)$ is not empty, two quadrilaterals; $v_1q_1r'p_4$ and $rp_1v_2A(r; v_2, p_2)$ are obtained. Let v be $A(r; v_4, p_3)$ or $A(r; r', v_4)$ if $C(r; r', v_4)$ is empty or not, respectively. Then we obtain $\triangle v_1p_4v_4$ and the pentagon $r'q_1p_1rv$ as desired.

(III) $|V(P)| = 3$: By Assumptions 1 and 3 there exists a characteristic domain, say T_1 containing exactly 1,2 or 3 points of P. If T_1 contains exactly one point, we are done by the same argument as for (II).

Let $p = A(v_2; p_1, p_3)$ and $q = A(v_3; p_3, p_1)$. Then if p is not in T_1, the empty quadrilateral $v_1p_1pA(p; v_1, p_3)$ can be separated from other points. Thus we can assume that both p and q are in T_1 by symmetry.

(1) T_1 contains exactly 2 points.

If $p = q$, either the empty quadrilateral v_1p_1rp or v_1prp_3 can be separated from the remaining point r in T_1. Thus suppose that q is in $C(v_3; p, p_3)$. If q is in $C(p_1; p, v_1)$ or in $C(p_3; p, p_1)$, the empty quadrilateral v_1p_1pq or v_1pqp_3 is obtained, respectively. Thus we can assume that $\{p, q, p_1, p_3\}$ is in convex position. Then if $C(v_2; p, q)$ is empty, we obtain $\triangle v_1p_3v_3$ and the pentagon $qpp_1v_2A(q; v_2, p_2)$. Let $s = A(v_2; p, q)$. We can assume that s is in $C(v_1; p, q)$ since, otherwise, the quadrilateral v_1p_1sp can be separated. If $C(p; v_2, s)$ is empty, the quadrilateral pp_1v_2s can be separated. If $C(p; v_2, s)$ is not empty, we obtain two desired polygons $\triangle v_1qp_3$ and the pentagon $spp_1v_2A(s; v_2, p_2)$.

(2) T_1 contains exactly 3 points.

Since, if $p = q$, we can easily find an empty quadrilateral to be separated, we suppose that $p \neq q$. Let r be the remaining point in T_1. If q is on the opposite side of v_1 with pp_3, the empty quadrilateral $p_3v_1qA(q; v_1, p)$ can be separated. Consider the case that q is on the same side of v_1 with pp_3. First, suppose that r is in $C(v_1; p, q)$. If r is in $C(p_1; p, v_1)$ or $C(p_3; q, v_1)$, the quadrilateral v_1p_1pr or v_1rqp_3 is obtained, respectively. If r is on the opposite side of v_1 with pq, the quadrilateral v_1prq is obtained. For otherwise, $\triangle v_1v_2p_1$ and the pentagon $p_3qrpA(p; p_3, p_2)$ can be separated. Thus suppose that r is in $C(v_1; p_1, p)$ without loss of generality. Then two quadrilaterals; v_1p_1rp and $v_3p_3qA(q; v_3, p_2)$ are obtained.   □

We need the next lemma proved in [1] to show the upper bound of Theorem 1.

Lemma C. *For any set of $2m + 4$ points in general position in the plane we can divide the plane into three disjoint convex regions such that one contains a convex quadrilateral and the others contain m points each.*

Combining Lemma C with Lemma 3, any set of 34 points can be partitioned into at most 9 empty polygons. Using a horizontal sweep, we can divide the plane into $\lceil n/34 \rceil$ disjoint strips so that $\lfloor n/34 \rfloor$ strips contain precisely 34 points of P each and one strip contains the remaining points S. Then S can be partitioned into at most $\lceil 9|S|/34 \rceil$ empty polygons by using Lemma 3 and Propositions A and B.
□

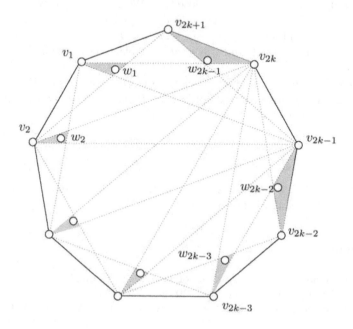

Fig. 2. The case $n = 4k$.

Next we show that Theorem 2 gives a stronger upper bound for an infinite sequence of integer n, using Lemma C. To prove this result, we will show the inequality, $G(19 \cdot 2^{k-1} - 4) \leq 5 \cdot 2^{k-1} - 1$ for $k \geq 1$.

For $k = 1$, the inequarity holds by Lemma 3. Let $\alpha(k) = 19 \cdot 2^{k-1} - 4$ and $\beta(k) = 5 \cdot 2^{k-1} - 1$. Suppose, by the induction hypothesis, $G(\alpha(k-1)) \leq \beta(k-1)$ holds. Since $\alpha(k) = 2\alpha(k-1) + 4$, every set of $\alpha(k)$ points can be divided on three disjoint convex regions such that one contains an empty quadrilateral and the others contain $\alpha(k - 1)$ points each. Thus we conclude that $G(\alpha(k)) = G(2\alpha(k-1) + 4) \leq 2\beta(k-1) + 1 = \beta(k)$.
□

3   Lower Bound

To prove the lower bound for $G(n)$, we construct a configuration P of n points satisfying $g(P) \geq \lceil (n+1)/4 \rceil$. In [2], it was shown $G(n) \geq \lceil (n-1)/4 \rceil$ for some configurations. Thus we have only to consider the cases $n = 4k$ and $4k + 1$ since $\lceil (n-1)/4 \rceil = \lceil (n+1)/4 \rceil$ holds for $n = 4k + 2, 4k + 3$.

We may assume $k \geq 2$ since the cases $k = 0, 1$ are trivial. Draw a regular $(2k + 1)$-gon C and label its vertices $v_1, v_2, \cdots v_{2k+1}$. Next denote the point of intersection of two line segments $\overline{v_{2k-3}v_{2k-1}}$ and $\overline{v_{2k-2}v_{2k}}$, and $\overline{v_{2k-1}v_{2k+1}}$ and $\overline{v_{2k}v_1}$ by x and y, respectively. Then we place two points w_{2k-2} in $\triangle v_{2k-2}v_{2k-1}x$ close to x, and w_{2k-1} in $\triangle v_{2k}v_{2k+1}y$ close to y. We choose other $2k - 3$ points w_i inside $\triangle v_{2k-1}v_{2k}v_i \cap \triangle v_{i-1}v_iv_{i+1}$ close to $\overline{v_{i-1}v_{i+1}}$ for $i = 1, 2, \cdots, 2k - 3$. Now we obtain the $4k$ point set $P = \{v_1, v_2, \cdots, v_{2k+1}, w_1, w_2, \cdots, w_{2k-1}\}$ in general position, as shown in Fig. 2, and if $n = 4k + 1$, place an additional point close to the center of C.

Any triangle on C is not empty since it inevitably contains an interior point of P. That is, any empty convex polygon of P has at most two points of C, implying that $g(P) \geq \lceil (2k + 1)/2 \rceil = \lceil (n + 1)/4 \rceil$. $\square$

4 Remark

We expect that $G(4m+7) \leq m+2$ also holds for $m \geq 3$. Therefore we conjecture that $G(n) = \lceil (n + 1)/4 \rceil$ by Theorem 1.

References

1. Hosono, K., Urabe, M.: On the number of disjoint convex quadrilaterals for a planar point set. Comp. Geom. Theory Appl. **20** (2001) 97–104
2. Urabe, M.: On a partition into convex polygons. Discr. Appl. Math. **64** (1996) 179–191

Relaxed Scheduling
in Dynamic Skin Triangulation*

Herbert Edelsbrunner[1] and Alper Üngör[2]

[1] Department of Computer Science, Duke University, Durham, NC 27708
and Raindrop Geomagic, Research Triangle Park, NC 27709
edels@cs.duke.edu
[2] Department of Computer Science, Duke University, Durham, NC 27708
ungor@cs.duke.edu

Abstract. We introduce relaxed scheduling as a paradigm for mesh maintenance and demonstrate its applicability to triangulating a skin surface in $\mathbb{R}^3$.

Keywords: Computational geometry, adaptive meshing, deformation, scheduling.

1   Introduction

In this paper, we describe a relaxed scheduling paradigm for operations that maintain the mesh of a deforming surface. We prove the correctness of this paradigm for skin surfaces.

Background. In 1999, Edelsbrunner [5] showed how a finite collection of spheres or weighted points can be used to construct a C^1-continuous surface in $\mathbb{R}^3$. It is referred to as the *skin* or the *skin surface* of the collection. If the spheres represent the atoms of a molecule then the appearance of that surface is similar to the molecular surface used in structural biology [2,8]. The two differ in a number of details, one being that the former uses hyperboloids to blend between sphere patches while the latter uses tori. The skin surface is not C^2-continuous, but its maximum normal curvature, κ, is continuous. This property is exploited by Cheng *et al.* [1], who describe an algorithm that constructs a triangular mesh representing the skin surface. In this mesh, the sizes of edges and triangles are inversely proportional to the maximum normal curvature. The main idea of the algorithm is to maintain the mesh while gradually growing the skin surface to the desired shape, as illustrated in Figure 1. The algorithm thus reduces the construction to a sequence of restructuring operations. There are *edge flips*, which maintain the mesh as the restricted Delaunay triangulation of its vertices, *edge contractions* and *vertex insertions*, which maintain a sampling whose local density is proportional to the maximum normal curvature, and *metamorphoses*, which adjust the mesh connectivity to reflect changes in the surface topology.

* Research of the two authors is supported by NSF under grant CCR-00-86013.

J. Akiyama and M. Kano (Eds.): JCDCG 2002, LNCS 2866, pp. 135–151, 2003.
© Springer-Verlag Berlin Heidelberg 2003

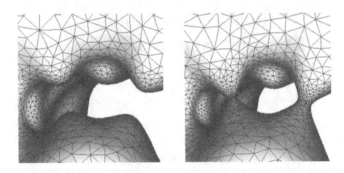

Fig. 1. The mesh is maintained as the surface on the left grows into that on the right.

Some of these operations are easier to schedule than others, and the most difficult ones are the edge contractions and vertex insertions. They depend on how the sampled points move with the surface as it deforms. The quality of the mesh is guaranteed by maintaining size constraints for all edges and triangles. When an edge gets too short we contract it, and when a triangle gets too large we insert a point near its circumcenter. Both events can be recognized by finding roots of fairly involved functions. Scheduling edge contractions and vertex insertions thus becomes a bottleneck, both in terms of the robustness and the running time of the algorithm.

Result. In this paper, we study how fast edges and triangles vary their size, and we use that knowledge to schedule these elements in a relaxed fashion. In other words, we do not determine when exactly an element violates its size constraint, but we catch it before the violation happens. Of course, the danger is now that we either update perfectly well-shaped elements or we waste time by checking elements unnecessarily often. To avoid the former, we introduce intervals or gray zones in which the shapes of the elements are neither good nor unacceptably bad. To avoid unnecessarily frequent checking, we prove lower bounds on how long an element stays in the gray zone before its shape becomes unacceptably bad. These bounds are different for edges and for triangles. Consider first an edge uv. Let $R = \|u - v\|/2$ be its half-length and $\varrho = 1/\max\{\kappa(u), \kappa(v)\}$ the smaller radius of curvature at its endpoints. We use judiciously chosen constants C, Q_0 and Q_1 and call the edge

$$\left.\begin{array}{r}\textit{acceptable}\\ \textit{borderline}\\ \textit{unacceptable}\end{array}\right\} \quad \text{if} \quad \left\{\begin{array}{r}C/Q_0 < R/\varrho,\\ C/Q_1 < R/\varrho \le C/Q_0,\\ R/\varrho \le C/Q_1.\end{array}\right.$$

The middle interval is what we called the gray zone above. Assuming uv is acceptable, we prove it will not become unacceptable within a time interval of duration $\Delta t = (2\theta - \theta^2)\varrho^2$, where

$$\theta = \frac{RQ_1 - C\varrho}{RQ_1 + C\varrho}.$$

In the worst case, R is barely larger than $C\varrho/Q_0$, so we have $\theta > (Q_1 - Q_0)/(Q_1 + Q_0)$ as a worst case bound. We will see that $C = 0.06$, $Q_0 = 1.6$ and $Q_1 = 2.3$ are feasible choices for the constants, and that for these we get $\theta > 0.179\ldots$ and $\Delta t/\varrho^2 > 0.326\ldots$. Consider next a triangle uvw. Let R be the radius of its circumcircle, and $\varrho = 1/\max\{\kappa(u), \kappa(v), \kappa(w)\}$ the smallest radius of curvature at its vertices. We call uvw

$$\left.\begin{array}{r} acceptable \\ borderline \\ unacceptable \end{array}\right\} \quad \text{if} \quad \left\{\begin{array}{c} R/\varrho < CQ_0, \\ CQ_0 \le R/\varrho < CQ_1, \\ CQ_1 \le R/\varrho. \end{array}\right.$$

Assuming uvw is acceptable, we prove it will not become unacceptable within a time interval of duration $\Delta t = (2\theta - \theta^2)\varrho^2$, where

$$\theta = 1 - \sqrt[4]{R/(CQ_1\varrho)}.$$

In the worst case, R is barely smaller than $CQ_0\varrho$, so we have $\theta > 1 - \sqrt[4]{Q_0/Q_1}$. For the above values of C, Q_0 and Q_1, this gives $\theta > 0.086\ldots$ and $\Delta t/\varrho^2 > 0.165\ldots$. It seems that triangles can get out of shape about twice as fast as edges, but we do not know whether this is really the case because our bounds are not tight.

Outline. Section 2 reviews skin surfaces and the dynamic triangulation algorithm. Section 3 introduces relaxed scheduling as a paradigm to keep track of moving or deforming data. Section 4 analyzes the local distortion within the mesh and derives the formulas needed for the relaxed scheduling paradigm. Section 5 concludes the paper.

2   Preliminaries

In this section, we introduce the necessary background from [5], where skin surfaces were originally defined, and from [1], where the meshing algorithm for deforming skin surfaces was described.

Skin Surfaces. We write $S_i = (z_i, r_i)$ for the sphere with center $z_i \in \mathbb{R}^3$ and radius r_i and think of it as the zero-set of the weighted square distance function $f_i : \mathbb{R}^3 \to \mathbb{R}$ defined by $f_i(x) = \|x - z_i\|^2 - r_i^2$. The square radius is a real number and the radius is either a non-negative real or a non-negative multiple of the imaginary unit. We know how to add functions and how to multiply them by scalars. For example, if we have a finite collection of spheres S_i and scalars $\sum \gamma_i = 1$ then $\sum \gamma_i f_i$ is again a weighted square distance function, and we denote by $S = \sum \gamma_i S_i$ the sphere that defines it. The *convex hull* of the S_i is the set of such spheres obtained using only non-negative scalars:

$$\mathcal{F} = \left\{\sum \gamma_i S_i \mid \sum \gamma_i = 1 \text{ and } \gamma_i \ge 0, \forall i\right\}.$$

We also shrink spheres and write $\sqrt{S} = (z, r/\sqrt{2})$, which is the zero-set of $2f - f(z)$. The *skin surface* defined by the S_i is then the envelope of the spheres in the convex hull, all scaled down by a factor $1/\sqrt{2}$, and we write this as $F = \text{env} \sqrt{\mathcal{F}}$. Equivalently, it is the zero-set of the point-wise minimum over all functions $2f - f(z)$, over all $S \in \mathcal{F}$, where f is the weighted square distance function defined by S. At first glance, this might seem like an unwieldy surface, but we can completely describe it as a collection of quadratic patches obtained by decomposing the surface with what we call the mixed complex. Its cells are Minkowski sums of Voronoi vertices, edges, polygons and polyhedra with their dually corresponding Delaunay tetrahedra, triangles, edges and vertices, all scaled down by a factor $1/2$. Instead of formally describing this construction, we illustrate it with a two-dimensional example in Figure 2. Depending on the di-

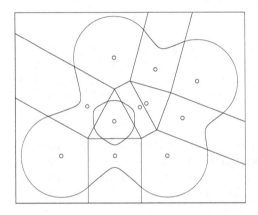

Fig. 2. The mixed complex decomposes the skin curve and the area it bounds.

mension of the contributing Delaunay simplex, we have four types of mixed cells. Because of symmetry, we have only two types of surface patches, namely pieces of spheres and of hyperboloids of revolution, which we frequently put in Standard Form:

$$\xi_1^2 + \xi_2^2 + \xi_3^2 = R^2, \tag{1}$$
$$\xi_1^2 + \xi_2^2 - \xi_3^2 = \pm R^2, \tag{2}$$

where the plus sign gives the one-sheeted hyperboloid and the minus sign gives the two-sheeted hyperboloid.

Meshing. The meshing algorithm triangulates the skin surface using edges and triangles whose sizes adapt to the local curvature. Let us be more specific. At any point $x \in F$, let $\kappa(x)$ be the maximum normal curvature at x. In contrast to other notions of curvature, κ is continuous over the skin surface and thus amenable to controlling the local size of the mesh. Call $\varrho(x) = 1/\kappa(x)$ the *local length scale* at x. The vertices of the mesh are points on the surface. For an edge

uv, let $R_{uv} = \|u - v\|/2$ be half its length, and for a triangle uvw, let R_{uvw} be the radius of its circumcircle. The algorithm obeys the Lower and Upper Size Bounds that require edges not be too short and triangles not be too large:

[L] $R_{uv}/\varrho_{uv} > C/Q$ for every edge uv, and
[U] $R_{uvw}/\varrho_{uvw} < CQ$ for every triangle uvw,

where ϱ_{uv} is the larger of $\varrho(u)$ and $\varrho(v)$, ϱ_{uvw} is the minimum of $\varrho(u)$, $\varrho(v)$ and $\varrho(w)$, and C and Q are judiciously chosen positive constants.

The particular algorithm we consider in this paper is dynamic, in the sense that it maintains the mesh while the surface deforms. We can use this algorithm to construct a mesh by starting with the empty surface and growing it into the desired shape. This is precisely the scenario in which our results apply. To model the growth process, we use a time parameter and let $S_i(t) = (z_i, \sqrt{r_i^2 + t})$ be the i-th sphere at time $t \in \mathbb{R}$. We start at $t = -\infty$, at which time all radii are imaginary and the surface is empty, and we end at $t = 0$, at which time the surface has the desired shape. This particular growth model is amenable to efficient computation because it does not affect the mixed complex, which stays the same at all times. Each patch of the surface sweeps out its mixed cell. At any moment, we have a collection of points sampled on the surface, and the mesh is the restricted Delaunay triangulation of these points, as defined in [4,7]. Given the surface and the points, this triangulation is unique. As the surface deforms, we move the points with it and update the mesh as required. From global and less frequent to local and more frequent these operations are:

1. topology changes that affect the local and global connectivity of the surface and the mesh,
2. edge contractions and vertex insertions that locally remove or add points to coarsen or refine the mesh, and
3. edge flips that locally adjust the mesh without affecting the point distribution or the surface topology.

For the particular growth model introduced above, the topology changes are easily predicted using the filtration of alpha complexes as described in [6]. To predict where and when we need to coarsen or refine the mesh is more difficult and depends on how the points move to follow the deforming surface. This is the topic of this paper and will be discussed in detail in the subsequent sections. Finally, edge flips are relatively robust operations, which can be performed in a lazy manner, without any sophisticated scheduling mechanism.

Point Motion. To describe the motion of the points sampled on the skin surface, it is convenient to consider the trajectory of the surface over time. Note that the i-th sphere at time t is $S_i(t) = f_i^{-1}(t)$. Similarly, the convex combination defined by coefficients γ_i at time t is $S(t) = f^{-1}(t)$, where $f = \sum \gamma_i f_i$. We can represent the skin surface in the same manner by introducing the function $g : \mathbb{R}^3 \to \mathbb{R}$ defined as the point-wise minimum of the functions representing the shrunken spheres. More formally, $g(x) = \min\{2f(x) - f(z)\}$, where the minimum is taken over all spheres $S \in \mathcal{F}$ and z is the center of S. The skin surface at time t is then $F(t) = g^{-1}(t)$, so it is appropriate to call the graph of g the *trajectory* of the skin surface. We see that growing the surface in time is equivalent to sweeping out its trajectory with a three-dimensional space that moves through time. It is natural to let the points sampled on $F(t)$ move normal to the surface. For a point $x = [\xi_1, \xi_2, \xi_3]^T$ on a sphere or hyperboloid in Standard Form $\xi_1^2 + \xi_2^2 \pm \xi_3^2 = \pm R^2$, the gradient is $\nabla g_x = 2[\xi_1, \xi_2, \pm\xi_3]^T$. The point x moves in the direction of the gradient with a speed that is inversely proportional to the length. In other words, the velocity vector at a point x is

$$\dot{x} = \frac{\mathrm{d}x}{\mathrm{d}t} = \frac{\nabla g_x}{\|\nabla g_x\|^2} = \frac{\nabla g_x}{4\|x\|^2}.$$

The speed of x is therefore $\|\dot{x}\| = 1/(2\|x\|)$. The implementation of the relaxed scheduling paradigm crucially depends on the properties of this motion. We use the remainder of this section to describe a symmetry property of the velocity vectors that is instrumental in the analysis of the motion. Consider two mixed cells that share a common face. The Standard Forms of the two corresponding surface patches differ by a single sign, and so do the gradients. If we reflect points in one cell across the plane of the common face into the other cell then we preserve the velocity vector, as illustrated in Figure 3. We use this observation

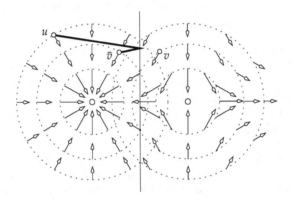

Fig. 3. Velocity vectors of a shrinking circle on the left and of a hyperbola on the right. The right portion of the edge uv is reflected across face shared by the two mixed cells.

about adjacent mixed cells to relate the velocity vectors of points in possibly non-adjacent cells. Consider points u and v and let $x_1, x_2, \ldots, x_k$ be the intersection points with faces of mixed cells encountered as we travel along the edge from u to v. Starting at $i = k$, we work backward and reflect the portion of the edge beyond x_i across the face that contains x_i. In the general case, this portion is a polygonal path that leads from x_i to the possibly multiply reflected image $\overline{v}$ of v. After k reflections we have a polygonal path from u to the final $\overline{v}$. The length of the path is equal to the length of the initial edge, and hence $\|u - \overline{v}\| \leq \|u - v\|$.

We note that $\bar{v}$ does not necessarily lie in the mixed cell of u, but its velocity vector — which is the same as that of v — is consistent with the family of spheres or hyperboloids that sweeps out that mixed cell. In other words, the motion of u and $\bar{v}$ is determined by the same quadratic function.

3 Relaxed Scheduling

In this section, we introduce relaxed scheduling as a paradigm for maintaining moving or deforming data. It is designed to cope with situations in which the precise moment for an update is either not known or too expensive to compute.

Correctness Constraints. In the context of maintaining the triangle mesh of a skin surface, we use relaxed scheduling to determine when to contract an edge and when to insert a new vertex. Since determining when the size of an edge or triangle stops to be acceptable is expensive, we introduce a gray zone between acceptability and unacceptability and update an element when we catch it inside that gray zone. That this course of action is even conceivable is based on the correctness proof of the dynamic skin triangulation algorithm for a range of its controlling parameters. The first three conditions defining that range refer to ε, C and Q. We have seen the latter two before in the formulation of the two Size Bounds [L] and [U]: C controls how well the mesh approximates the surface, and Q controls the quality of the mesh. Both are related to ε, which quantifies the sampling density.

(I) We require $0 < \varepsilon \leq \varepsilon_0$, where $\varepsilon_0 = 0.279\ldots$ is a root of $2\cos(\arcsin\frac{2\varepsilon}{1-\varepsilon} + \arcsin\varepsilon) - \frac{2\varepsilon}{1-\varepsilon} = 0$.

(II) $Q^2 - 4CQ > 2$.

(III) $\frac{\delta^2}{1+\delta^2} - \frac{\delta^2}{4} > C^2Q^2$, where $\delta = \varepsilon - \frac{2C(\varepsilon+1)}{Q+2C}$.

It is computationally efficient to select the loosest possible bound for the sampling density: $\varepsilon = \varepsilon_0$. Then we get $\delta = 0.166\ldots$ and, as noted in [1], we may choose $C = 0.08$ and $Q = 1.65$ to satisfy Conditions (I) to (III). Alternatively, we may lower C to 0.06 and are then free to pick Q anywhere inside the interval from 1.6 to 2.3. The two choices of parameters are marked by a hollow dot and a white bar in Figure 5. The last two conditions refer to h, ℓ and m. All three parameters control how metamorphoses that add or remove a handle are implemented. Since the curvature blows up at the point and time of a topology change, we use a special and relatively coarse sampling inside spherical neighborhoods of such points. Assuming a unit radius of such neighborhoods, we turn the special sampling strategy on and off when the skin surface enters and leaves the smaller spherical neighborhood of radius $h < 1.0$. If the skin enters as a two-sheeted hyperboloid we triangulate it using two ℓ-sided pyramids inside the unit sphere neighborhood. If it enters as a one-sheeted hyperboloid we triangulate it as an m-sided drum with a waist. The conditions are stated in terms of the edges ab, bc and wx and the triangles abc and vwx, as defined in Figure 4. Their sizes can all be expressed in terms of h, ℓ and m, and we refer to [1, Section 10] for the formulas.

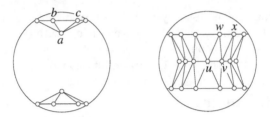

Fig. 4. The triangulation of a two-sheeted and a one-sheeted hyperboloid inside a unit neighborhood sphere around their apices.

(IV) $R_{ab}, R_{bc}, R_{wx} > C/Q$.

(V) $R_{abc}, R_{vwx} < \min\{Q, 2/Q\}Ch$.

Quality Buffer. The key technical insight about the dynamic skin triangulation algorithm is that we can find constants ε, C, h, ℓ, m and $Q_0 < Q_1$ such that Conditions (I) to (V) are satisfied for all $Q \in [Q_0, Q_1]$. This is illustrated in Figure 5, which shows the feasible region of points (C, Q) assuming fixed values for ε, h, ℓ and m. Instead of fixing Q and contracting an edge when its size-

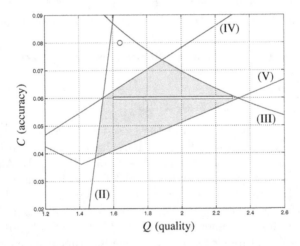

Fig. 5. The shaded feasible region of parameter pairs (C, Q) for $\varepsilon = \varepsilon_0$, $h = 0.993$, $\ell = 6$ and $m = 80$. For $C = 0.06$ this region contains the interval $Q \in [1.6, 2.3]$. The bounding curves are labeled by the corresponding constraints. Redundant constraints are not shown.

scale ratio reaches C/Q, we suggest to contract the edge any time its ratio is in the interval $(C/Q_1, C/Q_0]$. After the ratio enters this interval at C/Q_0 it can either leave again at C/Q_0 or it can get contracted, but it is not allowed to reach C/Q_1. Vertex insertions are treated symmetrically. Specifically, a triangle is removed by adding a vertex near its circumcenter, and this can happen at any moment its size-scale ratio is in $[CQ_0, CQ_1)$. The ratio can enter and leave the interval at CQ_0, but it is not allowed to reach CQ_1. We call $(C/Q_1, C/Q_0]$ and $[CQ_0, CQ_1)$ the *lower* and *upper size buffers*. The quality of the mesh is guaranteed because all edges and triangles satisfy the two Size Bounds [L] and [U] for $Q = Q_1$. Symmetrically, the correctness of the triangulation is guaranteed because edge contractions and vertex insertions are executed only if the same bounds are violated for $Q = Q_0$.

Early Warning. Recall that an edge is borderline iff its size-scale ratio is contained in the lower size buffer, and it becomes unacceptable at the moment it reaches C/Q_1. Similarly, a triangle is borderline iff its size-scale ratio is contained in the upper size buffer, and it becomes unacceptable at the moment it reaches CQ_1. The relaxed scheduling paradigm depends on an early warning algorithm that reports an element before it becomes unacceptable. That algorithm might err and produce false positives, but it may not let any element slip by and become unacceptable. False positives cost time but do not cause any harm, while unacceptable elements compromise the correctness of the meshing algorithm. In Figure 6, false positives are marked by hollow dots and deletions are marked by filled black dots. All false positive tests of edges are represented by dots above

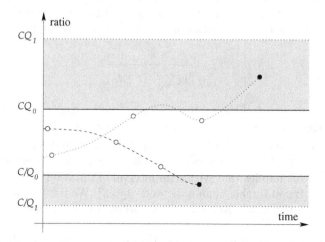

Fig. 6. The two buffers are shaded and the two curves are possible developments of size-scale ratios for an edge (dashed) and a triangle (dotted). The dots indicate moments at which the elements are tested and finally removed.

the lower size buffer. To get a correct early warning algorithm we just need to test each edge often enough so that its size-scale ratio cannot cross the entire lower size buffer between two contiguous tests. The symmetric rule applies to triangles. Bounds on the amount of time it takes to cross the size buffers will be given in Section 4.

Note that we have selected the parameters to obtain a fairly long interval $[Q_0, Q_1]$. It is not clear whether or not this is a good idea or whether a shorter interval would lead to a more efficient algorithm. An argument *for* a long interval is that the implied large size buffers let us get by with less frequent and therefore fewer tests. An argument *against* a long interval is that large size buffers are more likely to cause the deletion of elements that are on their way to better health but did not recover fast enough and get caught before they could leave the buffers. It might be useful to optimize the length of the intervals through experimentations after implementing the relaxed schedule as part of the skin triangulation algorithm.

4 Analysis

In this section, we derive lower bounds on the amount of time it takes for an edge or triangle to change its size by more than some threshold value. From these we will derive lower bounds on the time it takes an element to pass through the entire size buffer. We begin by studying the motion of a single point.

Traveling Point. We recall that the speed of a point u on the skin surface is $\|\dot{u}\| = 1/(2\|u\|)$, assuming we write the patch that contains it in Standard Form. The distance traveled by u in a small time interval is therefore maximized if it heads straight toward the origin, which for example happens if u lies on a shrinking sphere. Starting the motion at point u_0, which is the point u at time t_0, we get

$$\|u\| = \sqrt{\|u_0\|^2 - (t - t_0)}, \tag{3}$$

for the point u at time t. This implies $t - t_0 = \|u_0\|^2 - \|u\|^2$, so we see that u reaches the origin at time $t = t_0 + \|u_0\|^2$. More generally, we reach the point $u_1 = (1 - \theta)u_0$ between u_0 and the origin at time $t_1 = t_0 + \|u_0\|^2 - \|u_1\|^2 = t_0 + (2\theta - \theta^2)\|u_0\|^2$. Since the above analysis assumes the fastest way u can possibly travel, this implies that within an interval of duration $\Delta t = t_1 - t_0$, the point u_0 cannot travel further than a distance $\theta\varrho(u_0)$. We use θ as a convenient intermediate quantity that gives us indirect access to the important quantity, which is Δt.

Recall from the Curvature Variation Lemma of [1] that the difference in length scale between two points is at most the Euclidean distance. If that distance is $\|u_0 - u_1\| \leq \theta\varrho(u_0)$ then the length scale at u_1 is between $1 - \theta$ and $1 + \theta$ times the length scale at u_0. It follows that if we travel for a duration $\Delta t = (2\theta - \theta^2)\varrho^2(u_0)$, we can change the length scale only by a factor

$$1 - \theta \leq \frac{\varrho(u_1)}{\varrho(u_0)} < 1 + \theta. \tag{4}$$

The lower bound is tight, and the upper bound cannot be reached because the distance $\theta\varrho^2(u_0)$ from u_0 can only be achieved if the length scale shrinks. We will also be interested in the integral of $1/(2\|u\|^2)$, which is again maximized if u moves straight toward the origin:

$$
\int_{t_0}^{t_1} \frac{dt}{2\|u\|^2} \leq \int_{t_0}^{t_1} \frac{dt}{2\|u_0\|^2 - (2t - 2t_0)}
$$

$$
= \left(-\frac{1}{2}\right) \ln \frac{\|u_0\|^2 - (t_1 - t_0)}{\|u_0\|^2}
$$

$$
= \ln \frac{\|u_0\|}{\|u_1\|}.
$$

Denoting the above integral by X and choosing $t_1 - t_0 = (2\theta - \theta^2)\|u_0\|^2$, as before, we have

$$
e^X \leq \frac{\|u_0\|}{\|u_1\|} = \frac{\varrho(u_0)}{\varrho(u_1)} \leq \frac{1}{1 - \theta}. \tag{5}
$$

Edge Length Variation. Consider two points u and v on the skin surface during a time interval $[t_0, t_1]$. We assume that both points follow their trajectories undisturbed by any mesh maintenance operations. Let u_0 and u_1 be the point u at times t_0 and t_1 and, similarly, let v_0 and v_1 be the point v at these two moments. We prove that if the time interval is short relative to the length scale of the points then the distance between them cannot shrink or grow by much.

LENGTH LEMMA. Let $\varrho_0 = \min\{\varrho(u_0), \varrho(v_0)\}$ and $\Delta t = t_1 - t_0 = (2\theta - \theta^2)\varrho_0^2$, for some $0 \leq \theta \leq 1$. Then

$$
1 - \theta \leq \frac{\|u_1 - v_1\|}{\|u_0 - v_0\|} < \frac{1}{1 - \theta}.
$$

Proof. The derivative of the distance between points u and v with respect to time is

$$
\frac{d\|u - v\|}{dt} = \frac{d\|u - v\|}{du} \frac{du}{dt} + \frac{d\|u - v\|}{dv} \frac{dv}{dt}
$$

$$
= \frac{(u - v)^T}{\|u - v\|} (\dot{u} - \dot{v}). \tag{6}
$$

For example if u and v lie on a common sphere patch then $\varrho = \varrho(u) = \varrho(v)$, $\dot{u} = \pm u/(2\varrho^2)$ and $\dot{v} = \pm v/(2\varrho^2)$, which implies

$$
\frac{d\|u - v\|}{dt} = \pm \frac{(u - v)^T}{\|u - v\|} \frac{(u - v)}{2\varrho^2} = \pm \frac{\|u - v\|}{2\varrho^2}.
$$

We prove below that in the general case, the distance derivative stays between these two extremes:

$$
-\frac{\|u - v\|}{2\varrho^2} \leq \frac{d\|u - v\|}{dt} \leq \frac{\|u - v\|}{2\varrho^2}, \tag{7}
$$

where $\varrho = \min\{\varrho(u), \varrho(v)\}$. To get the final result from (6), we divide by $\|u - v\|$, multiply by dt, and use $d\ln x = dx/x$ to get

$$-\frac{dt}{2\varrho^2} \leq d(\ln\|u - v\|) \leq \frac{dt}{2\varrho^2}.$$

Next we integrate over $[t_0, t_1]$ and exponentiate to eliminate the natural logarithm:

$$e^{-X} \leq \frac{\|u_1 - v_1\|}{\|u_0 - v_0\|} \leq e^X.$$

The claimed pair of inequalities follows from (5) and the observation that the upper bound for X cannot be realized when the distance derivative is positive. To prove (7) for general points u and v, it suffices to show that the length of $\dot{u} - \dot{v}$ is at most $\|u - v\|/(2\varrho^2)$. We have seen that this is true if u and v belong to a common sphere patch. It is also true if u and v belong to a common hyperboloid patch because

$$\|\dot{u} - \dot{v}\| = \left\|\frac{u'}{2\varrho^2(u)} - \frac{v'}{2\varrho^2(v)}\right\| \leq \frac{\|u - v\|}{2\varrho^2},$$

where the primed and unprimed vectors are the same, except that they have a different sign in the third coordinate. We need a slightly more elaborate argument if u and v do not belong to the same mixed cell. We then reflect v across the faces of mixed cells that intersect the edge uv. As described in Section 2, such a sequence of reflections does not affect the velocity vector. The distance between u and the image $\bar{v}$ of v under the composition of reflections is at most that between u and v. Hence,

$$\|\dot{u} - \dot{v}\| = \|\dot{u} - \dot{\bar{v}}\| \leq \left\|\frac{u - v}{2\varrho^2}\right\|,$$

as required.   ⬜

The lower bound in the Length Lemma is tight and realized by points u and v on a common sphere patch.

Shrinking Edge. Consider an edge uv, whose half-length at time t_0 is R_0. As before, let u_0 and v_0 be the points u and v at time t_0. Let $\varrho_0 = \min\{\varrho(u_0), \varrho(v_0)\}$. We follow the two points during the time interval $[t_0, t_1]$, whose duration is $\Delta t = t_1 - t_0 = (2\theta - \theta^2)\varrho_0^2$. The Length Lemma implies that at time t_1, the length of the edge satisfies

$$\frac{\|u_1 - v_1\|}{\|u_0 - v_0\|} = \frac{R_1}{R_0} \geq 1 - \theta. \tag{8}$$

Our goal is to choose θ such that the edge at time t_1 is guaranteed to satisfy the Lower Size Bound for $Q = Q_1$. Using $R_1 \geq (1-\theta)R_0$ from (8) and $\varrho_1 < (1+\theta)\varrho_0$

from (4), we note that $R_1/\varrho_1 > C/Q_1$ is implied by $(1-\theta)R_0/(1+\theta) \geq C\varrho_0/Q_1$. In other words,

$$\theta = \frac{R_0 Q_1 - C\varrho_0}{R_0 Q_1 + C\varrho_0} \tag{9}$$

is sufficiently small. The corresponding time interval during which we can be sure that the edge uv does not become unacceptably short has duration $\Delta t = (2\theta - \theta^2)\varrho_0^2$. To get a better feeling for what these results mean, let us write the half-length of $u_0 v_0$ as a multiple of the lower bound in [L] for $Q = Q_0$: $R_0 = AC\varrho_0/Q_0$ with $A > 1.0$. We then get $\theta = (AQ_1 - Q_0)/(AQ_1 + Q_0)$ and Δt from θ as before. Table 1 shows the values of θ and Δt for a few values of A.

Table 1. For edges, the values of θ and Δt for $Q_0 = 1.6$, $Q_1 = 2.3$ and a few typical values of A.

A	θ	$\Delta t/\varrho_0^2$
1.0	0.179...	0.326...
1.5	0.366...	0.598...
2.0	0.483...	0.733...
2.5	0.564...	0.810...
3.0	0.623...	0.858...
3.5	0.668...	0.890...
4.0	0.703...	0.912...

Height Variation. Consider a triangle uvw during a time interval $[t_0, t_1]$. We assume that all three points follow their trajectories undisturbed by any mesh maintenance operations. Each vertex has a distance to the line spanned by the other two vertices, and the *height* H of uvw is the smallest of the three distances. If uv is the longest edge then $H = \|w - w'\|$, where w' is the orthogonal projection of w onto uv. We prove if the time interval is short relative to the length scale at the points then the height cannot shrink or grow by much. To state the claim we use indices 0 and 1 for points and heights at times t_0 and t_1.

HEIGHT LEMMA. Let $\varrho_0 = \min\{\varrho(u_0), \varrho(v_0), \varrho(w_0)\}$ and $\Delta t = t_1 - t_0 = (2\theta - \theta^2)\varrho_0^2$, for some $0 \leq \theta \leq 1$. Then

$$1 - \theta \leq \frac{H_1}{H_0} < \frac{1}{1-\theta}.$$

Proof. We prove that (7) is also true if we substitute the height H for the length of the edge uv:

$$-\frac{H}{2\varrho^2} \leq \frac{dH}{dt} \leq \frac{H}{2\varrho^2}, \tag{10}$$

where $\varrho = \min\{\varrho(u), \varrho(v), \varrho(w)\}$. The claimed pair of inequalities follows as explained in the proof of the Length Lemma. To see (10) note first that the height of the triangle is always determined by a vertex and a point on the opposite edge, eg. $H = \|w - w'\|$. Let $w' = (1 - \lambda)u + \lambda v$. If u and v belong to the same mixed cell then $\nabla g_{w'} = (1 - \lambda)\nabla g_u + \lambda \nabla g_v$ because the gradient varies linearly. Along a moving line segment uv the velocity vectors vary linearly, hence $\dot{w}' = (1 - \lambda)\dot{u} + \lambda\dot{v}$. Since the gradients and the velocity vectors at u and v point in the same directions, they do the same at w'. The length of the velocity vector at w' is at most that of the longer velocity vector at u and v. If w belongs to the same mixed cell as w', this implies

$$\|\dot{w} - \dot{w}'\| \leq \frac{\|w - w'\|}{2\varrho^2} = \frac{H}{2\varrho^2},$$

from which (10) follows. If u, v and w do not belong to the same mixed cell then we perform reflections, as in the proof of the Length Lemma, and get (10) because reflections do not affect velocity vectors.   □

In the following, we only need the lower bound in the Height Lemma, which is tight and is realized points u, v and w on a common sphere patch.

Expanding Triangle. We use both the Length Lemma and the Height Lemma to derive a lower bound on the length of time during which a triangle that initially satisfied the Upper Size Bound [U] for $Q = Q_0$ is guaranteed to satisfy the same for $Q = Q_1$. We begin by establishing a relation between the circumradius $R = R_{uvw}$ of a triangle uvw and its height and edge lengths. Referring to Figure 7, we let z denote the center of the circumcircle. Assuming uv is the longest of the three edges, the height is $H = \|w - w'\|$ and v and z lie on the same side of the line passing through u and w. Let z' be the midpoint of uw and note that the angle at z is twice that at v: $\angle uzw = 2\angle z'zw = 2\angle uvw$. This implies that the triangles $ww'v$ and $wz'z$ are similar, and therefore $\|z' - w\|/R = H/\|v - w\|$. It follows that the circumradius of uvw is

$$R = \frac{\|u - w\|\,\|v - w\|}{2H}.$$

There are three ways to write twice the area as the product of an edge length and the distance of the third vertex from the line of that edge: $\|u - v\|\,H = \|u - w\|\,\|v - v'\| = \|v - w\|\,\|u - u'\|$. Hence, the circumradius is also

$$R = \frac{\|u - v\|\,\|u - w\|}{2\|u - u'\|} = \frac{\|u - v\|\,\|v - w\|}{2\|v - v'\|}.$$

For the remainder of this section, we use indices 0 and 1 for points, heights and radii at times t_0 and t_1. The above equations for the circumradius imply

$$\frac{R_1}{R_0} = \frac{\|u_1 - w_1\|\,\|v_1 - w_1\|\,\|w_0 - w_0'\|}{\|u_0 - w_0\|\,\|v_0 - w_0\|\,\|w_1 - w_1'\|}.$$

Assuming $H_0 = \|w_0 - w_0'\|$ is the height at time t_0, we have $H_1 \leq \|w_1 - w_1'\|$ at time t_1. We can therefore use the Length Lemma to bound the first two ratios and the Height Lemma to bound the third to get

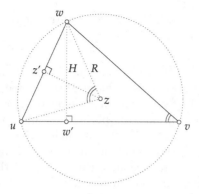

Fig. 7. The triangle uvw is similar to $wz'z$, which implies a relation between the height H and the circumradius R.

$$\frac{R_1}{R_0} < \frac{1}{(1-\theta)^3}. \tag{11}$$

We now choose θ such that a triangle that satisfies [U] for $Q = Q_0$ at time t_0 is guaranteed to satisfy [U] for $Q = Q_1$ at time t_1. Using $R_1 < R_0/(1-\theta)^3$ from (11) and $(1-\theta)\varrho_0 \le \varrho_1$ from (4), we note that $R_1/\varrho_1 < CQ_1$ is implied by $R_0/(1-\theta)^4 \le CQ_1\varrho_0$. In other words,

$$\theta = 1 - \sqrt[4]{R_0/(CQ_1\varrho_0)} \tag{12}$$

is sufficiently small. It is convenient to write the circumradius of the triangle $u_0v_0w_0$ as a fraction of the upper bound in [U]: $R_0 = CQ_0\varrho_0/A$ with $A > 1.0$. Then, $\theta = 1 - \sqrt[4]{Q_0/(AQ_1)}$. Table 2 shows the values of θ and Δt for a few values of A.

Table 2. For triangles, the values of θ and Δt for $Q_0 = 1.6$, $Q_1 = 2.3$ and a few typical values of A.

A	θ	$\Delta t/\varrho_0^2$
1.0	0.086...	0.165...
1.5	0.174...	0.319...
2.0	0.232...	0.410...
2.5	0.273...	0.472...
3.0	0.306...	0.518...
3.5	0.332...	0.554...
4.0	0.354...	0.583...

5   Discussion

The main contribution of this paper is the introduction of relaxed scheduling as a paradigm for maintaining moving or deforming data, and the demonstrations of its applicability to scheduling edge contractions and vertex insertions maintaining skin surfaces.

Algorithm Design. We view the dynamic skin triangulation algorithm, of which relaxed scheduling is now a part, as an interesting exercise in rational algorithm design. What are the limits for proving meshing algorithms correct? This design exercise gives us a glimpse on how complicated meshing problems can be. Perhaps more importantly, it illustrates what it might take to prove other meshing algorithms correct. We especially highlight the role of constant parameters in the algorithm and how they control the algorithm as well as the constructed mesh. In our example, the important parameters are C, which controls how closely the mesh approximates the surface, and Q, which controls the quality of the mesh. The effort of proving the various pieces of the algorithm correct has lead to inequalities for these parameters. In other words, we have identified a feasible region which is necessary for our proofs and sufficient for the correctness of the algorithm. The detailed knowledge of this feasible region has inspired the idea of relaxed scheduling, and it was necessary to formulate it in detail and to prove its correctness. Many meshing algorithms are based on parameters that are fine-tuned in the experimental phase of software design. We suggest that in the absence of detailed knowledge of limits, fine-tuning is a necessary activity that gropes for a place in the feasible region where correctness is implied. Of course, it might happen that this region is empty, but this is usually difficult to determine.

Future Work. It is not our intention to criticize work in mesh generation for the lack of correctness proofs. Indeed, it would be more appropriate to criticize our own work for the lack of generality. Although we laid out a complete algorithm for maintaining the mesh of a deforming surface, we are a far cry from being able to prove its correctness for any surface other than the skin surface introduced in [5]. We have also not been able to extend the algorithm beyond the deformations implied by growing the spheres that define the surface. For example, it would be desirable to maintain the mesh for deformations used for morphing as described in [3]. Generalizing the algorithm to include this application and proving it correct may be within reach.

Another worthwhile task is the implementation of relaxed scheduling as part of the dynamic skin algorithm. Are our lower bounds for the necessary Δt sufficient to eliminate edge contractions and vertex insertions as a bottleneck of the algorithm? Can these lower bounds be improved in any significant manner? Can we improve the performance by fine-tuning the parameters, in particular Q_0 and Q_1, while staying within the proved feasible region?

Acknowledgments

The authors thank Robert Bryant for helpful discussions concerning the proof of the Length Lemma.

References

1. H.-L. CHENG, T. K. DEY, H. EDELSBRUNNER AND J. SULLIVAN. Dynamic skin triangulation. *Discrete Comput. Geom.* **25** (2001), 525–568.
2. M. L. CONNOLLY. Analytic molecular surface calculation. *J. Appl. Crystallogr.* **6** (1983), 548–558.
3. H.-L. CHENG, H. EDELSBRUNNER AND P. FU. Shape space from deformation. *Comput. Geom. Theory Appl.* **19** (2001), 191–204.
4. L. P. CHEW. Guaranteed-quality mesh generation for curved surfaces. *In* Proc. 9th Ann. Sympos. Comput. Geom., 1993, 274–280.
5. H. EDELSBRUNNER. Deformable smooth surface design. *Discrete Comput. Geom.* **21** (1999), 87–115.
6. H. EDELSBRUNNER AND E. P. MÜCKE. Three-dimensional alpha shapes. *ACM Trans. Graphics* **13** (1994), 43–72.
7. H. EDELSBRUNNER AND N. R. SHAH. Triangulating topological spaces. *Internat. J. Comput. Geom. Appl.* **7** (1997), 365–378.
8. B. LEE AND F. M. RICHARDS. The interpretation of protein structures: estimation of static accessibility. *J. Mol. Biol.* **55** (1971), 379–400.

Appendix

Table 3. Notation for important geometric concepts, functions, variables, and constants.

$f_i : \mathbb{R}^3 \to \mathbb{R}$	weighted (square) distance function
$S_i = (z_i, r_i)$	zero-set of f_i; sphere with center z_i and radius r_i
$\mathcal{F}$	convex hull of spheres S_i
$F = \text{env } \sqrt{\mathcal{F}}$	skin surface
$\kappa, \varrho = 1/\kappa$	maximum curvature, length scale
$Q_0 \leq Q \leq Q_1$	constant controlling quality
$\varepsilon, C, h, \ell, m$	additional constants
$g : \mathbb{R}^3 \to \mathbb{R}$	point-wise min of the $2f - f(z)$
$F(t) = g^{-1}(t)$	skin surface at time t
t, θ	time parameter, relative travel distance
$[t_0, t_1]$	time interval
$\Delta t = t_1 - t_0$	duration
$u, u', \overline{u}$	point, projection, reflection
$\nabla g_u, \dot{u}$	gradient, velocity vector
uv, uvw, H, R	edge, triangle, height, radius

A Note on Point Subsets
with a Specified Number of Interior Points

Thomas Fevens

Department of Computer Science, Concordia University
Montréal, Québec, Canada H3G 1M8
fevens@cs.concordia.ca
http://www.cs.concordia.ca/~faculty/fevens

Abstract. An *interior* point of a finite point set is a point of the set that is not on the boundary of the convex hull of the set. For any integer $k \geq 1$, let $g(k)$ be the smallest integer such that every set of points in the plane, no three collinear, containing at least $g(k)$ interior points has a subset of points containing exactly k interior points. Similarly, for any integer $k \geq 3$, let $h(k)$ be the smallest integer such that every set of points in the plane, no three collinear, containing at least $h(k)$ interior points has a subset of points containing exactly k or $k+1$ interior points. In this note, we show that $g(k) \geq 3k - 1$ for $k \geq 3$. We also show that $h(k) \geq 2k + 1$ for $5 \leq k \leq 8$, and $h(k) \geq 3k - 7$ for $k \geq 8$.

1 Introduction

Let P be a set of points in the plane in general position, that is, with no three points in a line. For P, define its *vertices* to be those points on the boundary of the convex hull of P with the remaining points being *interior* points. In 1935, Erdős and Szekeres [3] proved that for any $n \geq 3$, there is an integer $f(n)$ such that any set P of at least $f(n)$ points contains a subset of points whose convex hull contains exactly n vertices. Horton [4] showed that, for $n \geq 7$, there does not exist a finite $b(n)$ such that any point set P of at least $b(n)$ points contains a subset with n vertices and no interior points. Avis, Hosono and Urabe [2] investigated the question as to when P contains a subset of points whose convex hull contains precisely k interior points of P. For any integer $k \geq 1$, let $g(k)$ be the smallest integer such that every set of points P containing at least $g(k)$ interior points has a subset of points containing exactly k interior points. Avis, Hosono and Urabe showed that $g(1) = 1$, $g(2) = 4$, $g(3) \geq 8$, and $g(k) \geq k + 2$, $k \geq 4$.

With regards to an upper bound for $g(k)$, it is not known if $g(k)$ is finite for $k \geq 3$. Define $g_\triangle(k)$ to denote the smallest integer such that every set of points P whose convex hull is a triangle and which has at least $g(k)$ interior points also contains a subset with exactly k interior points. Hosono, Károlyi and Urabe [5] show that if $g_\triangle(k)$ is finite then $g(k)$ is also finite.

For any integer $k \geq 1$, let $h(k)$ be the smallest integer such that every set of points P containing at least $h(k)$ interior points has a subset of points containing

J. Akiyama and M. Kano (Eds.): JCDCG 2002, LNCS 2866, pp. 152–158, 2003.

exactly k or $k+1$ interior points. Avis, Hosono and Urabe [1] showed that $h(3) = 3$, $h(4) = 7$ and $h(k) \geq \lceil (3k+3)/2 \rceil$ for $k \geq 5$. It is not known if $h(k)$ is finite for $k \geq 5$. Note that these lower bounds for $h(k)$ also directly imply that $g(4) \geq 7$ and $g(k) \geq \lceil (3k+3)/2 \rceil$ for $k \geq 5$.

To investigate the lower bounds on $g(k)$, we consider *deficient* point sets $P = P(m, s, k)$, $s \geq k \geq 1$, which are point sets with m vertices and s interior points which do not contain a subset of points with k interior points[1]. The existence of a deficient point set $P(m, s, k)$ implies that $g(k) \geq s+1$. Indeed, the existence of a deficient point set $P(3, s, k)$ is sufficient, since such examples can be extended to $P(m, s, k)$, for all $m \geq 3$ by using the following lemma, proved in [2].

Extension Lemma *Every deficient point set $P(m, s, k)$ can be extended to a deficient point set $P(m+1, s, k)$.*

Similarly, to investigate the lower bounds on $h(k)$, we consider *doubly deficient* point sets $P = P'(m, s, k)$, $s > k \geq 1$, which are point sets with m vertices and s interior points which do not contain a subset of points with k or $k+1$ interior points. Again, by using a lemma [1] analogous to the above Extension Lemma, the existence of a doubly deficient point set $P'(3, s, k)$ implies that $h(k) \geq s+1$.

2   Improved Lower Bounds

In this note, we will present deficient and doubly deficient point sets leading to the lower bounds for $g(k)$ and $h(k)$ presented in table 1.

Table 1. Known exact values and lower bounds for $g(k)$ and $h(k)$.

k	$g(k)$	Reference	k	$h(k)$	Reference
1	1	ref. [2]	3	3	ref. [1]
2	4	ref. [2]	4	7	ref. [1]
3	≥ 8	ref. [2]	$5 \leq k \leq 8$	$\geq 2k+1$	this note.
$k \geq 3$	$\geq 3k-1$	this note.	$k \geq 8$	$\geq 3k-7$	this note.

All the deficient and doubly deficient point sets we construct have the following properties. The convex hull of the point set forms an equilateral triangle, T, with vertices at $(0, 100)$, $(-100 \sin(\pi/3), -100 \cos(\pi/3))$ and $(100 \sin(\pi/3), -100 \cos(\pi/3))$. Label the three vertices on the boundary of T as $A = \{A_1, A_2, A_3\}$, in counterclockwise order, starting with the topmost vertex. One of the points from the point set is located at the center, $(0, 0)$, of the equilateral triangle (we will label this point as x). Each of the point sets presented in this note has *threefold rotational symmetry*: if you rotate a point about x by $2\pi/3$ or $4\pi/3$, the point will line up exactly with another point from the point set. The point sets, minus x and A, can be viewed as being composed of three disjoint subsets,

---

[1] We are following the notation used in [2].

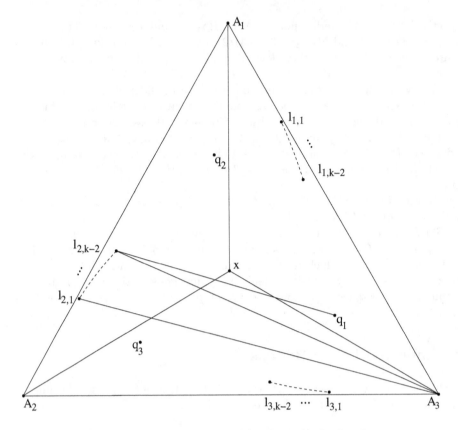

Fig. 1. Deficient point sets, $P(3, 3k - 2, k)$, for $k \geq 3$.

identical under threefold rotational symmetry, where each subset is bounded by the triangle defined by x and two vertices from A. All the points in a subset have the same index value (same first index if two indices) and any three points related by threefold rotational symmetry will have the same label (e.g., m_1, m_2, and m_3). The numbering of the indices of the points in the three subsets is counterclockwise cyclic over $\{1, 2, 3\}$ (the subset with index 1 is contained in the triangle defined by A_1, x and A_3, etc.).

2.1   Proof that $g(k) \geq 3k - 1$ for $k \geq 3$

Figure 1 shows deficient point sets with $3k - 2$ interior points that do not contain a convex subset with k interior points for $k \geq 3$. In these point sets, in addition to A and x, there are three points, labeled q_i, at $(43, -18)$ for $i = 1$, plus two other points from threefold rotational symmetry. In addition to these, there are $3(k - 2)$ points: $k - 2$ points, labeled $l_{i,j}$, $j = 1, \ldots, k - 2$ at (x_j, y_j) for $i = 1$ where x_j are $k - 2$ equally spaced values in the range from 22 to 34 and $y_j = -0.06x_j^2 + 0.5x_j + 78.0$ (if $k = 3$, then let $x_1 = 28$); plus $2(k - 2)$ other points from threefold rotational symmetry. These additional points are in convex position such that if the three vertices A are removed from the point set, all the $l_{i,j}$ and q_i points would be on the boundary of the convex hull of the remaining points (indeed, the only interior point for this convex subset would be x). Further, the triangle defined by $A_i, l_{i,j}$, and A_{i+2} has the points $l_{i,1}, l_{i,2}, \ldots, l_{i,j-1}$ as its only interior points.

We have to show that these are deficient point sets, $P(3, 3k - 2, k)$, $k \geq 3$. First note that if we remove any two of A from the point set, then the remaining points are in convex position with $k - 1$ interior points. Therefore, any possible subset of points in convex position with k interior points would have to include two of A on the boundary of its convex hull. Without loss of generality, assume that A_1 is removed from the point set, and that A_2 and A_3 are members in any subset. For a subset in convex position, there are three possible cases: i) x is on the convex hull of the subset; iii) x is outside of the convex hull of the subset; or iii) x is an interior point of the convex hull of the subset. Note that the triangle defined by x, A_2 and A_3 has $k - 1$ interior points. We will call this triangle T_3.

i) *x is on the convex hull of the subset*
   From figure 1, it can be seen that any subset that includes x on its convex hull has the interior points of T_3 as its interior points. In particular, we could only add q_1 and/or any subset of $l_{2,1}, \ldots, l_{2,k-2}$ to the subset defining T_3. Thus the maximum number of interior points for a subset that includes x on its convex hull is $k - 1$.

ii) *x is outside the convex hull of the subset*
   If x is outside the convex hull of a subset, then it could be added to a subset from case i) while either increasing or unchanging the number of interior points. But we have already seen that the maximum number of interior points for a subset that includes x on its convex hull is $k - 1$. Thus this case is equivalent to the previous case of having x on the convex hull of the point subset.

iii) *x is an interior point of the subset*
   In this case, any subset that contains x as an interior point must have either one of $l_{1,k-2}$ or q_2 as either an interior point or as a point on the boundary of the convex hull of the subset. But any triangle defined by one of these points and A_2 and A_3 has $k + 1$ interior points. Thus any convex subset with x as an interior point would have at least $k + 1$ interior points.

Therefore, there does not exist a subset in convex position with k interior points.

2.2   Proof that $h(k) \geq 2k + 1$ for $5 \leq k \leq 8$

Figure 2 shows that there exist point sets with $2k$ interior points that do not contain a subset in convex position with k or $k + 1$ interior points for $k = 5$ and $k = 8$. In the point set for $k = 8$, in addition to A and x, there are fifteen points, labeled m_i, n_i, o_i, p_i, and q_i, at $(1.5, 50.5)$, $(28, 44.96)$, $(35.5, -8)$, $(43, -13)$, and $(43, -18)$, respectively, for $i = 1$, plus ten other points from threefold rotational symmetry. To obtain the doubly deficient point sets for $k = 7$, 6, or 5 from the doubly deficient point set for $k = 8$, remove one or more pairs of points labeled o_i and n_{i+1} (one pair for $k = 7$; two pairs for $k = 6$; or all three pairs for $k = 5$).

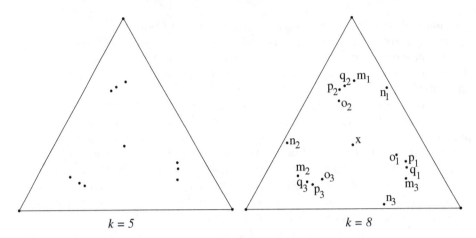

Fig. 2. Doubly deficient point sets, $P'(3, 2k, k)$, for $k = 5$ and $k = 8$, respectively.

2.3   Proof that $h(k) \geq 3k - 7$ for $k \geq 8$

Figure 3 shows that there exist point sets with $3k - 8$ interior points that do not contain a convex subset with k or $k + 1$ interior points for $k \geq 8$. In the point set, in addition to A and x, there are twelve points, labeled m_i, o_i, p_i, and q_i, at $(1.5, 50.5)$, $(35.5, -8)$, $(43, -13)$, and $(43, -18)$, respectively, for $i = 1$, plus eight other points from threefold rotational symmetry. In addition to these, there are $3(k - 7)$ points: $k - 7$ points, labeled $l_{i,j}$, $j = 1, \ldots, k - 7$ at (x_j, y_j) for $i = 1$ where x_j are $k - 7$ equally spaced values in the range from 22 to 34 and $y_j = -0.06x_j^2 + 0.5x_j + 78.0$ (if $k = 8$, then let $x_1 = 28$ and we have $P'(3, 16, 8)$ from figure 2); plus $2(k - 7)$ other points from threefold rotational symmetry. Note that these are the same as the deficient point sets $P(3, 3(k - 5) - 2, k - 5)$ shown in figure 3 with the addition of the nine points m_i, o_i, and p_i.

We have to show that such a point set does not contain a convex subset with k or $k + 1$ interior points. First note that if we remove two of A from the point set, then the remaining points are in convex position with $k - 1$ interior points. Therefore, any subset of points in convex position with k or $k + 1$ interior points would have to include two of A on the boundary of its convex hull. Without loss of generality, assume that A_1 is removed from the point set, and that A_2 and A_3 are members in any subset. Again, let us consider three possible cases: i) x is on the convex hull of subset in convex position; iii) x is outside of the convex hull of subset in convex position; iii) x is an interior point of the convex subset. Note that the triangle defined by x, A_2 and A_3 has $k - 3$ interior points. We will call this triangle T_3.

i) *x is on the convex hull of the subset*

From figure 3, we can see that any convex subset that includes x on its convex hull will include at most two interior points in addition to those in the interior of T_3 (namely, the points m_2 and q_1). Thus the maximum number of interior points for a subset that includes x on its convex hull is $k - 1$.

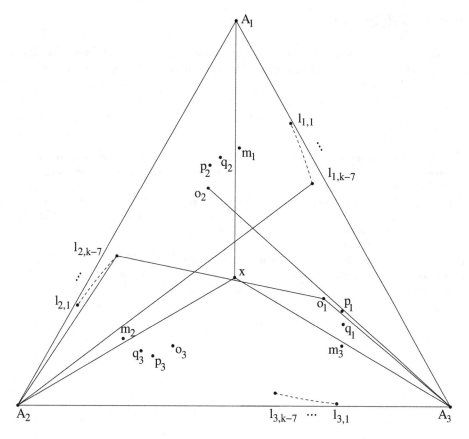

Fig. 3. Doubly deficient point sets, $P'(3, 3k - 8, k)$, for $k \geq 8$.

ii) *x is outside the convex hull of the subset*

Obviously, if x is outside the convex hull of a subset, then it could be added to a subset from case i) while either increasing or unchanging the number of interior points. But we have already seen that the maximum number of interior points for a subset that includes x on its convex hull is $k - 1$. Thus this is equivalent to the previous case of having x on the convex hull of the subset.

iii) *x is an interior point of the subset*

In this case, any subset that contains x as an interior point must have either o_2 or $l_{1,k-7}$ as either an interior point or a point on the boundary on the convex hull of the subset. But any triangle defined by either point and A_2 and A_3 has $k+2$ interior points. Thus any convex subset with x as an interior point would have at least $k + 2$ interior points.

Therefore, there does not exist a convex subset with k or $k + 1$ interior points.

Acknowledgement

I would like to thank David Avis for introducing me to this problem and for several useful discussions. Also, I would like to thank the referees for their careful reading of this manuscript.

References

1. D. Avis, K. Hosono, and M. Urabe. On the existance of a point subset with 4 or 5 interior points. volume 1763 of *Lecture Notes Comput. Sci.*, pages 57–64. Springer-Verlag, 2000.
2. D. Avis, K. Hosono, and M. Urabe. On the existence of a point subset with a specified number of interior points. *Discrete Mathematics*, 241(1-3):33–40, 2001.
3. P. Erdős and G. Szekeres. A combinatorial problem in geometry. *Compositio Mathematica*, 2:463–470, 1935.
4. J. Horton. Sets with no empty 7-gons. *Canad. Math. Bull.*, 26:482–484, 1983.
5. K. Hosono, G. Károlyi, and M. Urabe. Constructions from empty polygons. In A. Bezdek, editor, *Discrete Geometry: in Honor of W. Kuperberg's 60th Birthday*, pages 351–358. Marcel Dekker, 2003.

Piano-Hinged Dissections: Now Let's Fold!

Greg N. Frederickson

Department of Computer Science
Purdue University, West Lafayette, IN 47907, USA
gnf@cs.purdue.edu
http://www.cs.purdue.edu/people/gnf

Abstract. A new type of hinging is explored for geometric dissections of two-dimensional figures. The figures are represented by pieces on two adjacent *levels*. *Piano hinges* are used to rotate a piece B from being next to a piece A on one level to being above or below piece A on another level. Techniques are presented and analyzed for designing piano-hinged dissections. These include the use of polygon structure, the conversion from twisted-hinged dissections, the folding analogue of a P-slide, the folding analogue of a step dissection, and the use of tessellations. Properties of piano-hinged dissections are explored. An open problem relating to the possible universality of such hingings is posed.

1 Introduction

A *geometric dissection* is a cutting of a geometric figure into pieces that can be rearranged to form another figure [11,17]. When applied to two-dimensional figures, they are striking demonstrations of the equivalence of area and have intrigued people over the ages. Dissections date back to Arabic-Islamic mathematicians a millennium ago [1,24,27] and Greek mathematicians more than two millennia ago [3,6]. Their swift climb in popularity over the last century dates from their appearance in newspaper and magazine columns written by Sam Loyd [18] and Henry Ernest Dudeney [7].

Some dissections have a remarkable property, based on connecting the pieces with hinges in just the right way. We can then form one figure by swinging the pieces one way on the hinges, and the other figure when we swing them the other way. The earliest was described by Philip Kelland [15]. Almost a century ago, Dudeney [8] introduced perhaps the best-known *swing-hinged* dissection, of an equilateral triangle to a square. This hinged model (Figure 1) has captured the imagination of countless people and has been described in at least a dozen books. Fewer than fifty hinged dissections were known prior to the systematic discovery of a great many by the present author [12,13], and simultaneously by Jin Akiyama and Gisaku Nakamura[2]. This led naturally to the fundamental question:

Open Problem I [12,13]: For any two figures of equal area and bounded by straight line segments, is a swing-hingeable dissection possible?

J. Akiyama and M. Kano (Eds.): JCDCG 2002, LNCS 2866, pp. 159–171, 2003.
© Springer-Verlag Berlin Heidelberg 2003

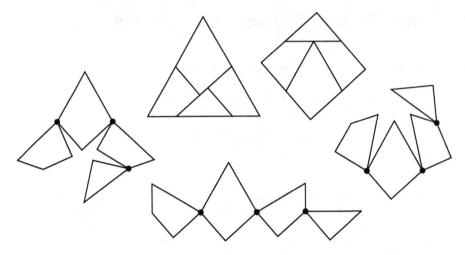

Fig. 1. Swing-hinged dissection of a triangle to a square.

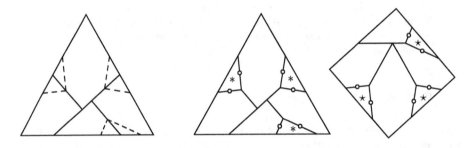

Fig. 2. Derivation, and twist-hinged dissection of a triangle to a square.

A second way to hinge dissections seems to be relatively recent. A *twist hinge* has a point of rotation on the interior of the line segment along which two pieces touch edge-to-edge. This allows one piece to be flipped over relative to the other, using rotation by 180° through the third dimension. A few isolated dissections [26,10,20] were the only examples prior to the discovery of great many by the current author [13]. A lovely example from that source, of an equilateral triangle to a square, is given on the right in Figure 2. Like many other twist-hinged dissections, it can be derived from swing-hinged dissection by a method from [13] that is hinted at on the left. Similarly, we have the question:

Open Problem II [12,13]: For any two figures of equal area and bounded by straight line segments, is a twist-hinged dissection possible?

There is a third way to hinge two pieces: with a piano hinge. (The term refers to the long hinge that attaches the top of a grand piano to the box containing the strings.) The model of a 2-dimensional figure must necessarily change, so that it consists of two *levels*. Consider two pieces that are *piano-hinged* or *fold-hinged*, together. They can rotate from being next to each other on the same level to being on different levels with one on top of the other. This type of movement has appeared a few times as a curiosity [5,9,25,28]. In addition to the piano-hinged dissections introduced here, there are a wealth more that are described in my book-in-progress [14]. Indeed, there are enough to motivate yet another fundamental question:

> **Open Problem III**: For any two figures of equal area and bounded by straight line segments, is a piano-hinged dissection possible?

We pose the interesting and challenging problem of identifying general techniques to produce piano-hinged dissections. We show how to exploit the structure of regular polygons and stars for these dissections. We introduce a technique to convert a twist-hinged dissection to be piano-hinged, a technique to change the length (and thus the height) of a rectangle, and a step-like technique that produces two (infinite) families of rectangle-to-rectangle dissections. We also show how to derive them through the use of tessellations. As in [11,13], we adopt the natural goal of minimizing the number of pieces, subject to the dissection being piano-hinged.

We also explore interesting properties of some fold-hinged transformations. A dissection is *cyclicly hinged* if removing one of the hinges does not disconnect the pieces. The over-abundance of hinges forces the motion of various pieces to be coordinated in fascinating ways.

This paper is excerpted from [14].

2   Definitions

We consider dissections of regular polygons and regular star polygons. Let $\{p\}$ be a regular polygon with p sides. Let $\{p/q\}$ be a star polygon with p points (vertices), where each point is connected to the q-th points clockwise and counterclockwise from it.

We take care when defining our piano-hinging of two-dimensional figures. If we were to assume that each level has thickness 0, then two pieces connected by a piano hinge would share points in common when they are folded against each other in 3-space. To avoid this problem, we assume that the pieces are open sets that at their simplest are prisms of thickness ϵ for some sufficiently small $\epsilon > 0$. A piano hinge will be a common edge of the bounding prisms of two connected pieces in each of the two figures that the pieces form.

If we assume a positive thickness for the pieces, we will rule out certain dissections, because the pieces can obstruct each other by a tiny amount (a function of ϵ) when they are rotated about the hinges. We can modify such dissections by rounding over (rounding the non-hinged edges by ϵ). The amount lost becomes negligible as ϵ approaches 0. We call such dissections *rounded*.

3   Cyclic Piano-Hingings

Let's first consider the dissection of a Greek cross to a square [16,19]. We present our 10-piece piano-hinged dissection in Figure 3, where both top and bottom

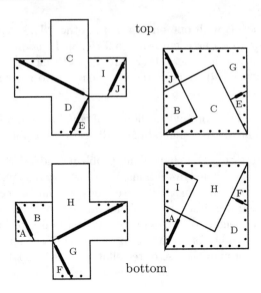

Fig. 3. Piano-hinged dissection of a Greek cross to a square.

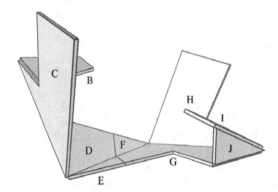

Fig. 4. Projected view of a Greek cross to a square.

levels are shown. We indicate a piano hinge that connects a piece on one level to a piece on the other with a line of dots next to the hinge line on each of the two levels. A thickened line segment indicates a piano hinge between two pieces on the same level. Figure 4 displays a perspective view.

A lovely property of some fold-hinged dissections is that of having a cyclic hinging. A dissection is *cyclicly hinged* when removing one of the hinges does not disconnect the pieces. This over-abundance of hinges forces the motion of various pieces to be coordinated. There are different ways that the pieces can be cyclicly hinged. The first, a *vertex-cyclic hinging*, is when four or more pieces touch at a vertex and each piece is hinged with its predecessor and successor on the cycle.

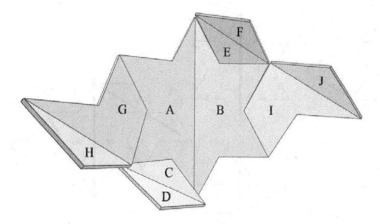

Fig. 5. Perspective view of assemblage for a hexagram to a triangle.

If the angles that meet at the vertex sum to less than 360°, then the vertex-cyclic hinging is a *cap-cyclic hinging*. There are two examples of cap-cycles in Figures 3 and 4: Pieces A, B, C, and D form one cap-cycle, and pieces G, H, I, and J form the other. If the angles that meet at the vertex sum to exactly 360°, then the vertex-cyclic hinging is a *flat-cyclic hinging*. There is one example of a flat-cycle in Figure 3, consisting of pieces D, E, F, and G.

If the angles that meet at the vertex sum to more than 360°, then the vertex-cyclic hinging is a *saddle-cyclic hinging*. This cycle would be at a concave vertex in a dissection. Its name derives from the shape of the surface as it moves from one configuration to the other. It opens from a concave angle to a surface with a saddle point and then closes to a concave angle on the other side of the surface. An example of a saddle-cyclic hinging is in a dissection of two equal Greek crosses to one. For other types of cyclic hingings that are non-vertex-cyclic, see [14].

4   Polygon and Star Structure

An unhinged dissection of a hexagram to an equilateral triangle was first described in [4]. We adapt the unhinged dissection in [23] to get the 10-piece piano-hinged dissection is in Figures 5 and 6. There is a 9-piece rounded piano-hinged dissection in [14].

Lindgren [17] identified a special trigonometric relationship between a $\{(4n+2)/(2n-1)\}$ and a $\{(2n+1)/n\}$, for any positive integer n. It leads to a $(4n+2)$-piece dissection of a $\{(4n+2)/(2n-1)\}$ to two $\{(2n+1)/n\}$s. These lovely dissections are swing-hingeable. However, the real surprise is that we can adapt them to be piano-hinged, For $n = 2$, we get the case of two pentagrams to a $\{10/3\}$, in Figure 8. We hinge the ten identical pieces from a pentagram alternately on a pentagram side and then a side of the $\{10/3\}$. As we see in Figure 7, the hinging gives the model an extraordinary pleated appearance.

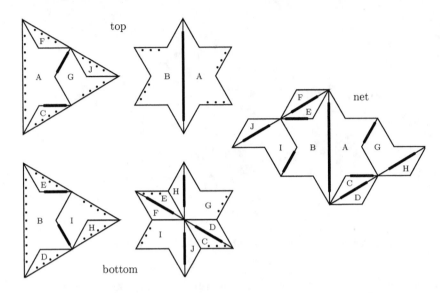

Fig. 6. Piano-hinged dissection of a hexagram to a triangle.

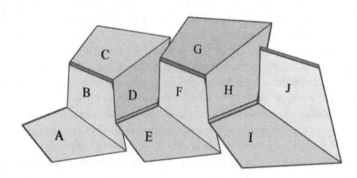

Fig. 7. Perspective view of one assemblage for two $\{5/2\}$s to a $\{10/3\}$.

Theorem 1. *For any natural number n, there is an $(8n+4)$-piece piano-hinged dissection of a $\{(4n+2)/(2n-1)\}$ to two $\{(2n+1)/n\}$s.*

Proof (Idea). Suggested by Figures 8 and 7. □

5   Conversion from Twist-Hinged

There are two general techniques for converting twist-hinged dissections to be piano-hinged, one of which is rounded.

Theorem 2. *Let $\mathcal{D}$ be an n-piece twist-hinged dissection. There is a $(4n-3)$-piece piano-hinged dissection $\mathcal{D}'$ in which each twist hinge is simulated by three piano hinges.*

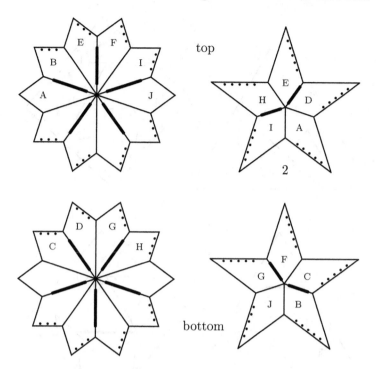

Fig. 8. Piano-hinged dissection of two {5/2}s to a {10/3}.

Proof (Idea). Duplicate the dissection on each of two levels. For each twist hinge (on pieces A and B), cut a rectangle R out of the bottom level of A, with one edge on the axis of the twist-hinge, and glue it to B. Place a piano-hinge along this edge. Cut out two more rectangles and piano-hinge them so that pieces A and B do not obstruct when rotated on the first piano hinge. Cut R in half and piano-hinge it, so that when folded up, it avoids the edge obstruction. □

If a piece in a twist-hinged dissection has more than one twist hinge incident on it, then we may be able to introduce just one piece that when flipped out of the way, avoids all obstructions of that piece with others. If we can do this for all pieces in the twist-hinged dissection, then our conversion will just double the number of pieces.

This is what happens when we convert the twist-hinged dissection of a triangle to a square in Figure 2. to get the rounded piano-hinged dissection of Figure 9, We start with the dissection of Figure 2 on both the top and the bottom. For each twist hinge, we identify a small right triangle with one leg on the axis of the twist hinge and the other leg against the edge of the piece. If we remove these small right triangles from the large pieces, then we will be able to rotate the pieces around on axes corresponding to the legs of the small right triangles that are perpendicular to legs of the original right triangles.

Theorem 3. *Let $\mathcal{D}$ be an n-piece twist-hinged dissection of two figures. There is a 2n-piece rounded piano-hinged dissection $\mathcal{D}'$ of the same two figures.*

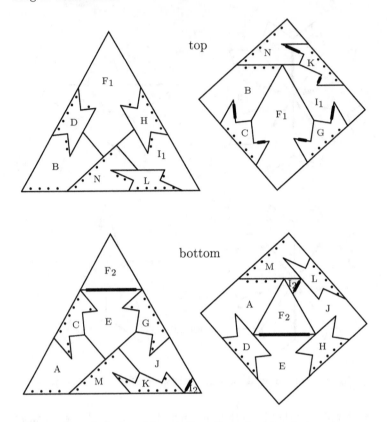

top

bottom

Fig. 9. Rounded piano-hinged dissection of a triangle to a square.

Proof (Idea). See the above discussion and Figure 9. □

6   Rectangle to Rectangle

There is a general technique that transforms a rectangle to another whose length is at most doubled. This gives the 8-piece piano-hinged dissection in Figure 10. This dissection is analogous to the 3-piece dissection of unhinged figures called a P-slide. (See [11,17].) By eliminating piece H and a corner of piece A, and then extending pieces B and C into the other level with corresponding projections, we produce the 7-piece rounded piano-hinged dissection in Figure 11.

Theorem 4. *Whenever $1 < \alpha < 2$, there is a 7-piece rounded piano-hinged dissection of an $(l \times w)$-rectangle to an $(\alpha l \times w/\alpha)$-rectangle.* □

When the ratio of the lengths of the rectangles are $(k+1)/k$, an unhinged P-slide dissection can be converted into a step dissection that uses one fewer piece. Similarly, the dissections in Figures 10 and 11 can be converted. The resulting number of pieces is 5, though there are two cases, with the second resulting in a rounded piano-hinged dissection. Examples are shown in Figures 12 and 13.

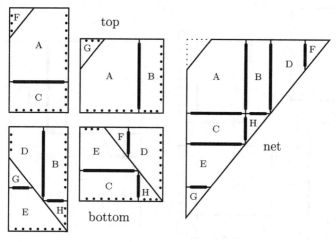

Fig. 10. Piano-hinged dissection of one rectangle to another.

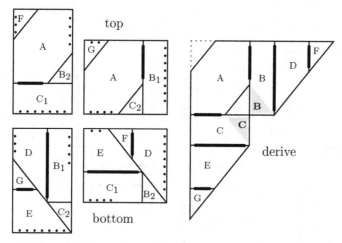

Fig. 11. Rounded piano-hinged dissection of one rectangle to another.

Theorem 5. *Whenever $\alpha = (2k+1)/2k$ for some natural number k, there is a 5-piece piano-hinged dissection of an $(l \times w)$-rectangle to an $(\alpha l \times w/\alpha)$-rectangle.* □

Theorem 6. *Whenever $\alpha = (2k+2)/(2k+1)$ for some natural number k, there is a 5-piece rounded piano-hinged dissection of an $(l \times w)$-rectangle to an $(\alpha l \times w/\alpha)$-rectangle.*

7   Derivations from Tessellations

A *tessellation of the plane* is a covering of the plane with copies of a figure in a repeating pattern such that the copies do not overlap. Here we shall see one example of how superposing tessellations can lead to piano-hinged dissections.

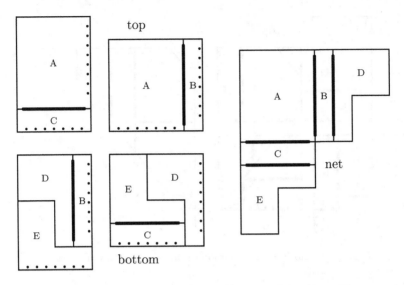

Fig. 12. Piano-hinged dissection of $(4a \times 5b)$-rectangle to $(5a \times 4b)$-rectangle.

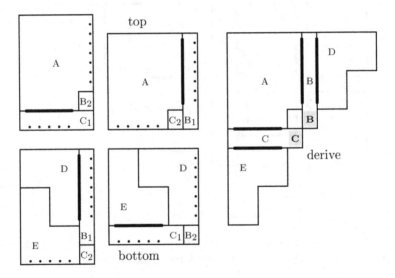

Fig. 13. Rounded piano-hinged dissection of $(5a \times 6b)$-rectangle to $(6a \times 5b)$-rectangle.

One of the oldest known dissections is Thābit ibn Qurra's 5-piece dissection of two unequal squares to one. It results from superposing two tessellations, one consisting of pairs of the unequal squares and the other consisting of the large squares [22,21]. Thābit's dissection is swing-hingeable and can be derived from a different superposition of tessellations [13].

We adapt Thābit's dissection to make it piano-hinged. Indeed, we start with the two tessellations from the superposition that gives rise to a swing-hinged

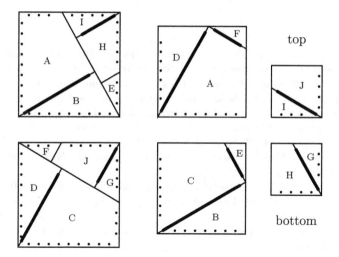

Fig. 14. Piano-hinged dissection of two unequal squares to one.

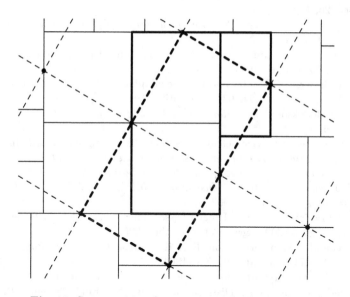

Fig. 15. Superposition for two unequal squares to one.

dissection, and then produce the piano-hinged dissection (Figure 14). The superposition of the tessellations is in Figure 15, where a pair of small squares and a pair of medium-sized squares are indicated by thicker solid lines. Similarly, a pair of large squares, is indicated by thicker dashed lines.

Many more examples of piano-hinged dissections derived from tessellations are given in [14].

References

1. Abu'l-Wafā' al-Būzjānī. Kitāb fīmā yahtāju al-sāni' min a' māl al-handasa (On the Geometric Constructions Necessary for the Artisan). Mashdad: Imam Riza 37, copied in the late 10th or the early 11th century. Persian manuscript.

2. Jin Akiyama and Gisaku Nakamura. Dudeney dissection of polygons. In Jin Akiyama, Mikio Kano, and Masatsugu Urabe, editors, *Discrete and Computational Geometry, Japanese Conference, JCDCG'98, Lecture Notes in Computer Science*, volume 1763, pages 14–29. Springer Verlag, 2000.

3. George Johnston Allman. *Greek Geometry from Thales to Euclid*. Hodges, Figgis & Co., Dublin, 1889.

4. Anonymous. Fī tadākhul al-ashkāl al-mutashābiha aw al-mutawāfiqa (Interlocks of Similar or Complementary Figures). Paris: Bibliothèque Nationale, ancien fonds. Persan 169, ff. 180r–199v.

5. Donald L. Bruyr. *Geometrical Models and Demonstrations*. J. Weston Walch, Portland, Maine, 1963.

6. Moritz Cantor. *Vorlesungen über Geschichte der Mathematik*, volume 1. B. G. Teubner, Stuttgart, third edition, 1907.

7. Henry E. Dudeney. Puzzles and prizes. Column in *Weekly Dispatch*, April 19, 1896–Dec. 26, 1903.

8. Henry Ernest Dudeney. *The Canterbury Puzzles and Other Curious Problems*. W. Heinemann, London, 1907.

9. Jan Essebaggers and Ivan Moscovich. Triangle hinged puzzle. European Patent EP0584883, 1994. Filed 1993.

10. William L. Esser, III. Jewelry and the like adapted to define a plurality of objects or shapes. U.S. Patent 4,542,631, 1985. Filed 1983.

11. Greg N. Frederickson. *Dissections Plane & Fancy*. Cambridge University Press, New York, 1997.

12. Greg N. Frederickson. Geometric dissections that swing and twist. In *Discrete and Computational Geometry, Japanese Conference, JCDCG'00, Lecture Notes in Computer Science, LNCS 2098*, pages 137–148. Springer Verlag, 2000.

13. Greg N. Frederickson. *Hinged Dissections: Swinging and Twisting*. Cambridge University Press, New York, 2002.

14. Greg N. Frederickson. "Piano-Hinged Dissections: Time to Fold". Rough draft of a book in progress, 200 pages as of November, 2002.

15. Philip Kelland. On superposition. Part II. *Transactions of the Royal Society of Edinburgh*, 33:471–473 and plate XX, 1864.

16. Don Lemon. *The Illustrated Book of Puzzles*. Saxon, London, 1890.

17. Harry Lindgren. *Geometric Dissections*. D. Van Nostrand Company, Princeton, New Jersey, 1964.

18. Sam Loyd. Mental Gymnastics. Puzzle column in Sunday edition of *Philadelphia Inquirer*, October 23, 1898–1901.

19. Sam Loyd. Weekly puzzle column in *Tit-Bits*, 1897. April 3, p. 3,.

20. Ernst Lurker. Heart pill. 7 inch tall model in nickel-plated aluminum, limited edition of 80 produced by Bayer, in Germany, 1984.

21. Percy A. MacMahon. Pythagoras's theorem as a repeating pattern. *Nature*, 109:479, 1922.

22. Paul Mahlo. *Topologische Untersuchungen über Zerlegung in ebene und sphaerische Polygone*. C. A. Kaemmerer, Halle, Germany, 1908. See pp. 13, 14 and Fig. 7.

23. Geoffrey Mott-Smith. *Mathematical Puzzles for Beginners and Enthusiasts*. Blakiston Co., Philadelphia, 1946.

24. Alpay Özdural. Mathematics and arts: Connections between theory and practice in the medieval Islamic world. *Historia Mathematica*, 27:171–201, 2000.

25. T. Sundara Row. *Geometric Exercises in Paper Folding*. Open Court, Chicago, 1901.

26. Erno Rubik. Toy with turnable elements for forming geometric shapes. U.S. Patent 4,392,323, 1983. Filed 1981; filed for a Hungarian patent in 1980.

27. Aydin Sayili. Thâbit ibn Qurra's generalization of the Pythagorean theorem. *Isis*, 51:35–37, 1960.

28. Kenneth V. Stevens. Folding puzzle using triangular pieces. U.S. Patent 5,299,804, 1994. Filed 1993.

The Convex Hull for Random Lines in the Plane

Mordecai Golin[1], Stefan Langerman[2,*], and William Steiger[3]

[1] Computer Science, University of Science and Technology, Hong Kong
[2] Département d'Informatique, Université Libre de Bruxelles
[3] Computer Science, Rutgers University

Abstract. An arrangement of n lines chosen at random from R^2 has a vertex set whose convex hull has constant (expected) size.

1 Introduction and Summary

Let $L = \{\ell_1, \ldots, \ell_n\}$ be a set of lines in general position in R^2. The vertex set $V = \{\ell_i \cap \ell_j, i < j\}$ of this arrangement has size $O(n^2)$ and we are interested in $|\mathrm{Conv}(V)|$, the number of extreme points of its convex hull. As observed by Atallah [1],

$$|\mathrm{conv}(V)| \leq 2n,$$

a fact that sparked algorithmic interest in the hull of line arrangements [2], [3], [4].

Suppose the lines are *chosen uniformly at random*. The specific model we use is that the lines in L are the duals of n points chosen uniformly and independently from $[0,1]^2$, under the familiar duality that maps a point $P = (x,y)$ to the line $TP = \{(u,v) : v = xu + y\}$ and maps the non-vertical line $\ell = \{(x,y) : y = mx+b\}$ to the point $T\ell = (-m, b)$. To get n randomly chosen lines $\ell_1, \ldots, \ell_n$, we start with points $P_i = (x_i, y_i)$, $i = 1, \ldots, n$ chosen uniformly and independently from $[0,1]^2$ and then take

$$\ell_i = \{(u,v) : v = x_i u + y_i\}, i = 1, \ldots, n.$$

We give a simple proof of the following statement.

Theorem 1 *Let L be a set of n lines chosen uniformly at random. There is a constant $c > 0$ so that*

$$E(|\mathrm{Conv}(V)|) < c; \qquad (1)$$

A similar statement holds when the lines are dual to points chosen uniformly from other convex polygons. We have not tried to estimate c carefully, but we believe it is smaller than 10.

Devroye and Toussaint [4] proved the same result when the lines are polar duals to points chosen at random from a wide range of radially symmetric distributions. The two models for random lines are quite different, and both are natural. Our proof is simple and elementary. Much more is needed to establish the statement in [4].

* Chargé de recherches du FNRS

J. Akiyama and M. Kano (Eds.): JCDCG 2002, LNCS 2866, pp. 172–175, 2003.
© Springer-Verlag Berlin Heidelberg 2003

2   The Proof

Choose $P_1 = (x_1, y_1), \ldots, P_n = (x_n, y_n)$ uniformly and independently from the unit square and numbered so $x_i < x_{i+1}$. We may assume that the points chosen are in general position in the sense that no three points lie on a common line and no two points have the same x coordinate, because these degeneracies occur with zero probability. The random lines are $\ell_i = \{(u, v) : v = x_i u + y_i\}$, $i = 1, \ldots, n$, and the vertex set is $V = \{\ell_i \cap \ell_j, i < j\}$. It is better to consider Conv(V) in the primal. A vertex $\ell_i \cap \ell_j \in V$ is an extreme point of Conv(V) only if $j = (i \bmod n) + 1$, so we seek the convex hull of the n vertices formed by the lines in L with successive slopes (in the radial ordering of the lines by slope). In the primal we seek lines through successive points P_i, P_{i+1} which are part of the upper or lower envelope of these lines. Specifically let r_i be the line joining P_i and P_{i+1}, $i = 1, \ldots, n-1$, and $r_i(t)$ the y-coordinate of the point on r_i with x-coordinate t. Write $U(t) = \max_i r_i(t)$ and $L(t) = \min_i r_i(t)$ for the upper and lower envelopes of the r_i. Then

$$|\mathrm{Conv(V)}| = |\mathbf{UP}| + |\mathbf{DN}|$$

where we write **UP** for the set $\{i : r_i \text{ has a segment in } U(t)\}$ and **DN** for the set $\{i : r_i \text{ has a segment in } L(t) \text{ but not in } U(t)\}$. We only show how to bound the expected size of **UP**, the argument for **DN** being similar.

Let A_i be the event that r_i meets $U(t)$ and Z_i, its indicator. We will show that $E(Z_i) \le c/n$ for an appropriate constant $c > 0$.

We cover A_i by simpler events whose probability is easier to estimate. Let λ be the $r_i, i > 3n/4$, of minimum slope, let ρ be the $r_i, i \le n/4$, of max slope, and

$$\sigma = \min\left[|slope(\lambda)|, |slope(\rho)|\right]$$

Let L be the line through $(x_{3n/4}, 0)$ of slope $-\sigma$, R the line through $(x_{n/4}, 0)$ of slope σ, and $U^*(t) = \max_t [L(t), R(t)]$. Clearly $U^*(t) \le U(t)$. Write $Q = (x, y) = L \cap R$ for the intersection of L and R (see figure).

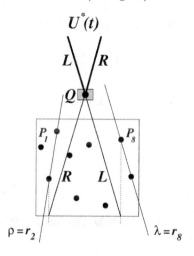

Let $E_1, \ldots, E_{n+1}$ be i.i.d. standard exponential random variables with partial sums

$$S_i = E_1 + \cdots + E_i,$$

and let $y_1, \ldots, y_n$ be i.i.d. uniforms on $[0, 1]$. It is familiar that the joint distribution of $S_1/S_{n+1}, \ldots, S_n/S_{n+1}$ is the same as that of $x_{(1)}, \ldots, x_{(n)}$, the order statistics of a sample $x_1, \ldots, x_n$ of i.i.d. uniforms. We think of the points $P_1, \ldots, P_n$ ordered by x-coordinate, with $P_i = (S_i, y_i)$, $i = 1, \ldots, n$. In this way the unit square is replaced by the random rectangle with corners at $(0,0)$ and $(S_{n+1}, 1)$. The law of large numbers implies that as $j \to \infty$

$$\text{Prob}[|S_j - j| < \varepsilon j] \geq 1 - 1/j$$

for any $\varepsilon > 0$, and that the point $Q = (x, y)$ where L and R meet satisfies $|x - n/2| < \varepsilon n$ and $|y - \sigma n/4| < \varepsilon n$ with probability at least $1 - 1/n$. We therefore assume that x and y, the coordinates of Q, satisfy those inequalities.

For $i \leq n/4$, $A_i \subseteq B_i \cup C_i$, where B_i is the event that r_i has slope less than $-\sigma$ and C_i the event that r_i is above Q; if neither B_i nor C_i occur, r_i does not meet $U^*(t)$, so it can't meet $U(t)$. To estimate the probabilities of B_i and C_i, note that the line r_i joining P_i and P_{i+1} has slope $s_i = (y_{i+1} - y_i)/E_{i+1}$. The numerator has density $f(t) = 1 - |t|$, $t \in [-1, 1]$. Therefore for $t > 0$

$$P[s_i \leq -t] = P[s_i \geq t] = \int_0^{1/t} P[y_{i+1} - y_i \geq ts]e^{-s}ds$$

$$= \int_0^{1/t} \frac{(1 - ts)^2}{2}e^{-s}ds = t^2(1 - e^{-1/t}) + \frac{1}{2} - t \qquad (2)$$

$$= \frac{1}{6t} - \frac{1}{24t^2} + \frac{1}{120t^3} + \cdots$$

Denote this function by $g(t)$.

Write $M = \max(s_2, s_4, \ldots, s_{n/4})$ and note that $\rho = \max_{i \leq n/4} s_i \geq M$. Also, because the random variables $s_{2j}, j = 1, \ldots, n/8$ are independent, we have

$$P[M \leq t] = P[s_{2i} \leq t]^{n/8} = (1 - P[s_{2i} \geq t])^{n/8} = [1 - g(t)]^{n/8}.$$

Similarly, writing $m = \min(s_{3n/4+2}, s_{3n/4+4}, \ldots, s_n)$, $\lambda = \min_{i > 3n/4} s_i \leq m$, and because the even slopes are independent,

$$P[m \geq -t] = P[s_{2i} \geq -t]^{n/8} = (1 - P[s_{2i} \leq -t])^{n/8} = [1 - g(t)]^{n/8}.$$

These combine to show

$$P[\sigma \leq t] \leq P[\min(M, |m|) \leq t] = 2[1 - g(t)]^{n/8} - [1 - g(t)]^{n/4}. \qquad (3)$$

The intuition from (2) and (3) is that M has median $\Theta(n)$ and s_i exceeds this with probability $O(1/n)$.

More formally, $C_i = \{r_i \text{ above } Q\} \subset \{s_i \geq \sigma/2 - \epsilon\}$ for some small $\epsilon > 0$, an event with probability at most

$$P[\sigma \leq K] + \int_K^\infty P[t/2 - \epsilon \leq s_i \leq t]h(t)dt,$$

for any positive K, where we write $h(t)$ for the density of $\min(M, |m|)$ obtained by differentiating the right hand side of (3). Therefore

$$P[C_i] \leq P[\sigma \leq K] + \int_{K-2\epsilon}^\infty P[s_i \geq t/2]h(t)dt.$$

Note that for large t, $1/(7t) < g(t) < 1/(5t)$ and for $K = an/\log n$, $P[\sigma \leq K] \leq 2e^{-ng(K)/8} \leq 1/n$. Applying these estimates,

$$P[C_i] \leq \frac{1}{n} + \int_K^\infty g(t/2)ne^{-ng(t)/8}/(6t^2)dt$$

$$\leq \frac{1}{n} + \int_K^\infty \frac{n}{120t^3 e^{n/(40t)}}dt,$$

an expression bounded by c/n.

We also have $P[B_i] = P[s_i < -\sigma] = P[s_i > \sigma] < P[C_i]$, and so $P[A_i] < 2P[C_i] < 2c/n$, for $i = 1, \ldots, n/4$. The same is true for $i = 3n/4 + 1, \ldots, n$ by symmetry. Finally, when $|i - n/2| \leq n/4$, r_i meets $U(t)$ only if $s_i > \sigma - \epsilon$ or $s_i < -\sigma + \epsilon$, and both these events have probability less than $P[C_i] < c/n$.   □

Acknowledgements

We thank an anonymous referee for valuable suggestions.

References

1. M. Atallah. Computing the Convex Hull of Line Intersections. *J. Algorithms 7*, 285-288 (1986).
2. B. Battacharya, H. Everett, and G. Toussaint. A Counter-Example to a Dynamic Algorithm for Convex Hulls of Line Arrangements. *Pattern Rec. Letters 12*, 145-147 (1991).
3. J. Boreddy. An Incremental Computation of Convex Hull of Planar Line Intersections. *Pattern Rec. Letters 11*, 541-543 (1990).
4. L. Devroye and G. Toussaint. Convex Hulls for Random Lines. *J. Algorithms 14*, 381-394 (1993).

Comparing Hypergraphs by Areas of Hyperedges Drawn on a Convex Polygon

Hiro Ito[1] and Hiroshi Nagamochi[2]

[1] Department of Communications and Computer Engineering, School of Informatics
Kyoto University, Kyoto 606-8501, Japan
itohiro@i.kyoto-u.ac.jp
[2] Department of Information and Computer Sciences
Toyohashi University of Technology, Aichi 441-8580, Japan
naga@ics.tut.ac.jp

Abstract. Let $H = (N, E, w)$ be a hypergraph with a node set $N = \{0, 1, \ldots, n-1\}$, a hyperedge set $E \subseteq 2^N$, and real edge-weights $w(e)$ for $e \in E$. Given a convex n-gon P in the plane with vertices $x_0, x_1, \ldots, x_{n-1}$ which are arranged in this order clockwisely, we let each node $i \in N$ correspond to the vertex x_i and define the area $A_P(H)$ of H on P by the sum of weighted areas of convex hulls for all hyperedges in H. For $0 \leq i < j < k \leq n-1$, a convex three-cut $C(i, j, k)$ of N is $\{\{i, \ldots, j-1\}, \{j, \ldots, k-1\}, \{k, \ldots, n-1, 0, \ldots, i-1\}\}$ and its size $c_H(i, j, k)$ in H is defined as the sum of weights of edges $e \in E$ such that e contains at least one node from each of $\{i, \ldots, j-1\}$, $\{j, \ldots, k-1\}$ and $\{k, \ldots, n-1, 0, \ldots, i-1\}$. We show that for two hypergraphs H and H' on N, the following two conditions are equivalent.
  - $A_P(H) \leq A_P(H')$ for all convex n-gons P.
  - $c_H(i, j, k) \leq c_{H'}(i, j, k)$ for all convex three-cuts $C(i, j, k)$.

1   Introduction

Every hypergraph H treated in this paper has a node set which consists of numbered nodes, and the set of n nodes is denoted by $N = \{0, 1, \ldots, n-1\}$. A hypergraph H has a set $E_H \subseteq 2^N$ of (hyper)edges $e \in E$; e also denotes the set of end nodes of e, and hence $|e|$ denotes the number of nodes incident to e.

Projection. Let $x_0, x_1, \ldots, x_{n-1}$ be vertices of a convex n-gon P in the plane, where $(x_i, x_{i+1 \bmod n})$ is an edge of P for $i = 0, 1, \ldots, n-1$. Each internal angle of P may be equal to π and some of vertices of P may be located at the same point in the plane. Let $\mathcal{P}$ denote the set of all such n-gons. For a subset $I \subseteq \{0, 1, \ldots, n-1\}$, we say that the convex hull of the vertices x_i with $i \in I$ is the $|I|$-gon induced by I, and is denoted by $P(I)$. We define a *projection* (H, P) of a hypergraph H of n nodes on an n-gon P by mapping each node $i \in N$ to the vertex x_i and each hyperedge e in H to the $|e|$-gon $P(e)$. For example, Fig. 1 illustrates two projections for a hypergraph H with a node set $N = \{0, 1, \ldots, 5\}$ and three hyperedges $\{0, 1, 2, 3\}$, $\{2, 3, 4, 5\}$ and $\{0, 1, 4, 5\}$.

J. Akiyama and M. Kano (Eds.): JCDCG 2002, LNCS 2866, pp. 176–181, 2003.

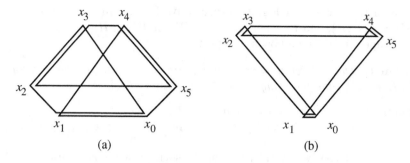

Fig. 1. Examples of projection (H, P).

Let $H = (N, E_H, w)$ be an *edge-weighted hypergraph* with a set N of n nodes, a set $E_H \subseteq 2^N$ of hyperedges, and a weight function $w : E_H \to \mathbf{R}$, where $w(e)$ means the number of copies of hyperedges e. Note that $w(e)$ may be negative. Let $a_P(e)$ denote the area of $P(e)$. Define the area of $H = (N, E, w)$ with respect to a convex n-gon P as the sum of the areas of $P(e)$ over all hyperedges $e \in E$, i.e.,

$$A_P(H) := \sum_{e \in E} w(e) \, a_P(e).$$

For $i, j \in N$, let

$$N[i, j] := \begin{cases} \{i, \ i+1, \ \ldots, \ j\}, & \text{if } i \leq j, \\ \{i, \ i+1, \ \ldots, \ n-1, \ 0, \ 1, \ \ldots, \ j\}, & \text{if } i > j. \end{cases}$$

We adopt the cyclic order for treating integers (or numbered nodes) in N. Thus for $i, j, k \in N$, $i \leq j \leq k$ means $j \in N[i, k]$, and $i \pm j$ means $i' \in N$ such that $i' \equiv i \pm j \pmod{n}$.

For $h_1, h_2, h_3 \in N$ ($h_1 < h_2 < h_3$), a *convex three-cut* $C(h_1, h_2, h_3)$ denotes a three-partition $\{N[h_1, h_2 - 1], \ N[h_2, h_3 - 1], \ N[h_3, h_1 - 1]\}$ of N. The size of a convex three-cut $C(h_1, h_2, h_3)$ of a hypergraph H is defined as

$$c_H(h_1, h_2, h_3) := \sum \{w(e) \mid e \cap N[h_1, h_2 - 1] \neq \emptyset, \ e \cap N[h_2, h_3 - 1] \neq \emptyset,$$
$$e \cap N[h_3, h_1 - 1] \neq \emptyset\}.$$

For example, for the hypergraph H of Fig. 1 with $w(e) = 1$ for all edges e of H, $c_H(0, 2, 4) = 3$, and $c_H(0, 1, 2) = 2$.

We introduce two relations $\preceq_a$ and $\preceq_c$ between hypergraphs as follows.

Definition 1. *Let H and H' be two hypergraphs. $H \preceq_a H'$ means $A_P(H) \leq A_P(H')$ for every convex n-gon P. $H \preceq_c H'$ means $c_H(h_1, h_2, h_3) \leq c_{H'}(h_1, h_2, h_3)$ for all convex three-cuts $C(h_1, h_2, h_3)$.* $\square$

For example, we compare the hypergraph of Fig. 1, say H, and $H' = (\{0, 1, \ldots, 5\}, \{\{0, 1, 3, 4\}\}, w)$ ($w(e) = 1$ for all edges e of H and H'). Let P and P' be polygons of (a) and (b) of Fig. 1, respectively. Then, $A_P(H) > A_P(H')$ and $A_{P'}(H) < A_{P'}(H')$, and thus neither $H \preceq_a H'$ nor $H' \preceq_a H$.

This paper shows the following property.

Theorem 1. *The two relations $\preceq_a$ and $\preceq_c$ are equivalent, i.e., for two hypergraphs H and H', the following two conditions are equivalent.*

- $A_P(H) \leq A_P(H')$ *for all convex n-gons P.*
- $c_H(h_1, h_2, h_3) \leq c_{H'}(h_1, h_2, h_3)$ *for all convex three-cuts $C(h_1, h_2, h_3)$.*   □

Related work. Recently, several problems in such projections of graphs or hypergraphs on convex polygons were studied. In the articles [1,3], given an edge-weighted graph $G = (N, E, w)$ and an n-gon P, the weighted sum $S_P(G)$ of lengths of line segments in the projection (G, P) were studied. Note that a graph is a hypergraph in which every edge consists of two nodes and thereby a convex hull $P(e)$ for an edge e with $|e| = 2$ becomes a line segment in a projection. The articles proved equivalence of two relations $\preceq_l$ and $\preceq_c$ that are defined respectively by $G \preceq_l G'$ if and only if $S_P(G) \leq S_P(G')$ for all convex n-gons P, and $G \preceq_c G'$ if and only if $c_G(h_1, h_2) \leq c_{G'}(h_1, h_2)$ for all linear cuts $C(h_1, h_2)$, where $c_G(h_1, h_2) := \sum \{w(e) \mid e \cap N[h_1, h_2 - 1] \neq \emptyset, \ e \cap N[h_2, h_1 - 1] \neq \emptyset\}$ is defined to be the size of a linear-cut $C(h_1, h_2) := \{N[h_1, h_2 - 1], N[h_2, h_1 - 1]\}$ in a graph $G = (N, E, w)$. Definition 1 provides a hypergraph version of these relations, and Theorem 1 shows that an analogous property holds in hypergraphs.

The articles [5,6] investigated problems on graph projection on convex polygons with minimizing the size of linear cuts. Our problem is different from them in that node i of the graph or the hypergraph must be corresponded to vertex x_i of the polygon.

Skiena [7] showed that a graph is uniquely determined if the size of all linear-cuts are specified. Theorem 1 of this paper is an extension of the theorem.

For hypergraph projections, the article [2] considered a problem of determining whether or not a given hypergraph H covers all convex n-gons, where we call that H covers P if every point of the internal region of P is included in at least one $P(e)$ of a hyperedge e of H. It presented a polynomial time algorithm for the problem.

2   Proof

In this section, we give a proof of Theorem 1. Consider a hypergraph $H_\emptyset = (N, \emptyset)$ with an empty edge set and no edge-weight function. Note that $c_{H_\emptyset}(h_1, h_2, h_3) = 0$ for any convex three-cut $C(h_1, h_2, h_3)$, and $A_P(H_\emptyset) = 0$ for any convex n-gon P. For any pair of $H = (N, E, w)$ and $H' = (N, E', w')$, we define $H - H' = (N, E'', w'')$ as

$$E'' := \{e \subset E \cup E' \mid w(e) - w'(e) \neq 0\}$$

and

$$w''(e) := w(e) - w'(e),$$

where $w(e) = 0$ for $e \notin E$ and where $w'(e) = 0$ for $e \notin E'$.

Observe that $H \preceq_a H$ (resp., $H \preceq_c H$) is equivalent to $H - H' \preceq_a H_\emptyset$ (resp., $H - H' \preceq_c H_\emptyset$). Therefore, it is enough to consider $H' = H_\emptyset$ for proving Theorem 1. Our proof of Theorem 1 consists of two parts:

(1) $H \preceq_a H_\emptyset \Rightarrow H \preceq_c H_\emptyset$ (Lemma 1) and
(2) $H \preceq_c H_\emptyset \Rightarrow H \preceq_a H_\emptyset$ (Lemma 3).

Lemma 1. *For a hypergraphs H, if $H \preceq_a H_\emptyset$, then $H \preceq_c H_\emptyset$.*

Proof. Assume that $H \preceq_c H_\emptyset$ does not hold, i.e., there is a convex three-cut $C(h_1, h_2, h_3)$ such that $c_H(h_1, h_2, h_3) > c_{H_\emptyset}(h_1, h_2, h_3) = 0$. We show a convex n-gon P that satisfies $A_P(H) > 0 = A_P(H_\emptyset)$ as follows.

In the plane with xy-coordinate, we consider three circles O_0, O_1, and O_2 with radius $r > 0$ whose centers are $(0,0)$, $(1,0)$, and $(0,1)$, respectively. $X_0 = \{x_i \mid i \in N[h_1, h_2 - 1]\}$, $X_1 = \{x_i \mid i \in N[h_2, h_3 - 1]\}$, $X_2 = \{x_i \mid i \in N[h_3, h_1 - 1]\}$. It is a simple matter to see that, for any small $r > 0$ all vertices of X_i ($i = 0, 1, 2$) can be put in the interior region of O_i so that they form a convex n-gon P. For a sufficiently small r, $A_P(H)$ becomes greater than zero.   □

Before proving the converse, we show the following lemma. For two points a and b, we denote by $\overline{ab}$ the line (or line segment between a and b) containing a and b.

Lemma 2. *Let P be a convex n-gon with $n > 3$. Let l be a line that contains a vertex x_i of P, but no point in the interior of P. Assume that there is a cross point a (resp., b) of l and line $\overline{x_{i-2}x_{i-1}}$ (resp., line $\overline{x_{i+1}x_{i+2}}$) such that a, x_{i-1}, x_{i-2} and b, x_{i+1}, x_{i+2} appear in these orders in the lines (see, Fig. 2). Construct a polygon P_a (resp., P_b) from P by replacing vertex x_i with a new vertex a (resp., b). Then for any hypergraph H,*

$$A_{P_a}(H) \geq A_P(H) \geq A_{P_b}(H) \quad or \quad A_{P_a}(H) \leq A_P(H) \leq A_{P_b}(H).$$

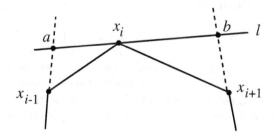

Fig. 2. Two new possible vertices a and b for a given polygon P and line l.

Proof. For a point y on the line segment $\overline{ab}$, let P_y be a polygon constructed from P by replacing x_i with y. The area of any triangle $x_j x_k y$ becomes a linear function of the distance $|ay|$. Since the area $a_{P_y}(e)$ for a hyperedge e is the sum of areas of triangles, $A_{P_y}(H)$ is also a linear function of the distance $|ay|$. Therefore the desired inequality holds.   □

Lemma 3. *For a hypergraph H, if $H \preceq_c H_\emptyset$, then $H \preceq_a H_\emptyset$.*

Proof: Assume that $H \preceq_c H_\emptyset$. Let P be a convex n-gon. We derive that $A_P(H) \leq 0 = A_P(H_\emptyset)$ from the assumption. Choose three distinct vertices x_{i-1}, x_i, x_{i+1} which are consecutively arranged along P. We apply Lemma 2 to the vertex x_i and the line l that contains x_i and is parallel to $\overline{x_{i-1}x_{i+1}}$ (note that $l = \overline{x_{i-1}x_{i+1}}$ iff the internal angle at x_i is π). W.o.l.g. we can assume $A_{P_a}(H) \geq A_P(H) \geq A_{P_b}(H)$. We replace P with P_a without changing the area of P. By this replacement, the internal angle at x_{i-1} becomes π while the area $A_P(H)$ never decreases.

If the internal angle at x_i is π, then a equals to x_{i-1} and hence the resulting polygon P_a is an $(n-1)$-gon.

By applying the above operation iteratively, P finally becomes a triangle $y_{N[h_1,h_2-1]}y_{N[h_2,h_3-1]}y_{N[h_3,h_1-1]}$, which we denote by P^*, where $y_{N[h_i,h_{i+1}-1]}$ $(i = 1, 2, 3)$ is a vertex made by contracting $x_{h_i}, x_{h_i+1}, \ldots, x_{h_{i+1}-1}$. In each step, the area $A_P(H)$ is not decreased. Hence

$$A_P(H) \leq A_{P^*}(H). \tag{1}$$

The unique convex three-cut separating the three vertices of the triangle P^* is $c_H(h_1, h_2, h_3)$, and from the assumption, we have

$$c_H(h_1, h_2, h_3) \leq 0,$$

which means

$$A_{P^*}(H) \leq 0. \tag{2}$$

From (1) and (2), $A_P(H) \leq 0$ is obtained.   □

Proof of Theorem 1: Follows immediately from Lemmas 1 and 3.   □

3   Concluding Remarks

We showed that for two hypergraphs H and H', "$A_P(H) \leq A_P(H')$ for all convex n-gons P" if and only if "$c_H(h_1, h_2, h_3) \leq c_{H'}(h_1, h_2, h_3)$ for all convex three-cuts $C(h_1, h_2, h_3)$". The number of convex three-cuts is $O(n^3)$, and thus the latter condition can be checked in polynomial time for given H and H'. This assures us that the former condition, which is stated on the basis of infinitely many number of convex n-gons, can be checked in polynomial time.

References

1. H. Ito, Sum of edge lengths of a multigraph drawn on a convex polygon, Computational Geometry, **24** (2003) 41–47.
2. H. Ito and H. Nagamochi, Can a hypergraph cover every convex polygon?, Proc. of the 3rd Hungarian-Japanese Symposium on Discrete Mathematics and Its Applications Sanjo Conference Hall, The Univ. of Tokyo, Jan. 21–24 (2003) 293–302.

3. H. Ito, H. Uehara, and M. Yokoyama, Lengths of tours and permutations on a vertex set of a convex polygon, Discrete Applied Mathematics, **115** (2001) 63–72.
4. H. Ito, H. Uehara, and M. Yokoyama, Sum of edge lengths of a graph drawn on a convex polygon, in: Discrete and Computational Geometry: Proc. of JCDCG2000, Lecture Notes in Computer Science, Vol. 2098 (Springer, Berlin, 2001) 160–166.
5. E. Mäkinen, On circular layouts, Intern. J. Computer Math., **24** (1988) 29–37.
6. H. Schröder, O. Sỳkora, and I. Vrt'o, Cyclic cutwidth of the mesh, in: Proc. of SOFSEM'99, Lecture Notes in Computer Science, Vol. 1725 (Springer, Berlin, 1999) 449–458.
7. S. S. Skiena, Reconstructing graphs from cut-set sizes, Information Processing Letters, **32** (1989) 123–127.

On Reconfiguring Radial Trees

Yoshiyuki Kusakari

Akita Prefectural University, Honjo Akita, 015-0055, Japan
Kusakari@akita-pu.ac.jp

Abstract. A linkage is a collection of line segments, called bars, possibly joined at their ends, called joints. Flattening a tree linkage is a continuous motion of their bars from an initial configuration to a "straight line segment," preserving the length of each bar and not crossing any two bars. In this paper, we introduce a new class of linkages, called "radial trees," and show that there exists a radial tree which can not be flattened.

1   Introduction

A *linkage* is a collection of line segments, called *bars*, possibly joined at their ends, called *joints*. A linkage is called *planar* if all bars are in the plane $\mathbb{R}^2$ with no self-intersection. A *reconfiguration* of a linkage is a continuous motion of their bars, or equivalently a continuous motion of their joints, that preserves the length of each bar. A reconfiguration of a linkage is called *planar* if all bars are in the plane during the motion, and is called *non-crossing* if every two bars do not cross each other during the motion. In this paper, we consider only a planar reconfiguration of a planar linkage, and we may omit the word "planar." Furthermore, we consider only a non-crossing reconfiguration, and we may omit the word "non-crossing."

For such planar reconfiguration problems, there is a fundamental question: whether any polygonal chain can be *straightend*. This problem had been open from the 1970's to the 1990's. However, Connelly et al. have answered this question affirmatively: they show that any polygonal chain can be straightened [3]. On the other hand, negative results are known for a non-crossing planar reconfiguration of a tree linkage: there exist trees which cannot be "flattened" [2,4]. Figure 1 illustrates a tree which cannot be straightened [2]. In [4], Connelly et al. gives a method for proving that some trees are locked. On the other hand, an affirmative result is reported for reconfiguring tree linkages: Kusakari et al. show that any "monotone tree" can be straightened, and give a method for flattening "monotone trees"[5]. Figure 2 illustrates a monotone tree [5]. Furthermore, recently, the complexity of flattening tree linkages has been proved: Alt et al. show that deciding lockability for trees is PSPACE-complete [1]. It is desired to characterize the class of trees which can be flattened.

In this paper, we define a new class of trees, called "radially monotone trees" or "radial trees," which is a natural modification of the class of monotone trees, and show that there exists a radial tree which cannot be flattened. The remainder of this paper is organized as follows. In Section 2, we give some preliminary

J. Akiyama and M. Kano (Eds.): JCDCG 2002, LNCS 2866, pp. 182–191, 2003.

Fig. 1. Locked tree.

definitions. In Section 3, we give a method to construct a locked radial tree. In Section 4, we show that the initial tree constructed in Section 3 is simple and radial. In Section 5, we show that the tree constructed in Section 3 can not be flattened. Finally, we conclude in Section 6.

2   Preliminaries

In this section, we define terms and formally describe our problem.

Let $L = (J, B)$ be a linkage consisting of a joint set J and a bar set B. A structural graph of a linkage L is denoted by $SG(L)$. An embedding of a structural graph $SG(L)$ is called a *configuration* of linkage L. A linkage L is called a *(rooted) tree linkage* or a *(rooted) tree* if the structural graph $SG(L)$ is a (rooted) tree. Let $T = (J, B)$ be such a rooted tree linkage, and $r \in J$ be the root of T. A bar $b \in B$ is denoted by (j_s, j_t) if $j_s \in J$ is the parent of $j_t \in J$. For any joint $j \in J$, an incident bar $b = (j, j')$ is called a *child bar of joint j*. A *leaf* is a joint having no child bar. For any joint $j \in J - \{r\}$, an incident bar $b = (j', j)$ is called a *parent bar of joint j*. For any joint $j \in J - \{r\}$, a parent bar of j is unique, and is denoted by $\bar{j}$. A joint $j \in J$ is *internal* if j is neither the root nor a leaf. A *flattened configuration of a rooted tree linkage* is one in which, for any internal joint j, the parent bar $\bar{j}$ of j makes angle π with each child bar of j, and the angle between each pair of child bars of j is zero. *Flattening a tree linkage T* is a reconfiguration of T from an initial configuration to a flattened configuration.

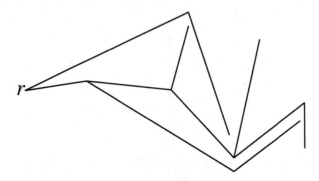

Fig. 2. Monotone tree.

For an initial configuration of a tree linkage, we first describe a definition of a monotone tree [5]. A polygonal chain P is *x-monotone* if the intersection of P and any vertical line is either a single point or a line segment if the intersection is not empty. A configured tree T is *x-monotone* if T is a rooted tree and the polygonal chain in T from the root r to any leaf is x-monotone. (See Figure 2.) Next, we define radial trees by slightly modifying the definition of monotone trees. A polygonal chain P is *radially monotone (for a point p)* or *radial (for a point p)* if the intersection of P and any circle with the same center p is either a single point or empty. A tree T is *radially monotone* or *radial* if T is a rooted tree and the directed polygonal chain in T from the root r to any leaf is radially monotone for the root r. A radial tree is illustrated in Figure 3. Note that an x-monotone tree may not be radial, and a radial tree may not be x-monotone.

For three points $p_1, p_2, p_3 \in \mathbb{R}^2$, the angle $\angle p_1 p_2 p_3$ is measured counterclockwise from the direction of $\overrightarrow{p_2 p_1}$ to the direction of $\overrightarrow{p_2 p_3}$, and ranges in $[0, 2\pi)$. For two bars $\bar{j}_1 = (j_0, j_1), \bar{j}_2 = (j_1, j_2) \in B$ joined with joint j_1, the angle $\angle j_0 j_1 j_2$ is denoted by $\theta(\bar{j}_1 \bar{j}_2)$. The *slope* $s(\bar{j}_1)$ of bar $\bar{j}_1 = (j_0, j_1)$ is the angle measured counterclockwise at the parent joint j_0 from $+x$ direction to the direction $\overrightarrow{j_0 j_1}$, and ranges in $[0, 2\pi)$. The length of bar b is denoted by $|b|$. For two points $p_1, p_2 \in \mathbb{R}^2$, the ray starting from p_1 and passing through p_2 is denoted by $R(p_1, p_2)$. For a point $p \in \mathbb{R}^2$ and a direction $d \in [0, 2\pi)$, the ray starting from p and going in the direction d is denoted by $R_p(d)$.

3   Constructing a Locked Radial Tree

In this section, we construct a radial tree which can not be flattened, i.e., we construct a *locked* radial tree. Figure 3 illustrates such a locked radial tree.

3.1   Overview

The locked tree T in Figure 3 contains six congruent components $C_0, C_1, \cdots,$ C_5, all of which are joined at the root r of T. More generally, one can construct a locked radial tree by such $n(> 4)$ congruent components, each of which is

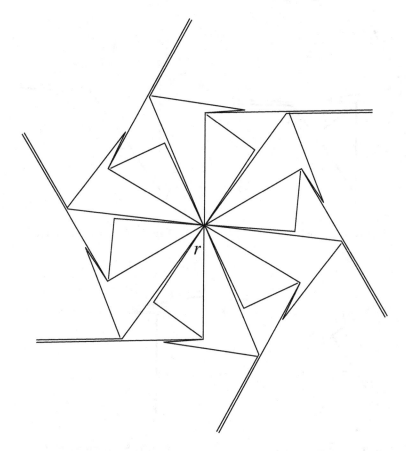

Fig. 3. Locked radial tree.

called a C_i-*component* and is often denoted by C_i, for i, $0 \leq i \leq n-1$. Each C_i-component consists of three subcomponents: a V_i-*component*, an L_i-*component* and a Γ_i-*component*. These V_i-component, L_i-component and Γ_i-component are often denoted by V_i, L_i and Γ_i, respectively. For each i, $0 \leq i \leq n-1$, these V_i, L_i and Γ_i are incident to the root r counterclockwise in this order. Furthermore, L_i is wrapped by V_i and Γ_i, as illustrated in Figure 4.

A V_i-component has two bars $\bar{v}_1 = (v_0, v_1)$, $\bar{v}_2 = (v_1, v_2)$ joined with the internal joint v_1 whose angle $\theta(\bar{v}_1\bar{v}_2)$ is nearly $\frac{\pi}{2} + \frac{\pi}{n}(= \frac{\pi}{2} + \frac{\pi}{6})$, and looks like the letter "V", as illustrated in Figure 5. An L_i-component has two bars $\bar{l}_1 = (l_0, l_1)$, $\bar{l}_2 = (l_1, l_2)$ joined with the internal joint l_1 whose angle $\theta(\bar{l}_1\bar{l}_2)$ is nearly $\frac{3\pi}{2}$, and looks like the letter "L", as illustrated in Figure 6. A Γ_i-component has four bars $\bar{\gamma}_1 = (\gamma_0, \gamma_1)$, $\bar{\gamma}_2 = (\gamma_1, \gamma_2)$, $\bar{\gamma}_3 = (\gamma_1, \gamma_3)$ and $\bar{\gamma}_4 = (\gamma_3, \gamma_4)$, and two internal joints γ_1, γ_3, and looks like the letter "Γ", as illustrated in Figure 7. The angles $\angle\gamma_0\gamma_1\gamma_2$, $\angle\gamma_0\gamma_1\gamma_3$ and $\angle\gamma_0\gamma_3\gamma_4$ are nearly $\frac{\pi}{2}$, $\frac{\pi}{n}$ and $\frac{3\pi}{2}$, respectively.

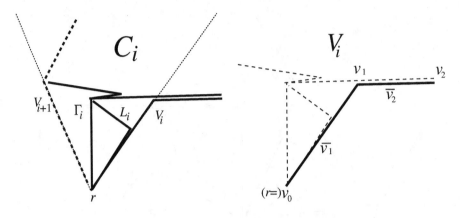

Fig. 4. Component C_i.

Fig. 5. Subcomponent V_i.

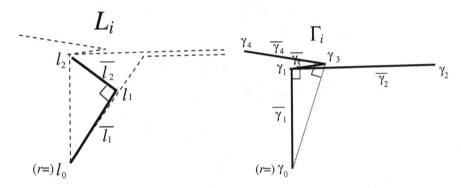

Fig. 6. Subcomponent L_i.

Fig. 7. Subcomponent Γ_i.

3.2   A Detail of the Construction

In this subsection, we focus on a single C_i-component, and may often omit the index i for simplification. Furthermore, subcomponents, bars, and joints in C_{i-1} or C_{i+1} are designated by the corresponding notation with symbol "–" or "+", respectively. For example, Γ_{i-1}, Γ_i and Γ_{i+1} are denoted by Γ^-, Γ and Γ^+, respectively. Moreover, the bar in Γ_{i+1} corresponding to the bar $\bar{\gamma}_1$ in Γ_i is denoted by $\bar{\gamma}_1{}^+$. We use similar notations for the others. For two points $p_1, p_2 \in \mathbb{R}^2$, we use $p_1 p_2$ to designate the line segment between p_1 and p_2, and $|p_1 p_2|$ to denote the length of the segment $p_1 p_2$.

We will draw a figure C_i^* containing the initial configuration of the C_i-component. In order to designate each points or segments in C_i^*, we use the notation adding symbol "*" to the corresponding notation of the joint $j \in J$ or the bar $b \in B$. Note that, for i, $0 \le i \le n-1$, we draw all figures C_i^* simultaneously, so that each pair of corresponding bars in consequent components makes angle $\frac{2\pi}{n}$, and the index i increases counterclockwise. Without loss of generality, we may assume that the length $|\bar{\gamma}_1{}^*| = 1$, and the slope $s(\bar{\gamma}_1{}^*) = \frac{\pi}{2}$. Renote that we first draw all bars corresponding to $\bar{\gamma}_1{}^*$ for all C_i^* simultaneously. Then, we draw a line segment $\bar{\gamma}_2{}^*$ from the point γ_1^* with the slope $s(\bar{\gamma}_2{}^*)$, so that the angle $\theta(\bar{\gamma}_1{}^*\bar{\gamma}_2{}^*) = \frac{\pi}{2}$. We choose the length $|\bar{\gamma}_2{}^*|$ long enough, so that the ray $R(r, \gamma_1^{-*})$ intersects $\bar{\gamma}_2{}^*$. Thus, we choose $|\bar{\gamma}_2{}^*| > \tan(\frac{2\pi}{n})$. Next, we choose a point γ_4^* on the bar $\bar{\gamma}_2{}^{+*}$, so that the following equation holds:

$$\tan(\frac{\pi}{n}) < |\gamma_1^{+*}\gamma_4^*| < \min\{\tan(\frac{2\pi}{n}), \frac{1}{\sin(\frac{2\pi}{n})}\}. \tag{1}$$

Then, we can find the point γ_3^* on the bar $\bar{\gamma}_2{}^*$ satisfying $\angle r\gamma_3^*\gamma_4^* = \frac{3\pi}{2}$. We draw two line segments γ_4^*r and $\gamma_4^*\gamma_3^*$. We finally drop a perpendicular from γ_1^* to $r\gamma_4^{-*}$ and the foot of the perpendicular is l_1^*.

We construct C_i on the figure C_i^*. The notation $j \approx p$ denote that joint j is configured sufficiently near point p, and the notation $b \approx s$ also denote that bar b is configured sufficiently near line segment s. The initial configuration of the C_i-component is obtained as follows: for the V_i-components, let $\bar{v}_1 \approx \bar{v}_1{}^* = r\gamma_4^{-*}$ and $\bar{v}_2 \approx \bar{v}_2{}^* = \gamma_4^{-*}\gamma_3^*$; for the L_i-components, let $\bar{l}_1 \approx \bar{l}_1{}^* = rl_1^*$ and $\bar{l}_2 \approx \bar{l}_2{}^* = l_1^*\gamma_1^*$; for the Γ_i-components, $\bar{\gamma}_1 \approx \bar{\gamma}_1{}^* = r\gamma_1^*$, $\bar{\gamma}_2 \approx \bar{\gamma}_2{}^* = \gamma_1^*\gamma_2^*$, $\bar{\gamma}_3 \approx \bar{\gamma}_3{}^* = \gamma_1^*\gamma_3^*$ and $\bar{\gamma}_4 \approx \bar{\gamma}_4{}^* = \gamma_3^*\gamma_4^*$. (See Figure 4.)

4   The Initial Configuration

In this section, we show that the tree constructed in the previous section is simple and radial. From now on, we often do not distinguish between the linkage and its configuration, and may often omit the symbol "$*$."

4.1   Simplicity

The slope $s(\bar{\gamma}_2{}^+) = \frac{2\pi}{n} < \frac{\pi}{2}$ if $n > 4$, and hence the ray $R_{\gamma_1}(\frac{\pi}{2})$ must cross $\bar{\gamma}_2{}^+$. Let X be the intersection point of the ray $R_{\gamma_1}(\frac{\pi}{2})$ with $\bar{\gamma}_2{}^+$, and let Y the intersection point of the ray $R_{\gamma_1}(\pi)$ with $\bar{\gamma}_2{}^+$, as illustrated in Figure 8. Since $|\bar{\gamma}_1{}^+| = 1$ and $\angle Xr\gamma_1^+ = \angle \gamma_1 r\gamma_1^+ = \frac{2\pi}{n}$, $|\gamma_1^+ X| = \tan(\frac{2\pi}{n})$. Furthermore, one can observe that $|\gamma_1^+ Y| = \tan(\frac{\pi}{n})$ as follows: since the hypotenuses are common and $|r\gamma_1| = |r\gamma_1^+| = 1$, two right triangles $\triangle rY\gamma_1$ and $\triangle rY\gamma_1^+$ are congruent, and hence $\angle \gamma_1 rY = \angle Yr\gamma_1^+ = \frac{\pi}{n}$. By equation (1), $\tan(\frac{\pi}{n}) < |\gamma_1^{+*}\gamma_4^*| < \tan(\frac{2\pi}{n})$, and hence the point γ_4^* is contained in the open line segment XY. Thus, one can observe that the two line segments $\gamma_4^*r(= \bar{v}_1{}^{+*})$ and $\gamma_4^*\gamma_3^*(= \bar{\gamma}_4{}^*)$ can be drawn without crossing any other line segments. Furthermore, one can observe that any pair of line segments in C_i^* can be drawn without crossing even if the pair contains neither $\bar{v}_1{}^{+*}$ nor $\bar{\gamma}_4{}^*$. Therefore, the tree constructed in Section 3 is simple.

4.2   Radial Monotonicity

For any joint $j \in J$, a subtree of T rooted at j is denoted by $T(j)$. For any configured joint $j \in \mathbb{R}^2$, the circle with the center o passing through j is denoted

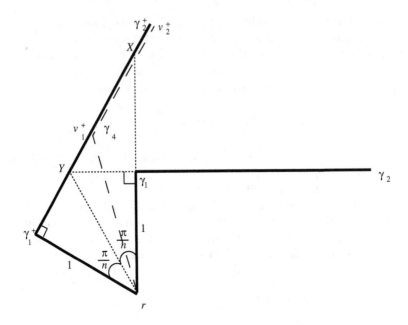

Fig. 8. Position of v_1^+.

by $C_o(j)$. A *wedge* p_1jp_2 is the set of points swept out by a ray starting j moving counterclockwise from the direction $\overrightarrow{jp_1}$ to the direction $\overrightarrow{jp_2}$, and contains points on both $R(j, p_1)$ and $R(j, p_2)$. A wedge p_1jp_2 may be denoted by $w_j[\theta_1, \theta_2]$, where $\theta_1 = \angle rjp_1$, $\theta_2 = \angle rjp_2$, and r is the root of the tree T.

The following lemmas hold.

Lemma 1. (i) *A tree $T = (J, B)$ is radial if and only if, for any joint $j \in J$, all joints j' (except j) in the subtree $T(j)$ are properly outside of $C_r(j)$.*

(ii) *A tree $T = (J, B)$ is radial if and only if, for any joint $j \in J - \{r\}$, a child bar $b = (j, j') \in B$ is contained in the wedge $w_j[\frac{\pi}{2}, \frac{3\pi}{2}]$.*

Proof. Both (i) and (ii) are obvious from the definition of radial trees.   □

Note that, for any bar $b = (r, j)$ incident to the root r, the slope $s(b)$ can be taken any values in $[0, 2\pi)$ even if T is radial.

Lemma 2. (i) *A V_i-component is radial for the root r.*

(ii) *An L_i-component is radial for the root r.*

(iii) *A Γ_i-component is radial monotone for the root r.*

Proof. (i)   One can easily observe that the angle $\theta(\bar{v}_1\bar{v}_2) \geq \frac{\pi}{2}$. (See Figure 5.) Thus, $\bar{v}_2 \subseteq w_{v_1}[\frac{\pi}{2}, \frac{3\pi}{2}]$, and hence the path $(r =)v_0v_1v_2$ is radial by Lemma 1 (ii).

(ii)   From the construction of a L_i-component, $\theta(\bar{l}_1\bar{l}_2) = \frac{3\pi}{2}$. Thus, $\bar{l}_2 \subseteq w_{l_1}[\frac{\pi}{2}, \frac{3\pi}{2}]$, and hence the path $(r =)l_0l_1l_2$ is radial by Lemma 1 (ii).

(iii)   Since a Γ_i-component has two leaves, and hence it is sufficient to show that both path $P_1 = \gamma_0\gamma_1\gamma_2$ and path $P_2 = \gamma_0\gamma_1\gamma_3\gamma_4$ are radial. From the construction, $\theta(\bar{\gamma}_1\bar{\gamma}_2) = \frac{\pi}{2}$, and hence P_1 is radial by Lemma 1 (ii). Furthermore, one can observe that $\bar{\gamma}_3 \subseteq w_{\gamma_1}[\frac{\pi}{2}, \frac{3\pi}{2}]$ and $\bar{\gamma}_4 \subseteq w_{\gamma_3}[\frac{\pi}{2}, \frac{3\pi}{2}]$ since $\theta(\bar{\gamma}_1\bar{\gamma}_3) = \frac{\pi}{2}$ and $\angle\gamma_0\gamma_3\gamma_4 = \frac{3\pi}{2}$. Thus, by Lemma 1 (ii), P_2 is radial. $\qquad\square$

Any C_i-component is radial since all subcomponents are radial by Lemma 2 above, and hence the tree constructed in Section 3 is radial.

5   Lockability

In this section, we show that the tree constructed in Section 3 can not be flattened. Note that the proving method described in [4] can not directly apply to our tree.

For the sake of simplicity, we assume that $|\bar{v}_1| = |\bar{v}_1{}^*|$, $|\bar{v}_2| = |\bar{v}_2{}^*|$, $|\bar{l}_1| = |\bar{l}_1{}^*|$, $|\bar{l}_2| = |\bar{l}_2{}^*|$, $|\bar{\gamma}_1| = |\bar{\gamma}_1{}^*|$, $|\bar{\gamma}_2| = |\bar{\gamma}_2{}^*|$, $|\bar{\gamma}_3| = |\bar{\gamma}_3{}^*|$ and $|\bar{\gamma}_4| = |\bar{\gamma}_4{}^*|$. Note that each pair of bars does not "properly cross" each other even if lengths of bars are chosen by above way and the initial configuration of C_i is embedded on C_i^*. Furthermore, all lengths $|\bar{\gamma}_3|$, $|\bar{v}_1|$, $|\bar{l}_1|$ and $|\bar{l}_2|$ are determined if $|\gamma_1^+\gamma_4|$ is determined. Thus, there are no choice of lengths for bars $\bar{\gamma}_3$, $\bar{v}_1$, $\bar{l}_1$ and $\bar{l}_2$.

For the initial configuration, the following lemma holds.

Lemma 3. *The following four equations hold:*

(i) $\angle r\gamma_1 l_1 < \frac{\pi}{2}$,
(ii) $\angle l_1\gamma_1\gamma_2 < \frac{\pi}{2}$,
(iii) $\angle r\gamma_4\gamma_3 < \frac{\pi}{2}$, *and*
(iv) $\angle\gamma_3\gamma_4 v_2^+ < \frac{\pi}{2}$.

Proof. Since $\angle r\gamma_1\gamma_2 = \angle r\gamma_1 l_1 + \angle l_1\gamma_1\gamma_2 = \frac{\pi}{2}$, both (i) and (ii) immediately hold. Moreover, since $\angle\gamma_4\gamma_3 r = \frac{\pi}{2}$ and $\angle\gamma_4\gamma_3 r + \angle\gamma_3 r\gamma_4 + \angle r\gamma_4\gamma_3 = \pi$, (iii) holds. Thus, we only prove (iv) below.

For the quadrangle $r\gamma_3\gamma_4\gamma_1^+$, $\angle\gamma_4\gamma_3 r = \angle r\gamma_1^+\gamma_4 = \frac{\pi}{2}$ from our construction of the tree, and hence $\angle\gamma_1^+\gamma_4\gamma_3 + \angle\gamma_3 r\gamma_1^+ = \pi$. Let Z and W be vertices of a rectangle $rZW\gamma_1^+$ such that the vertex Z is on $\bar{\gamma}_2$ and the vertex W is on γ_2^+, as illustrated in Figure 9. Then, $\angle\gamma_1 Zr = \angle\gamma_1 r\gamma_1^+ = \frac{2\pi}{n}$, since $\angle Zr\gamma_1 + \gamma_1 Zr = \angle Zr\gamma_1 + \angle\gamma_1 r\gamma_1^+ = \frac{\pi}{2}$. Therefore, $|\gamma_1^+ W| = |rZ| = \frac{1}{\sin(\frac{2\pi}{n})}$. From equation (1), $|\gamma_1^+ v_1^+| < \frac{1}{\sin(\frac{2\pi}{n})}$, and hence v_1^+ is on the open line segment $\gamma_1^+ W$. On the other hand, one can easily observe that the length $|\gamma_1^+ v_1^+|$ increases if and only if the angle $\angle\gamma_3 v_1^+ v_2^+$ increases. Thus, $\angle\gamma_3 v_1^+ v_2^+ < \frac{\pi}{2}$. $\qquad\square$

For each C_i, the angle $\angle v_1 r\gamma_4$ ($= \angle v_1 r v_1^+$) is called *the angle of* C_i and may be denoted by $\angle C_i$. A reconfiguration *widens* C_i if it makes the angle $\angle C_i$ increase, and *squeezes* C_i if it makes the angle $\angle C_i$ decrease.

The following lemmas hold.

Lemma 4. *There exists a widened C_i-component if and only if there exists a squeezed C_j-component, where $0 \leq i, j \leq n - 1$ and $i \neq j$.*

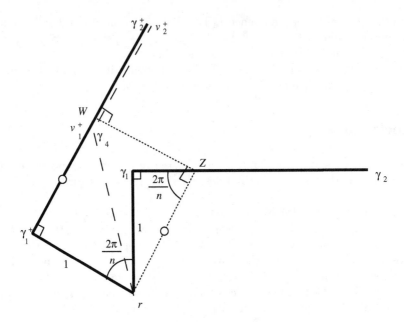

Fig. 9. Rectangle $rCD\gamma_1^+$.

Proof. Since $\sum_{i=0}^{n-1} \angle C_i = 2\pi$, the claim immediately holds. □

Lemma 5. (i) *No reconfigurations can squeeze any C_i-component.*
(ii) *No reconfigurations can widen any C_i-component.*

Proof. (sketch) By Lemma 4, it is sufficient to show only (i).

We may assume, without loss of generality, that the root r is located on the origin of the xy-plane, and the bar $\bar{v}_1$ in C_i is fixed during the reconfiguration. Furthermore, we assume, for a contradiction, that C_i is reconfigured to C_i' such that

$$\angle v_1' r v_1^{+'} < \frac{2\pi}{n} = \angle v_1 r v_1^+, \tag{2}$$

where the objects (subcomponents, bars and joints) in the C_i'-component are denoted by notations adding the symbol "'" to the corresponding notations of the objects in the C_i-component.

The bar $\bar{l}_2$ can not swing with the center l_1 both clockwise and counterclockwise since the angles $\angle l_1 l_2 r$ and $\angle l_1 l_2 \gamma_2$ is less than $\frac{\pi}{2}$ from Lemma 3 (i) and (ii). Furthermore, the bar $\bar{\gamma}_4$ can not swing with the center γ_3 both clockwise and counterclockwise since the angles $\angle r\gamma_4\gamma_3$ and $\angle \gamma_3\gamma_4 v_2^+$ is less than $\frac{\pi}{2}$ from Lemma 3 (iii) and (iv).

Thus, the only feasible motion is either expanding or reducing the diagonal $\gamma_1\gamma_4$ of the reflex quadrangle $r\gamma_1\gamma_3\gamma_4$. Thus, the following two cases may occur:

Case 1: The diagonal $\gamma_1'\gamma_4'$ of the reflex quadrangle $r\gamma_1'\gamma_3'\gamma_4'$ is longer than the diagonal $\gamma_1\gamma_4$ of the initial configuration of the reflex quadrangle $r\gamma_1\gamma_3\gamma_4$; and

Case 2: The diagonal $\gamma_1'\gamma_4'$ of the reflex quadrangle $r\gamma_1'\gamma_3'\gamma_4'$ is shorter than the diagonal $\gamma_1\gamma_4$ of the initial configuration of the reflex quadrangle $r\gamma_1\gamma_3\gamma_4$.

Case 1: Since $|\gamma_1'\gamma_4'| > |\gamma_1\gamma_4|$, then $\angle\gamma_1'r\gamma_4' > \angle\gamma_1r\gamma_4$ by the cosine rule for the triangle $\triangle r\gamma_1\gamma_4$. Therefore, by equation (2), $\angle v_1'r\gamma_1' < \angle v_1r\gamma_1$. This means that the distance between joint γ_1' and bar $\bar{j}_1'(=\bar{j}_1)$ is shorter than the distance between joint γ_1 and bar $\bar{j}_1$. However, the distance between joint γ_1 and bar $\bar{j}_1$ is equal to $|\bar{l}_2|$, and hence $\bar{l}_2$ must cross the path $r\gamma_1\gamma_2$, contradicting a condition of the planar reconfiguration.

Case 2: Since $|\gamma_1'\gamma_4'| < |\gamma_1\gamma_4|$, then $\angle\gamma_1'r\gamma_4' < \angle\gamma_1r\gamma_4$ by the cosine rule for the triangle $\triangle r\gamma_1\gamma_4$, and then $\angle\gamma_4'\gamma_3'\gamma_1' < \angle\gamma_4\gamma_3\gamma_1$ by the cosine rule for the triangle $\triangle\gamma_3\gamma_1\gamma_4$. Therefore, one can easily observe that $\angle\gamma_3'\gamma_1'\gamma_4' > \angle\gamma_3\gamma_1\gamma_4$ and $\angle\gamma_4'\gamma_1'r > \angle\gamma_4\gamma_1r$. Thus, $\angle r\gamma_1'\gamma_3' < r\gamma_1\gamma_3 = \frac{\pi}{2}$. However, the angle $\angle r\gamma_1\gamma_2$ can not decrease from $\frac{\pi}{2}$, since $\bar{\gamma}_2$ can not swing clockwise, and hence $\bar{\gamma}_3$ must pass through $\bar{\gamma}_2$, contradicting a condition of the planar reconfiguration.   □

Thus, the following theorem holds.

Theorem 1. *There exists a radial tree which can not be flattened.*

6   Conclusion

In this paper, we show that there exists a tree linkage which is radially monotone and can not be flattened. One of the future works is to find a method for flattening tree linkages in other classes.

References

1. H. Alt, C. Knauer, G. Rote, and S. Whitesides, The Complexity of (Un)folding, In Proceeding of 19th ACM Symposium on Computational Geometry, 2003, to appear.
2. T. Biedl, E. Demaine, M. Demaine, S. Lazard, A. Lubiw, J. O'Rourke, S. Robbins, I. Streinu, G. Toussaint, and S. Whitesides, A note on reconfiguring tree linkages: Trees can lock, Discrete Applied Mathematics, vol. 254, pp.19-32, 2002. A preliminary version appeared in Proceedings of 10th Canadian Conference on Computational Geometry, 1998.
3. R. Connelly, E. Demaine, and G. Rote, Straightening polygonal arcs and convexifying polygonal cycles, In Proceedings of 41st IEEE Symposium on Foundations of Computer Science, 2000.
4. R. Connelly, E. Demaine, and G. Rote, Infinitesimally Locked Self-Touching Linkages with Applications to Locked Trees, In Physical Knots: Knotting, Linking, and Folding Geometric Objects in R^3, American Mathematical Society, pp. 287-311,2002. A preliminary version appeared in in Proceedings of the 11th Annual Fall Workshop on Computational Geometry, 2001.
5. Y. Kusakari, M. Sato, and T. Nishizeki, Planar Reconfiguration of Monotone Trees, IEICE Trans. Fund., Vol.E85-A, No.5, pp.938-943, 2002.

Viewing Cube and Its Visual Angles

Yoichi Maeda

Tokai University, Hiratsuka, Kanagawa, 259-1292, Japan
maeda@keyaki.cc.u-tokai.ac.jp

Abstract. We discuss several relations among visual angles of a rectangle, those of orthogonal axes, and those of a cube. Depending on a viewpoint, a given angle changes its visual angle from 0 to π in radian. There are simple and beautiful relations between visual angles at the vertices of such specified shapes. The determination of the distance from a viewpoint to orthogonal axes is also discussed.

1   Introduction

We are always getting lots of information from outside. An angle in the 3-dimensional Euclidean space changes its visual angle as the observer moves. This visual angle also gives us some information. For example, the four visual angles of a rectangle tells us its field of vision (the area of the projected image on the unit sphere centered at the observer), that is, the sum of these visual angles minus 2π. Actually, many researches on the projected image of an object have been done in the fields of computer vision and machine intelligence ([1,2,3,5,6,7,8,9,10,11]), however, every projected image seems to be a central or parallel projection on a plane. We deal with the central projection on a sphere centered at the viewpoint or the given angle. In this paper, let us focus on right angles. After the definition of the visual angle in Section 2, we show a simple equation which four visual angles of a rectangle hold in Section 3. In Section 4, let us view orthogonal axes from various viewpoints. As an application, we can calculate the distance from the eyes of the observer to the origin of orthogonal axes by the pair of three visual angles of the axes. In addition, we study the projected image of a cube in Section 5. Several relations among 24 visual angles will be derived. Especially, a pair of three visual angles determines the projected image of a cube.

2   Definition of Visual Angle

Let us first define "visual angle".

Definition 1. *Let $\angle BAC$ be an angle in the 3-dimensional Euclidean space* $\mathbf{E}^3$. *When we view this angle $\angle BAC$ from a viewpoint $O \in \mathbf{E}^3$, its visual angle is defined as the dihedral angle of the two faces BAO and CAO of the tetrahedron (possibly degenerate) $OABC$, and this visual angle is occasionally denoted by A_v.*

J. Akiyama and M. Kano (Eds.): JCDCG 2002, LNCS 2866, pp. 192–199, 2003.

The following proposition will be clear.

Proposition 1. *(i) Let A', B', and C' be the projected points of A, B, and C from O onto the unit sphere centered at O. Then the visual angle A_v of $\angle BAC$ is equal to the angle $\angle A'$ of the spherical triangle $\triangle A'B'C'$.*
(ii) Let O', B'', and C'' be the projected points of O, B, and C from A onto the unit sphere centered at A. Then the visual angle A_v of $\angle BAC$ is equal to the angle $\angle O'$ of the spherical triangle $\triangle O'B''C''$.

3   Visual Angles of a Rectangle

In this section, we begin the search for visual angles of a rectangle. It is well known that the field of vision (the area of the projected image on the unit sphere centered at the viewpoint) of a rectangle $ABCD$ is given as

$$\text{(the field of vision of } ABCD) = A_v + B_v + C_v + D_v - 2\pi \ (> 0). \tag{1}$$

Theorem 1. *Let $ABCD$ be a rectangle in $\mathbf{E}^3$ and O be a viewpoint. Then the four visual angles A_v, B_v, C_v, and D_v from O satisfy the following equation:*

$$\cos A_v \cos C_v = \cos B_v \cos D_v \ . \tag{2}$$

Proof. Let S be the unit sphere centered at the viewpoint $O(0,0,0)$. Without loss of generality, the rectangle $ABCD$ is on the plane $z = 1$, and the line AB is parallel to the x-axis, and $A_x < B_x$. Projecting the rectangle $ABCD$ to the spherical quadrangle $A'B'C'D'$ on S, the common vanishing points of the lines AB and CD are projected to the two points $X'_\infty(1,0,0)$ and $X'_{-\infty}(-1,0,0)$, and the common vanishing points of the lines BC and DA are also projected to the two points $Y'_\infty(0,1,0)$ and $Y'_{-\infty}(0,-1,0)$. Note that

$$\overline{X'_\infty Y'_\infty} = \overline{Y'_\infty X'_{-\infty}} = \overline{X'_{-\infty} Y'_{-\infty}} = \overline{Y'_{-\infty} X'_\infty} = \pi/2$$

on S. Let

$$\alpha = \angle A' X'_{-\infty} Y'_{-\infty} = \angle B' X'_\infty Y'_{-\infty}, \quad \beta = \angle C' X'_\infty Y'_\infty = \angle D' X'_{-\infty} Y'_\infty,$$
$$\gamma = \angle D' Y'_\infty X'_{-\infty} = \angle A' Y'_{-\infty} X'_{-\infty}, \quad \delta = \angle B' Y'_{-\infty} X'_\infty = \angle C' Y'_\infty X'_\infty.$$

Then, by the law of cosine for angles [4] (pp 59)

$$\cos A_v = -\cos\alpha\cos\gamma, \quad \cos B_v = -\cos\delta\cos\alpha,$$
$$\cos C_v = -\cos\beta\cos\delta, \quad \cos D_v = -\cos\gamma\cos\beta.$$

Hence, Equation (2) is trivial.   □

Corollary 1. *The visual angles of a rectangle $ABCD$ have the following properties:*

(i) There is at least one obtuse visual angle.
(ii) If one of the visual angles is equal to $\pi/2$(right angle), the number of right visual angles is two or three.
(iii) If neither of the visual angles is equal to $\pi/2$, the number of obtuse visual angles is two or four.

Proof. If $A_v, B_v, C_v, D_v \leq \pi/2$, then $A_v + B_v + C_v + D_v \leq 2\pi$, which would contradict Equation (1). For part (ii), assume that $A_v = \pi/2$. By Theorem 1, $\cos B_v \cos D_v = 0$, hence $\cos B_v = 0$ or $\cos C_v = 0$. The case that $A_v = B_v = C_v = D_v = \pi/2$ is impossible because of part (i). Finally, suppose that all of the visual angles are acute or obtuse. If the sign of the value of the left(right) side of Equation (2) is negative, one of A_v and C_v is obtuse, and also one of B_v and D_v is obtuse. In this case, two of the visual angles are obtuse and the others are acute. On the other hand, if the sign is positive, then both A_v and C_v are obtuse or both B_v and D_v are obtuse. Therefore two or four of the visual angles are obtuse in the second case.   □

4   Visual Angles of Orthogonal Axes

In the rest of this paper, let us consider the visual angles as oriented. Let $\angle B'A'C'$ on S be the oriented angle which is measured in the counterclockwise direction from $A'B'$ to $A'C'$ looking from outside of S. Then the range of visual angle is $[0, 2\pi)$, for example, $\angle X'_\infty Z'_\infty Y'_\infty = \pi/2$ and $\angle Y'_\infty Z'_\infty X'_\infty = 3\pi/2$ where $X'_\infty = (1,0,0), Y'_\infty = (0,1,0)$ and $Z'_\infty = (0,0,1)$. With this oriented visual angle, let us look at 3-dimensional orthogonal axes $VXYZ(V(\alpha, \beta, \gamma), X(\alpha+1, \beta, \gamma), Y(\alpha, \beta+1, \gamma), Z(\alpha, \beta, \gamma+1), \alpha, \beta, \gamma \in \mathbf{R})$ from the viewpoint $O(0,0,0)$. There are three visual angles on S: let a be $\angle Y'V'Z'$ as the visual angle of $\angle YVZ$, b be $\angle Z'V'X'$ as that of $\angle ZVX$, and c be $\angle X'V'Y'$ as that of $\angle XVY$. Note that

$$a = \angle Y'_\infty V'Z'_\infty, \quad b = \angle Z'_\infty V'X'_\infty, \quad c = \angle X'_\infty V'Y'_\infty.$$

The following theorem shows that these visual angles almost surely identify the position of $V'(x, y, z)(x^2 + y^2 + z^2 = 1)$ which is the projected point of V, the origin of the axes.

Theorem 2. *The visual angles a, b and c around $V'(x, y, z)(x, y, z \neq \pm 1)$ are given as*

$$\tan a = -\frac{x}{yz}, \quad \tan b = -\frac{y}{zx}, \quad \tan c = -\frac{z}{xy}. \tag{3}$$

If neither a, b nor c are equal to 0 and π, then $V'(x, y, z)$ is identified as

$$\begin{pmatrix} x \\ y \\ z \end{pmatrix} = \begin{pmatrix} sign(\sin a)\sqrt{\cot b \cot c} \\ sign(\sin b)\sqrt{\cot c \cot a} \\ sign(\sin c)\sqrt{\cot a \cot b} \end{pmatrix},$$

where $sign(t) = \pm 1$ is the sign of t.

Proof. $x, y, z \neq \pm 1$ imply that eight vanishing points are omitted because two of the visual angles are indefinite at these points. Assume that $V'(x, y, z)$ is not at these vanishing points. First, it is easy to check that $x \sin a \geq 0$. Appling the cosine law for angles to $\triangle V'Y'_\infty Z'_\infty$,

$$\cos \frac{\pi}{2} = yz + \sin(\arccos y) \sin(\arccos z) \cos a,$$

$$\cos a = -\frac{yz}{\sqrt{1 - y^2}\sqrt{1 - z^2}}, \quad \sin a = \frac{x}{\sqrt{1 - y^2}\sqrt{1 - z^2}}$$

where we use the fact that $\sin a$ has the same sign as x. Hence $\tan a = -\frac{x}{yz}$ is derived. $\tan b$ and $\tan c$ are also led in the same way. Next, note that $a = 0, \pi$ if and only if $x = 0$. Therefore, $a, b, c \neq 0, \pi$ is equivalent to $xyz \neq 0$. Then, $\cot b \cot c = x^2 (> 0)$, so $x = sign(\sin a)\sqrt{\cot b \cot c}$.   □

Remark 1. The visual angles a, b and c satisfy the equation

$$\tan a + \tan b + \tan c = \tan a \tan b \tan c,$$

which is equivalent to $a + b + c = n\pi \ (n \in \mathbf{Z})$.

The following corollary will be derived from Equations (3).

Corollary 2. *The tangent values of the visual angles of orthogonal axes are classified into four cases:*
(i) All are positive,
(ii) All are negative,
(iii) One is equal to 0, and the others are equal to ∞,
(iv) One is equal to ∞ and the others are indefinite.

Proof. If $xyz \neq 0$, then one of case (i) or (ii) is satisfied from Equations (3). Case (iv) is the case that V' is at one of the eight vanishing points. Otherwise, $xyz = 0$ and $x, y, z \neq \pm 1$, that is case (iii).   □

Using Theorem 2, let us introduce an application which indicates that we can find out the distance from the two eyes of the observer to the origin of the orthogonal axes only with the pair of three visual angles.

Theorem 3. *Let $(a_i, b_i, c_i)(i = 1, 2)$ be the pair of three visual angles of orthogonal axes $VXYZ$ from left and right eyes, O_1 and O_2. If $\overline{O_1 V} = \overline{O_2 V}$ and neither of six visual angles is equal to 0 and π, then the distance from the eyes to the origin V of the axes is*

$$\overline{O_1 V} = \overline{O_2 V} = \frac{\overline{O_1 O_2}}{\sqrt{2}\sqrt{1 - (x_1 x_2 + y_1 y_2 + z_1 z_2)}}$$

where $(x_i, y_i, z_i)(i = 1, 2)$ is given as follows:

$$\begin{pmatrix} x_i \\ y_i \\ z_i \end{pmatrix} = \begin{pmatrix} sign(\sin a_i)\sqrt{\cot b_i \cot c_i} \\ sign(\sin b_i)\sqrt{\cot c_i \cot a_i} \\ sign(\sin c_i)\sqrt{\cot a_i \cot b_i} \end{pmatrix} \quad (i = 1, 2). \tag{4}$$

Proof. Without loss of generality, we set the axes $VXYZ$ as $V(0,0,0)$, $X(1,0,0)$, $Y(0,1,0)$, and $Z(0,0,1)$. Let O_1' and O_2' be the projected points of O_1 and O_2 to the unit sphere centered at V. By the symmetric property of the axes $VXYZ$, we can assume that a_i(resp. b_i and c_i) is the visual angle of $\angle YVZ$(resp. $\angle ZVX$ and $\angle XVY$). Then from part (ii) of Proposition 1, $a_i = \angle YO_i'Z$, $b_i = \angle ZO_i'X$, and $c_i = \angle XO_i'Y (i = 1, 2)$ on the sphere. With the similar argument as Theorem 2, we can identify the position of $O_1'(x_1, y_1, z_1)$ and $O_2'(x_2, y_2, z_2)$ as Equation (4). Let $2\theta = \angle O_1 V O_2$. Then $\cos 2\theta = x_1 x_2 + y_1 y_2 + z_1 z_2$ and $\overline{O_1 O_2} = 2\overline{O_1 V} \sin \theta$. Therefore,

$$\overline{O_1 V} = \frac{\overline{O_1 O_2}}{2\sin\theta} = \frac{\overline{O_1 O_2}}{\sqrt{2}\sqrt{2\sin^2\theta}} = \frac{\overline{O_1 O_2}}{\sqrt{2}\sqrt{1 - \cos 2\theta}}.$$

$\square$

5   Visual Angles of a Cube

In the last section, let us look at a cube from various viewpoints. We set the viewpoint at the origin $O(0,0,0)$. Without loss of generality, assume that the edge length of a cube has the constant value two, because we can not find out the exact length from the projected image of the cube. Then the eight vertices of a cube in $\mathbf{E}^3$ are parametrized by three parameters, α, β, and γ:

$$
\begin{aligned}
V_1 &: (\alpha_1, \beta_1, \gamma_1) = (\alpha_-, \beta_-, \gamma_-), & V_5 &: (\alpha_5, \beta_5, \gamma_5) = (\alpha_+, \beta_+, \gamma_+), \\
V_2 &: (\alpha_2, \beta_2, \gamma_2) = (\alpha_-, \beta_+, \gamma_+), & V_6 &: (\alpha_6, \beta_6, \gamma_6) = (\alpha_+, \beta_-, \gamma_-), \\
V_3 &: (\alpha_3, \beta_3, \gamma_3) = (\alpha_+, \beta_-, \gamma_+), & V_7 &: (\alpha_7, \beta_7, \gamma_7) = (\alpha_-, \beta_+, \gamma_-), \\
V_4 &: (\alpha_4, \beta_4, \gamma_4) = (\alpha_+, \beta_+, \gamma_-), & V_8 &: (\alpha_8, \beta_8, \gamma_8) = (\alpha_-, \beta_-, \gamma_+),
\end{aligned}
$$

where

$$
\begin{aligned}
\alpha_- &= \alpha - 1, & \beta_- &= \beta - 1, & \gamma_- &= \gamma - 1, \\
\alpha_+ &= \alpha + 1, & \beta_+ &= \beta + 1, & \gamma_+ &= \gamma + 1.
\end{aligned}
$$

The point $G(\alpha, \beta, \gamma)$ is the center of gravity of this cube. For simplicity, let us consider the case $\alpha, \beta, \gamma > 1$, i.e., $\alpha_i, \beta_i, \gamma_i > 0 (i = 1, 2, \cdots, 8)$. The projected points $V_i' (i = 1, 2, \cdots, 8)$ on the unit sphere S centered at O are given as

$$V_i' : (x_i, y_i, z_i) = \frac{1}{\sqrt{\alpha_i^2 + \beta_i^2 + \gamma_i^2}} (\alpha_i, \beta_i, \gamma_i) \quad (i = 1, 2, \cdots, 8). \tag{5}$$

Each vertex of the cube is considered as orthogonal axes with three visual angles. Let us define the visual angles as follows:

$$a_i = \angle Y_\infty' V_i' Z_\infty', \quad b_i = \angle Z_\infty' V_i' X_\infty', \quad c_i = \angle X_\infty' V_i' Y_\infty' \quad (i = 1, 2, \cdots, 8).$$

Then from Equations (3), these visual angles are represented by the three parameters α, β, and γ, in fact,

$$(\tan a_i, \tan b_i, \tan c_i) = -\frac{\sqrt{\alpha_i^2 + \beta_i^2 + \gamma_i^2}}{\alpha_i \beta_i \gamma_i}(\alpha_i^2, \beta_i^2, \gamma_i^2) \quad (i = 1, 2, \cdots, 8). \tag{6}$$

Note that $\pi/2 < a_i, b_i, c_i < \pi$, and also

$$x_i : y_i : z_i = \sqrt{-\tan a_i} : \sqrt{-\tan b_i} : \sqrt{-\tan c_i} , \tag{7}$$

$$\frac{\beta_i}{\alpha_i} = \sqrt{\cot a_i \tan b_i}, \quad \frac{\gamma_i}{\beta_i} = \sqrt{\cot b_i \tan c_i}, \quad \frac{\alpha_i}{\gamma_i} = \sqrt{\cot c_i \tan a_i} \tag{8}$$

for any $i = 1, 2, \cdots, 8$. For any pair of vertices, there must be a relation between their visual angles:

Theorem 4. *Let* $\vec{A} = \overrightarrow{V_i V_j}$, $\vec{B} = (\sqrt{-\tan a_i}, \sqrt{-\tan b_i}, \sqrt{-\tan c_i})$, *and* $\vec{C} = (\sqrt{-\tan a_j}, \sqrt{-\tan b_j}, \sqrt{-\tan c_j})$ *for* $i \neq j$. *Then*

$$\det[\vec{A}, \vec{B}, \vec{C}] = 0. \tag{9}$$

Proof. Let $V'_{i,j}$ be the projected point of one of the vanishing points of the line $V_i V_j$ on the unit sphere S. Then $\vec{A} = \overrightarrow{V_i V_j}$ is parallel to $\overrightarrow{OV''_{i,j}}$. These projected points V'_{ij}, V'_i and V'_j are on the same great circle of S. From Equation (7), $\vec{B} // \overrightarrow{OV'_i}$ and $\vec{C} // \overrightarrow{OV'_j}$ which derive Equation (9).   □

Example 1. Let us introduce several examples.
(i) V_1 and V_6 (end points of an edge).

$$\begin{vmatrix} 1 & 0 & 0 \\ \sqrt{-\tan a_1} & \sqrt{-\tan b_1} & \sqrt{-\tan c_1} \\ \sqrt{-\tan a_6} & \sqrt{-\tan b_6} & \sqrt{-\tan c_6} \end{vmatrix} = 0,$$

that is, $\tan b_1 : \tan c_1 = \tan b_6 : \tan c_6$.
(ii) V_1 and V_2 (diagonal vertices of a face).

$$\begin{vmatrix} 0 & 1 & 1 \\ \sqrt{-\tan a_1} & \sqrt{-\tan b_1} & \sqrt{-\tan c_1} \\ \sqrt{-\tan a_2} & \sqrt{-\tan b_2} & \sqrt{-\tan c_2} \end{vmatrix} = 0.$$

(iii) V_1 and V_5 (diagonal vertices of the cube).

$$\begin{vmatrix} 1 & 1 & 1 \\ \sqrt{-\tan a_1} & \sqrt{-\tan b_1} & \sqrt{-\tan c_1} \\ \sqrt{-\tan a_5} & \sqrt{-\tan b_5} & \sqrt{-\tan c_5} \end{vmatrix} = 0.$$

Conversely, we can determine the parameters from a pair of three visual angles.

Theorem 5. *Assume that* $V'_1 \neq V'_5$. *Then any pair of visual angles* (a_i, b_i, c_i) *and* $(a_j, b_j, c_j)(i \neq j)$ *determines the parameters* $\alpha, \beta,$ *and* γ *of the projected image of the cube.*

Proof. Under the assumption above, it is easy to check that $(x_i, y_i, z_i) \neq (x_j, y_j, z_j)$ and also $(a_i, b_i, c_i) \neq (a_j, b_j, c_j)$. Then from Equations (8),

$$\left(\frac{\beta_i}{\alpha_i}, \frac{\gamma_i}{\beta_i}, \frac{\alpha_i}{\gamma_i} \right) \neq \left(\frac{\beta_j}{\alpha_j}, \frac{\gamma_j}{\beta_j}, \frac{\alpha_j}{\gamma_j} \right).$$

In the case that $\beta_i/\alpha_i \neq \beta_j/\alpha_j$, we can determine α and β using the conditions $|\alpha_i - \alpha_j| = 0, 2$ and $|\beta_i - \beta_j| = 0, 2$. In the same way, we can determine these three parameters in the other cases. □

Example 2. Let us introduce several examples.
(i) V_1 and V_6 (end points of an edge).

$$\alpha = \frac{\sqrt{\cot a_1 \tan b_1} + \sqrt{\cot a_6 \tan b_6}}{\sqrt{\cot a_1 \tan b_1} - \sqrt{\cot a_6 \tan b_6}} = \frac{\sqrt{\cot c_6 \tan a_6} + \sqrt{\cot c_1 \tan a_1}}{\sqrt{\cot c_6 \tan a_6} - \sqrt{\cot c_1 \tan a_1}},$$

$$\beta = \frac{2}{\sqrt{\cot b_6 \tan a_6} - \sqrt{\cot b_1 \tan a_1}} + 1, \quad \gamma = \frac{2}{\sqrt{\cot c_6 \tan a_6} - \sqrt{\cot c_1 \tan a_1}} + 1.$$

(ii) V_1 and V_2 (diagonal vertices of a face).

$$\alpha = \frac{2}{\sqrt{\cot a_2 \tan b_2} - \sqrt{\cot a_1 \tan b_1}} + 1,$$

$$\beta = \frac{\sqrt{\cot a_2 \tan b_2} + \sqrt{\cot a_1 \tan b_1}}{\sqrt{\cot a_2 \tan b_2} - \sqrt{\cot a_1 \tan b_1}}, \quad \gamma = \frac{\sqrt{\cot c_1 \tan a_1} + \sqrt{\cot c_2 \tan a_2}}{\sqrt{\cot c_1 \tan a_1} - \sqrt{\cot c_2 \tan a_2}}.$$

(iii) V_1 and V_5 (diagonal vertices of the cube).

$$\alpha = \frac{\sqrt{\cot a_1 \tan b_1} + \sqrt{\cot a_5 \tan b_5} - 2}{\sqrt{\cot a_1 \tan b_1} - \sqrt{\cot a_5 \tan b_5}},$$

$$\beta = \frac{\sqrt{\cot b_1 \tan c_1} + \sqrt{\cot b_5 \tan c_5} - 2}{\sqrt{\cot b_1 \tan c_1} - \sqrt{\cot b_5 \tan c_5}},$$

$$\gamma = \frac{\sqrt{\cot c_1 \tan a_1} + \sqrt{\cot c_5 \tan a_5} - 2}{\sqrt{\cot c_1 \tan a_1} - \sqrt{\cot c_5 \tan a_5}}.$$

On the other hand, there is a symmetric property in the case of $V_1' = V_5'$.

Theorem 6. *If $V_1' = V_5'$, then*
(i) $a_1 = b_1 = c_1 = a_5 = b_5 = c_5 = 2\pi/3$,
(ii) $a_2 = b_3 = c_4$ and $a_6 = b_7 = c_8$,
(iii) $\cos a_2 + \cos a_6 = -1$.

Proof. From Equation (5), $(x_1, y_1, z_1) = (x_5, y_5, z_5)$ implies $\alpha_+/\alpha_- = \beta_+/\beta_- = \gamma_+/\gamma_-$, hence $\alpha = \beta = \gamma$. From Equation (6), $\tan a_1 = \tan b_1 = \tan c_1 = \tan a_5 = \tan b_5 = \tan c_5 = -\sqrt{3}$, that is equivalent to part (i). Part (ii) is also trivial from Equation (6). Finally, by the law of cosine for angles,

$$\cos a_2 = -\cos(\angle Z_\infty' Y_\infty' V_2') \cos(\angle V_2' Z_\infty' Y_\infty'),$$

$$\cos a_6 = -\cos(\angle Z_\infty' Y_\infty' V_6') \cos(\angle V_6' Z_\infty' Y_\infty').$$

Let θ be the angle $\angle Z_\infty' Y_\infty' V_2'$. From $\alpha = \beta = \gamma$, $\angle V_2' Z_\infty' Y_\infty' = \theta$ and $\angle Z_\infty' Y_\infty' V_6' = \angle V_6' Z_\infty' Y_\infty' = \pi/2 - \theta$. Therefore,

$$\cos a_2 + \cos a_6 = -\cos^2 \theta - \cos^2(\pi/2 - \theta) = -1.$$

□

Remark 2. The visual angles a_2 and a_6 converge to $2\pi/3$ as $\alpha \nearrow \infty$. Conversely, as $\alpha \searrow 1$, $a_2 \nearrow \pi$ and $a_6 \searrow \pi/2$.

References

1. Barnard, S.T.: Interpreting perspective images. Artificial Intelligence **21** (1983) 435-462
2. Barnard, S.T.: Choosing a basis for perceptual space. Computer Vision, Graphics, and Image Processing **29** (1985) 87-99
3. Ikeuchi, K.: Shape from regular patterns. Artificial Intelligence **22** (1984) 49-75
4. Jennings, G.: Modern Geometry with Applications. Springer-Verlag, New York. (1994)
5. Kanade, T.: Recovery of the three-dimensional shape of an object from a single view. Artificial Intelligence **17** (1981) 409-460
6. Kanatani, K.: The constraints on images of rectangular polyhedra. IEEE Transactions on Pattern Analysis and Machine Intelligence **PAMI-8** (1986) 456-463
7. Kanatani, K.: Constraints on length and angle. Computer Vision, Graphics, and Image Processing **41** (1988) 28-42
8. Lutton, E., Maître, H., Lopez-Krahe, J.: Contribution to the determination of vanishing points using Hough transform. IEEE Transactions on Pattern Analysis and Machine Intelligence **16** (1994) 430-438
9. Mackworth, A.K.: Interpreting pictures of polyhedral scenes. Artificial Intelligence **4** (1973) 121-137
10. Shakunaga, T., Kaneko, H.: Perspective angle transform: Principle of shape from angles. International Journal of Computer Vision **3** (1989) 239-254
11. Shufelt, J.A.: Performance evaluation and analysis of vanishing point detection techniques. IEEE Transactions on Pattern Analysis and Machine Intelligence **21** (1999) 282-288

Observing an Angle from Various Viewpoints

Yoichi Maeda[1] and Hiroshi Maehara[2]

[1] Tokai University, Hiratsuka, Kanagawa, 259-1292 Japan
[2] Ryukyu University, Nishihara, Okinawa, 903-0213 Japan

Abstract. Let AOB be a triangle in R^3. When we look at this triangle from various viewpoints, the angle $\angle AOB$ changes its appearance, and its 'visual size' is not constant. We prove, nevertheless, that the average visual size of $\angle AOB$ is equal to the true size of the angle when viewpoints are chosen at random on the surface of a sphere centered at O. We also present a formula to compute the variance of the visual size.

1 Introduction

Let $\angle AOB$ be a fixed angle determined by three points O, A, B in the three dimensional Euclidean space R^3. When we look at this angle, its appearance changes according to our viewpoint. Let us denote by

$$\angle_P AOB$$

the dihedral angle of the two faces OAP, OBP of the (possibly degenerate) tetrahedron $POAB$, see Figure 1. This angle $\angle_P AOB$ is called the *visual angle* of $\angle AOB$ from the viewpoint P, and its size (measure) is called the *visual size* of $\angle AOB$ from P. For an angle with fixed size ω $(0 < \omega < \pi)$, its visual size can

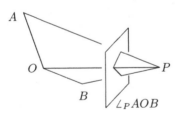

O P

A

B $\angle_P AOB$

Fig. 1. A visual angle.

vary from 0 to π depending on the viewpoint.

For a given angle $\angle AOB$ in R^3, take a random point P distributed uniformly on the unit sphere centered at O. Then the size of $\angle_P AOB$ is a random variable, which is called the *random visual size* of $\angle AOB$. It will be clear that this random visual size depends only on the size of $\angle AOB$. So, for an angle $\angle AOB$ of size $\omega(0 < \omega < \pi)$, we may denote its random visual size by Θ_ω.

J. Akiyama and M. Kano (Eds.): JCDCG 2002, LNCS 2866, pp. 200–203, 2003.

Theorem 1. *For any angle of size* ω $(0 < \omega < \pi)$, *the expected value of the random visual size* Θ_ω *is equal to* ω, *that is,* $\mathbf{E}(\Theta_\omega) = \omega$.

Thus, when we observe an angle from several viewpoints, each chosen at random, the average visual size is approximately equal to the true size.

For potential applications of Theorem 1, let us present a formula to compute the variance $\mathbf{V}(\Theta_\omega)$ of Θ_ω.

Theorem 2. $\mathbf{V}(\Theta_\omega) = \mathbf{V}(\Theta_{\pi-\omega})$ *and*

$$\mathbf{V}(\Theta_\omega) = \int_0^\pi \int_0^\pi \left(\frac{\pi}{2} - \tan^{-1} \left(\frac{\cot \omega \sin y + \cos x \cos y}{\sin x} \right) \right)^2 \frac{\sin y}{2\pi} dx dy - \omega^2.$$

Though this looks complicated, we can easily compute the variance from this double integral by computer. The following table shows the values of $\mathbf{V}(\Theta_\omega)$ for $\omega = k\pi/12$, $k = 1, 2, 3, 4, 5, 6$.

ω	$\pi/12$	$2\pi/12$	$3\pi/12$	$4\pi/12$	$5\pi/12$	$6\pi/12$
$\mathbf{V}(\Theta_\omega)$	0.0699	0.1874	0.3042	0.3988	0.4595	0.4804

2   Proof of Theorem 1

Let $\angle AOB$ be an angle of size ω, and let P be a random point on the unit sphere S^2 centered at O in R^3. We may suppose that A, B lie on S^2. Then the spherical distance $\widehat{AB}$ of A and B is equal to ω. (We denote the shortest geodesic connecting A, B and its length by the same notation $\widehat{AB}$.) Notice that $\angle_P AOB$ is equal to the interior angle $\angle P$ of the spherical triangle $\triangle APB$. Since P is a *random* point on S^2, we have $\Theta_\omega = \angle P$.

Now, consider the (polar) dual triangle $\triangle A^* B^* P^*$ of the spherical triangle $\triangle ABP$:

$$\triangle A^* B^* P^* = H(A) \cap H(B) \cap H(P),$$

where $H(A)$ denotes the hemisphere with pole A. Let $\angle P^* (=: \tau)$ denote the interior angle at P^* of this spherical triangle $\triangle A^* B^* P^*$, see Figure 2. Then, by the duality (see, e.g. [1] Chapter 2), we have

$$\widehat{AB} + \angle P^* = \pi, \qquad \widehat{A^* B^*} + \angle P = \pi.$$

Hence $\angle P^* = \pi - \omega$ and $\widehat{A^* B^*} = \pi - \Theta_\omega$. Let $\Lambda = H(A) \cap H(B)$. Then the angle of the lune Λ is equal to $\pi - \omega$, and its area is equal to $2(\pi - \omega)$. Note that since P is a random point on S^2, the boundary $\partial H(P)$ of $H(P)$ is a random great circle, and $\widehat{A^* B^*} = \partial H(P) \cap \Lambda$.

Here we recall Santaló's chord theorem:

Theorem[2] *Let* $\Omega \subset S^2$ *be a subset obtained as the intersection of a number of hemispheres. Let* G *be a random great circle, and let* φ *be the length of the arc* $G \cap \Omega$. $(G \cap \Omega = \emptyset$ *implies* $\varphi = 0$.) *Then* $\mathbf{E}(\varphi) = area(\Omega)/2$.

Applying this theorem, we have

$$\mathbf{E}(\widehat{A^*B^*}) = \frac{\text{area}(\varLambda)}{2} = \pi - \omega.$$

Therefore $\mathbf{E}(\pi - \Theta_\omega) = \pi - \omega$, and $\mathbf{E}(\Theta_\omega) = \omega$. This proves Theorem 1.

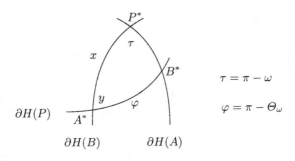

Fig. 2. The polar dual $\triangle P^* A^* B^*$.

3   Proof of Theorem 2

Since $\pi - \Theta_\omega$ can be regarded as $\Theta_{\pi-\omega}$, Θ_ω and $\Theta_{\pi-\omega}$ have the same variance.

Now, let $\angle AOB$ be an angle of size ω with $\overline{OA} = \overline{OB} = 1$, and let P be a random point on the unit sphere S^2 centered at O in R^3 as in the proof of Theorem 1. Let $\triangle A^* B^* P^*$ be the polar dual of $\triangle ABP$, and let $\varLambda = H(A) \cap H(B)$, $\tau = \pi - \omega$. Let φ denote the length of the arc $\widehat{A^*B^*} = \partial H(P) \cap \varLambda$. Then by Santalo's chord theorem, $\mathbf{E}(\varphi) = \tau$. Since $\varphi = \pi - \Theta_\omega$ by duality, we have $\mathbf{V}(\varphi) = \mathbf{V}(\pi - \Theta_\omega) = \mathbf{V}(\Theta_\omega)$. So, we consider the variance of φ. Let $y = \angle A^*$ and $x = \widehat{P^*A^*}$, see Figure 2. By applying the spherical cosine law for angles (see e.g. [1]) to $\triangle P^* A^* B^*$,

$$\cos \angle B^* = -\cos \tau \cos y + \sin \tau \sin y \cos x,$$
$$\cos \tau = -\cos y \cos \angle B^* + \sin y \sin \angle B^* \cos \varphi.$$

Hence

$$
\begin{aligned}
\cos \varphi &= \frac{\cos \tau + \cos y \cos \angle B^*}{\sin y \sin \angle B^*} \\
&= \frac{\cos \tau + \cos y(-\cos \tau \cos y + \sin \tau \sin y \cos x)}{\sin y \sin \angle B^*} \\
&= \frac{\cos \tau - \cos \tau \cos^2 y + \cos y \sin \tau \sin y \cos x}{\sin y \sin \angle B^*} \\
&= \frac{\cos \tau \sin y + \cos y \sin \tau \cos x}{\sin \angle B^*}
\end{aligned}
$$

On the other hand, by the spherical sine law (see [1]), we have

$$\frac{\sin \varphi}{\sin \tau} = \frac{\sin x}{\sin \angle B^*},$$

and hence

$$\sin \varphi = \frac{\sin \tau \sin x}{\sin \angle B^*}.$$

Therefore,

$$\cot \varphi = \frac{\cos \tau \sin y + \cos y \sin \tau \cos x}{\sin \tau \sin x} = \frac{\cot \tau \sin y + \cos x \cos y}{\sin x}.$$

Since $0 \le \varphi \le \pi$, we have

$$\varphi = \frac{\pi}{2} - \tan^{-1}\left(\frac{\cot \tau \sin y + \cos x \cos y}{\sin x}\right).$$

Notice that since $\partial H(P)$ is a random great circle, x and y are mutually *independent*. (Indeed, relative to the position of the random great circle $\partial H(P)$, the point P^* on the fixed great circle $\partial H(B)$ can be regarded as a *uniform random point* on this great circle.) Since $\Pr(y \le y_0) = \Pr(\widehat{PB} \ge \pi - y_0) = (1 - \cos y_0)/2$, the angle y is distributed on the interval $[0, \pi]$ with probability density $\frac{1}{2} \sin y$; and x is distributed uniformly on $[0, \pi]$. Therefore,

$$\mathbf{V}(\varphi) = \int_0^\pi \int_0^\pi \varphi^2 \frac{\sin y}{2\pi} dx dy - \mathbf{E}(\varphi)^2$$

$$= \int_0^\pi \int_0^\pi \left[\frac{\pi}{2} - \tan^{-1}\left(\frac{\cot \tau \sin y + \cos x \cos y}{\sin x}\right)\right]^2 \frac{\sin y}{2\pi} dx dy - \tau^2.$$

Since $\mathbf{V}(\Theta_{\pi-\omega}) = \mathbf{V}(\Theta_\omega)$, $\mathbf{V}(\varphi)$ is equal to $\mathbf{V}(\Theta_\tau)$. Now, by replacing τ with ω, we have the second formula of Theorem 2.

References

1. G. A. Jennings, *Modern Geometry with Applications*, Springer-Verlag, New York 1994.
2. L. A. Santaló, Integral formulas in Crofton's style on the sphere and some inequalities referring to spherical curves. *Duke Math. J.* **9**(1942) 707–722.

The Polyhedra of Maximal Volume Inscribed in the Unit Sphere and of Minimal Volume Circumscribed about the Unit Sphere

Nobuaki Mutoh

School of Administration and Informatics, University of Shizuoka
52-1, Yada, Shizuoka-shi, Shizuoka, 422-8002, Japan
muto@u-shizuoka-ken.ac.jp

Abstract. In this paper, we consider two classes of polyhedra. One is the class of polyhedra of maximal volume with n vertices that are inscribed in the unit sphere of R^3. The other class is polyhedra of minimal volume with n vertices that are circumscribed about the unit sphere of R^3. We construct such polyhedra for n up to 30 by a computer aided search and discuss some of their properties.

1 Introduction

In this paper, we consider two classes of polyhedra. One is the class of polyhedra of maximal volume with n vertices that are inscribed in the unit sphere of R^3. The other class is polyhedra of minimal volume with n vertices that are circumscribed about the unit sphere of R^3.

Berman and Hanes discussed polyhedra of the first class and formed the polyhedra of maximal volume for $n \leq 9$. They showed that for $n = 4$, the polyhedron of maximal volume is the regular tetrahedron. For $5 \leq n \leq 7$, the polyhedra consist of two pyramids. In particular, for $n = 5$ the polyhedron is a double tetrahedra. For $n = 6$, the polyhedron is the octahedron. For $n = 7$, the polyhedron consists of double five-sided pyramids. For $n = 8$, the polyhedron has 4 vertices of degree 4 and 4 vertices of degree 5. It is not the cube. Moreover, they proved that the polyhedra of maximal volume inscribed in the unit sphere are Euclidian simplexes, i.e., all faces of the polyhedra are triangles.

Grace conjectured that the degree of the vertices of the polyhedra are m or $m + 1$, where $m \leq 6 - 12/n < m + 1$. Grace also expected that the polyhedron of $n = 12$ is the icosahedron.

Goldberg referred to the relation of the polyhedra of maximal volume inscribed in the unit sphere and the polyhedra of minimal volume circumscribed about the unit sphere. He conjectured that for given n two such polyhedra are dual.

In this paper, we construct such polyhedra for n up to 30 by using a computer aided search. We discuss some properties of the polyhedra, and examine Grace and Goldberg's conjecture.

J. Akiyama and M. Kano (Eds.): JCDCG 2002, LNCS 2866, pp. 204–214, 2003.

Table 1. The polyhedra of maximal volume inscribed in the unit sphere.

N	V	F	degree	E_{min}	E_{max}	E_{min}/E_{max}
4	0.51320010	4	3×4	1.63261848	1.63335658	0.99954810
5	0.86602375	6	3×2 4×3	1.41273620	1.73244016	0.81546032
6	1.33333036	8	4×6	1.41301062	1.41573098	0.99807848
7	1.58508910	10	4×5 5×2	1.17439900	1.41629677	0.82920403
8	1.81571182	12	4×4 5×4	1.13754324	1.45682579	0.78083684
9	2.04374046	14	4×3 5×6	1.12352943	1.36344511	0.82403716
10	2.21872888	16	4×2 5×8	1.04153932	1.26202346	0.82529315
11	2.35462915	18	4×2 5×8 6×1	0.96536493	1.26366642	0.76393969
12	2.53614471	20	5×12	1.04956370	1.05406113	0.99573324
13	2.61282570	22	4×1 5×10 6×2	0.80234323	1.14003700	0.70378701
14	2.72096433	24	5×12 6×2	0.89290608	1.05849227	0.84356410
15	2.80436840	26	5×12 6×3	0.81809612	1.04523381	0.78269198
16	2.88644378	28	5×12 6×4	0.81890957	0.97608070	0.83897732
17	2.94750699	30	5×12 6×5	0.74657798	1.02119982	0.73107923
18	3.00958510	32	5×12 6×6	0.75499655	0.96805125	0.77991382
19	3.06319073	34	5×12 6×7	0.72816306	0.99849367	0.72926157
20	3.11851200	36	5×12 6×8	0.74113726	0.95901998	0.77280690
21	3.16440426	38	5×12 6×9	0.69438111	0.94733206	0.73298597
22	3.20820707	40	5×12 6×10	0.69345933	0.89581626	0.77410889
23	3.24694072	42	5×12 6×11	0.66928634	0.87244988	0.76713442
24	3.28399413	44	5×12 6×12	0.69163182	0.87601499	0.78952053
25	3.31626151	46	5×12 6×13	0.66118725	0.86554529	0.76389677
26	3.34935826	48	5×12 6×14	0.65046670	0.85140448	0.76399258
27	3.38027449	50	5×12 6×15	0.65839644	0.82328392	0.79971978
28	3.40577470	52	5×12 6×16	0.59817265	0.81912078	0.73026185
29	3.42990751	54	5×12 6×17	0.58296887	0.80547257	0.72376005
30	3.45322727	56	5×12 6×18	0.59082147	0.79788645	0.74048315

2   The Polyhedra of Maximal Volume Inscribed in the Unit Sphere

First, we show our result on the polyhedra of maximal volume inscribed in the unit sphere calculated by computer. Table 1 shows some properties of the polyhedra of maximal volume inscribed in the unit sphere for $4 \le n \le 30$. In Table 1, V is the volume of each polyhedron, F is the number of faces of each polyhedron. The number of edges is $3F/2$ because the polyhedra are simplex is as shown by Berman and Hanes. From Table 1, we can that $F = 2N - 4$. This is also an obvious consequencd of the fact that the polyhedra are simplexes. The "$m \times n$" entries in the column of the table denoted by "degree" designate that each polyhedron has n vertices of degree m. In Addition, E_{min} and E_{max} are the minimal length and the maximal length of the edges of each polyhedron. Figure 1–3 shows the polyhedra for $4 \le n \le 30$. Whew you see the left figure with your left eye and the right figure with your right eye, the figure can be solidly seen.

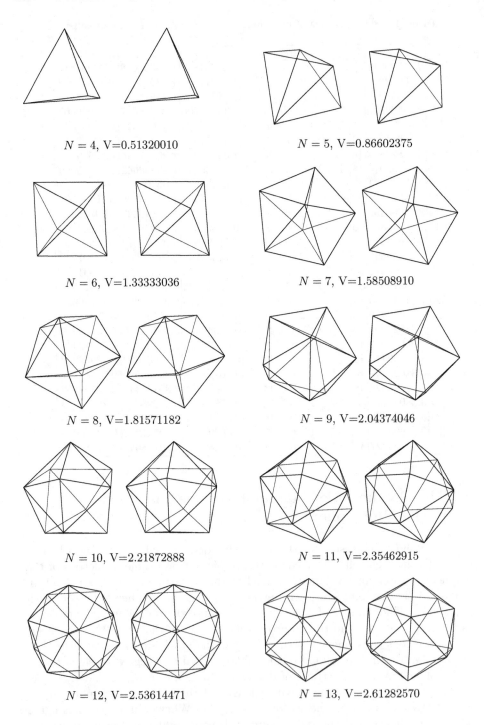

$N = 4$, V=0.51320010

$N = 5$, V=0.86602375

$N = 6$, V=1.33333036

$N = 7$, V=1.58508910

$N = 8$, V=1.81571182

$N = 9$, V=2.04374046

$N = 10$, V=2.21872888

$N = 11$, V=2.35462915

$N = 12$, V=2.53614471

$N = 13$, V=2.61282570

Fig. 1. The polyhedra of maximal volume inscribed in the unit sphere for $4 \leq n \leq 13$

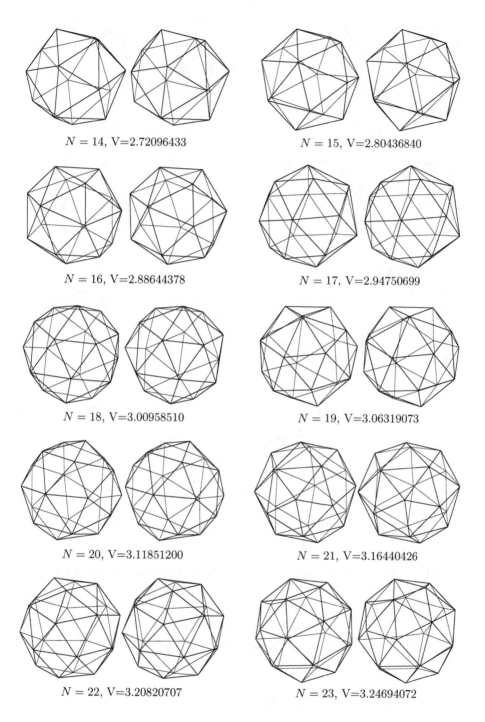

$N = 14$, V=2.72096433

$N = 15$, V=2.80436840

$N = 16$, V=2.88644378

$N = 17$, V=2.94750699

$N = 18$, V=3.00958510

$N = 19$, V=3.06319073

$N = 20$, V=3.11851200

$N = 21$, V=3.16440426

$N = 22$, V=3.20820707

$N = 23$, V=3.24694072

Fig. 2. The polyhedra of maximal volume inscribed in the unit sphere for $14 \leq n \leq 23$

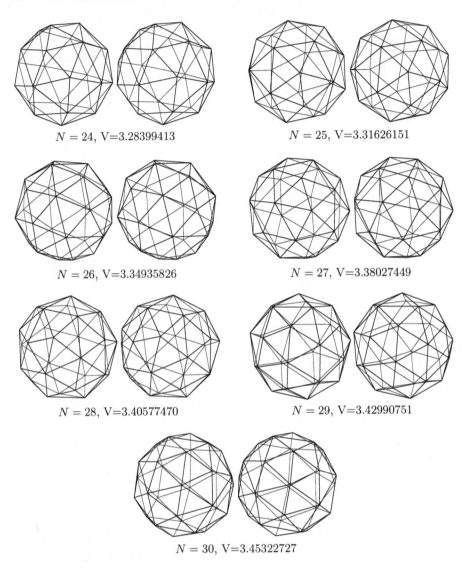

$N = 24$, V=3.28399413 $N = 25$, V=3.31626151

$N = 26$, V=3.34935826 $N = 27$, V=3.38027449

$N = 28$, V=3.40577470 $N = 29$, V=3.42990751

$N = 30$, V=3.45322727

Fig. 3. The polyhedra of maximal volume inscribed in the unit sphere for $24 \leq n \leq 30$

As mentioned in the section 1, Berman and Hanes showed the polyhedra of maximal volume inscribed in the unit sphere for $n \leq 9$. The volume of the polyhedra for $4 \leq n \leq 8$ is the following.

$$n = 4 \quad \frac{8}{9}\sqrt{3} = 0.51320024$$

$$n = 5 \quad \frac{1}{2}\sqrt{3} = 0.86602540$$

Table 2. Some local optimal solutions of the polyhedra of the maximal volume inscribed in the unit sphere for $n = 11$ and 13.

$N = 11$

V	F	degree		
2.35462915 (optimal)	18	4×2	5×8	6×1
2.32969914	18	4×3	5×6	6×2
2.32001635	18	4×4	5×4	6×3

$N = 13$

V	F	degree		
2.61282570 (optimal)	22	4×1	5×10	6×2
2.59084710	22	4×3	5×6	6×4
2.55057603	22	4×2	5×8	6×3

$$n = 6 \quad \frac{4}{3}$$

$$n = 7 \quad \frac{5}{3}\sqrt{\frac{5 + \sqrt{5}}{8}} = 1.58509419$$

$$n = 8 \quad \sqrt{\frac{475 + 29\sqrt{145}}{250}} = 1.81571610$$

The accuracy of our calculation, i. e., the relative error of volume in Table 1 against the theoretical value shown above is 1.996×10^{-6}.

Now we discuss the degree of the vertices of the polyhedra of maximal volume. Grace conjectured that the degree of the vertices of the polyhedra are m or $m+1$, where $m \leq 6 - 12/n < m + 1$. Our result supports Grace's conjecture except for $n = 11$ and 13. If Grace's conjecture is right, the polyhedron for $n = 11$ has vertices of degree 4 or 5 and the polyhedron for $n = 13$ has vertices of degree 5 or 6. For $n = 11$, there is a polyhedron with $E = 27$ and $F = 18$ which has 1 vertex of degree 4 and 10 vertices of degree 5. For $n = 13$, there is a polyhedron with $E = 33$ and $F = 22$ which has 12 vertices of degree 5 and 1 vertices of degree 6. However, the optimal solutions for $n = 11$ and 13 are polyhedra which have the vertices of degree 4, 5 and 6 (Table 1). We have some local optimal solutions for the polyhedra of maximal volume inscribed in the unit sphere (Table 2).

Grace also expected that the polyhedron of $n = 12$ is the icosahedron. This expectation is seems to be right. We find the icosahedron and our optimal solution are isomorphic. The relative error of the volume of the icosahedron that is $20\sqrt{\gamma}/(3 \cdot 5^{\frac{3}{4}}) = 2.53615071$, where $\gamma = (\sqrt{5} + 1)/2$, and the volume of our optimal solution is 2.367×10^{-6}.

In addition, for $n \geq 14$, we may predict from Table 1 that the polyhedron of $n \geq 14$ has 12 vertices of degree 5 and $n - 12$ vertices of degree 6.

Now we discuss the regularity of the polyhedra of maximal volume inscribed in the unit sphere. Table 1 shows the minimal length E_{min}, the maximal length

Table 3. The polyhedra of minimal volume circumscribed about the unit sphere.

N	V	v	face	E_{min}	E_{max}	E_{min}/E_{max}
4	13.85643934	4	3×4	4.89497665	4.90275065	0.99841436
5	10.39237187	6	3×2 4×3	1.99405215	3.46908051	0.57480711
6	8.00006475	8	4×6	1.99002944	2.00998481	0.99007188
7	7.26548656	10	4×5 5×2	1.44373739	2.00161826	0.72128508
8	6.70100546	12	4×4 5×4	0.83423494	1.86418974	0.44750538
9	6.26322791	14	4×3 5×6	0.70579749	1.38275551	0.51042826
10	5.97462020	16	4×2 5×8	0.71622425	1.28160081	0.55885128
11	5.77915420	18	4×2 5×8 6×1	0.65242786	1.34717683	0.48429267
12	5.55039837	20	5×12	0.88027917	0.91840096	0.95849113
13	5.45064989	22	4×1 5×10 6×2	0.49826743	1.14409212	0.43551339
14	5.33186370	24	5×12 6×2	0.64485254	1.02101518	0.63157977
15	5.24075498	26	5×12 6×3	0.55190785	1.04920489	0.52602485
16	5.15752220	28	5×12 6×4	0.62321255	0.91309493	0.68252767
17	5.09465493	30	5×12 6×5	0.43058955	0.94392164	0.45617086
18	5.03757656	32	5×12 6×6	0.23606084	0.90741900	0.26014536
19	4.98466185	34	5×12 6×7	0.31150863	0.88502416	0.35197755
20	4.93634204	36	5×12 6×8	0.31652420	0.73551928	0.43034113
21	4.89438272	38	5×12 6×9	0.30618086	0.75344682	0.40637355
22	4.85837764	40	5×12 6×10	0.15842315	0.81393479	0.19463863
23	4.82507072	42	5×12 6×11	0.32146783	0.72095208	0.44589347
24	4.79382934	44	5×12 6×12	0.14429001	0.71631496	0.20143375
25	4.76816757	46	5×12 6×13	0.17302622	0.73014192	0.23697615
26	4.74438404	48	5×12 6×14	0.10418232	0.74767040	0.13934257
27	4.71877747	50	5×12 6×15	0.01273378	0.72927387	0.01746090
28	4.70046819	52	5×12 6×16	0.13362517	0.69573664	0.19206286
29	4.67939498	54	5×12 6×17	0.10090586	0.67268554	0.15000450
30	4.66190180	56	5×12 6×18	0.02593854	0.69315995	0.03742071

E_{max} and the ratio E_{min}/E_{max} of each polyhedron. Except for $n = 4$, 6 and 12, E_{min}/E_{max} is not equal to 1. The mean of E_{min}/E_{max} is 0.80000285.

3   The Polyhedra of Minimal Volume Circumscribed about the Unit Sphere

Table 3 shows some properties of the polyhedra of minimal volume circumscribed about the unit sphere for $4 \leq n \leq 30$. In Table 3, V is the volume of each polyhedron and v is the number of its vertices. The entries "$m \times n$" in the column designated by "face" designate that each polyhedron has n faces with m vertices. E_{min} and E_{max} are the minimal length and the maximal length of the edges of each polyhedron. The mean of the ratio E_{min}/E_{max} for $4 \leq n \leq 30$ is 0.44592722. The polyhedra are farther from regular than the polyhedra of maximal volume inscribed in the unit sphere. From Table 3, for $n \geq 14$ we may predict that the polyhedra of minimal volume circumscribed about the unit sphere have 12 pentagon and $n - 12$ hexagon.

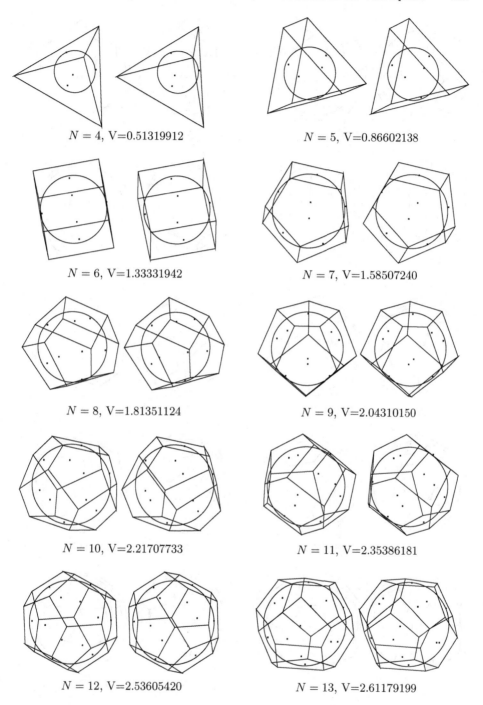

$N = 4$, V=0.51319912

$N = 5$, V=0.86602138

$N = 6$, V=1.33331942

$N = 7$, V=1.58507240

$N = 8$, V=1.81351124

$N = 9$, V=2.04310150

$N = 10$, V=2.21707733

$N = 11$, V=2.35386181

$N = 12$, V=2.53605420

$N = 13$, V=2.61179199

Fig. 4. The polyhedra of minimal volume circumscribed about the unit sphere for $4 \leq n \leq 13$

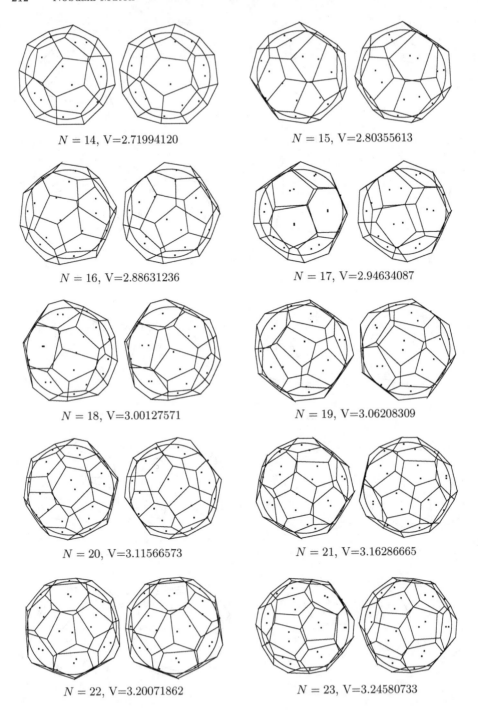

$N = 14$, V=2.71994120

$N = 15$, V=2.80355613

$N = 16$, V=2.88631236

$N = 17$, V=2.94634087

$N = 18$, V=3.00127571

$N = 19$, V=3.06208309

$N = 20$, V=3.11566573

$N = 21$, V=3.16286665

$N = 22$, V=3.20071862

$N = 23$, V=3.24580733

Fig. 5. The polyhedra of minimal volume circumscribed about the unit sphere for $14 \leq n \leq 23$

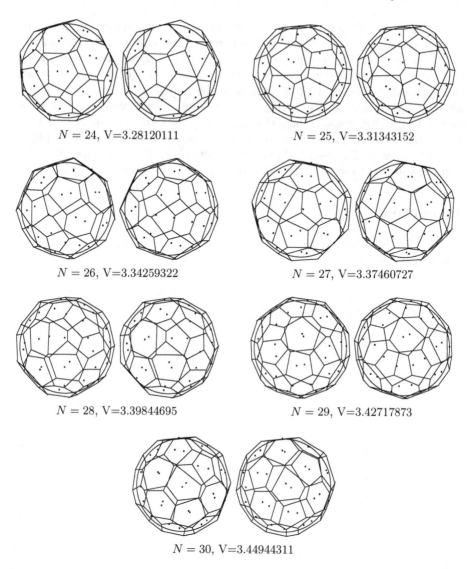

$N = 24$, V=3.28120111 $N = 25$, V=3.31343152

$N = 26$, V=3.34259322 $N = 27$, V=3.37460727

$N = 28$, V=3.39844695 $N = 29$, V=3.42717873

$N = 30$, V=3.44944311

Fig. 6. The polyhedra of minimal volume circumscribed about the unit sphere for $24 \leq n \leq 30$

The accuracy of the volume is 3.626×10^{-5}. We calculate this as the difference between our computational result and the volume of polyhedra with assuming for $n = 4$, the polyhedra is the tetrahedron, for $n = 5$ it is the regular triangular prism, for $n = 6$ it is cube and for $n = 12$ it is the dodecahedron.

Figure 4–6 shows the polyhedra for $4 \leq n \leq 30$. In Figure 5–8, black dots designate the contact points of the faces of polyhedra and the unit sphere.

Goldberg conjectured that the polyhedron of maximal volume inscribed to the unit sphere and the polyhedron of minimal volume circumscribed about the unit sphere are dual. A comparison of Table 1 and 3 shows that the number of vertices and the number of faces of the two class of polyhedra correspond with each other. The degrees of vertices of the polyhedra of maximal volume inscribed in the unit sphere correspond to the numbers of vertices of faces of the polyhedra of minimal volume circumscribed about the unit sphere. Indeed, the volume of polyhedra whose vertices are the contact points of the unit sphere and the polyhedra circumscribed about the unit sphere differs only by 0.07299% from the volume of the polyhedra inscribed in the unit sphere.

References

1. H. T. Croft, K. J. Falconer and R. K. Guy. *Unsolved Problems in Geometry*. Springer Verlag, 1995.

Maximal Number of Edges in Geometric Graphs without Convex Polygons

Chie Nara[1], Toshinori Sakai[1], and Jorge Urrutia[2,⋆]

[1] Research Institute of Educational Development, Tokai University
2-28-4 Tomigaya, Shibuya-ku, Tokyo 151-8677, Japan
{cnara,tsakai}@ried.tokai.ac.jp
[2] Instituto de Matemáticas, Ciudad Universitaria
Universidad Nacional Autónoma de México, México D.F., México
urrutia@matem.unam.mx

Abstract. A geometric graph G is a graph whose vertex set is a set P_n of n points on the plane in general position, and whose edges are straight line segments (which may cross) joining pairs of vertices of G. We say that G contains a convex r-gon if its vertex and edge sets contain, respectively, the vertices and edges of a convex polygon with r vertices. In this paper we study the following problem: Which is the largest number of edges that a geometric graph with n vertices may have in such a way that it does not contain a convex r-gon? We give sharp bounds for this problem. We also give some bounds for the following problem: Given a point set, how many edges can a geometric graph with vertex set P_n have such that it does not contain a convex r-gon?
A result of independent interest is also proved here, namely: Let P_n be a set of n points in general position. Then there are always three concurrent lines such that each of the six wedges defined by the lines contains exactly $\lfloor \frac{n}{6} \rfloor$ or $\lceil \frac{n}{6} \rceil$ elements of P_n.

1 Introduction

A geometric graph G is a graph whose vertex set is a set of P_n of n points on the plane in general position such that its edges are straight line segments joining some pairs of elements of P_n. Geometric graphs have received considerable attention lately, see for example a recent survey by J. Pach [6]. Some classical topics in Graph Theory have been studied for geometric graphs, e.g. Ramsey-type problems on geometric graphs have been studied in [4][5]. A classical problem in Graph Theory solved by Turán [7] is that of determining the largest number of edges that a graph has such that it does not contain a complete graph on r vertices. In this paper we study the corresponding problem for geometric graphs. We say that a geometric graph G *contains a convex r-gon* if its vertex and edge sets contain, respectively, the vertices and edges of a convex polygon with r vertices. We study the following problem: What is the largest number of edges a geometric graph may have in such a way that it does not contain a convex r-gon? In Section 2 we give tight bounds for our problem.

⋆ Supported by CONACYT of Mexico, Proyecto 37540-A.

J. Akiyama and M. Kano (Eds.): JCDCG 2002, LNCS 2866, pp. 215–220, 2003.
© Springer-Verlag Berlin Heidelberg 2003

In Section 3 we study the following related problem: Given a point set P_n what is the largest number of edges that a geometric graph containing no convex r-gons may have such that its vertex set is P_n? Sharp bounds are given for the case $r = 5$ and $r = 7$. In this section we also prove the following problem which is interesting on its own right: Let P_n be a set of n points in general position. Then there are always three concurrent lines such that each of the six wedges defined by the lines contains exactly $\lfloor \frac{n}{6} \rfloor$ or $\lceil \frac{n}{6} \rceil$ elements of P_n. This is the discrete version of a theorem by R. C. and E.F. Buck [1] which states that any convex set can be divided by three concurrent lines into six parts of equal area.

2   Geometric Graphs without Convex r-Gons

Let $T_{r-1}(n)$ be the Turán graph, that is the complete $(r-1)$-partite graph whose classes have sizes as equal as possible, and denote by $t_{r-1}(n)$ the number of edges in $T_{r-1}(n)$. We recall a result of Turán:

Theorem 1 *[7] The maximal number of edges in simple graphs of order n not containing a complete graph of order r is $t_{r-1}(n)$ and $T_{r-1}(n)$ is the unique graph of order n with $t_{r-1}(n)$ edges that does not contain a complete graph of r points.*

The next result follows immediately:

Theorem 2 *The maximum number of edges that a geometric r-graph whose vertices are in convex position is $t_{r-1}(n)$. The bound is tight.*

The following result of Erdös and Szekeres will be useful:

Theorem 3 *[2] Let k be a natural number. There exists a natural number $p(k)$ such that any set with at least $p(k)$ points on the plane, in general position, contains k points in convex position.*

It is well known that $p(4) = 5$, $p(5) = 9$, and that

$$2^{k-2} + 1 \le p(k) \le \binom{2k-5}{k-2} + 2$$

In fact it is conjectured that $p(k) = 2^{k-2} + 1$.

Our objective in this section is to prove the following result:

Theorem 4 *Let k be a natural number, $k \ge 3$, and let $r = p(k) - 1$. Then the maximal number of edges in a geometric graph on n points which does not contain a convex k-gon is $\lfloor \frac{r-1}{2r} n^2 \rfloor$. This bound is tight.*

Proof. Suppose that a geometric graph with n vertices contains more than $t_k(n)$ edges. Then by Turán's Theorem it contains a complete subgraph H with $p(k)$ vertices. By Theorem 3, the vertex set of H contains k elements in convex position, and thus G contains a convex k-gon.

To show that our bound is tight, take a point set S in general position with $p(k) - 1$ points labeled $q_1, \ldots, q_{p(k)-1}$ on the plane such that S contains no k points in convex position, and let m be an integer greater than or equal to 0. If we substitute each of the points q_i by a set S_i with $r_i = m$, or $r_i = m + 1$ points within a sufficiently small neighborhood of q_i, and join all pairs of points u, v with $u \in S_i$, $v \in S_j$, $i \neq j$, $i = 1, \ldots, p(k) - 1$ we obtain a geometric graph with $n = r_1 + \ldots + r_{p(k)-1}$ vertices and $t_{p(k)-1}(n)$ edges which does not contain a convex k-gon. Our result follows.   □

For the cases $r = 4$, and $r = 5$ it follows that any geometric graph containing no convex quadrilateral (respectively convex pentagons) contains at most $\lfloor \frac{3}{8}n^2 \rfloor$ (respectively $\lfloor \frac{7}{16}n^2 \rfloor$) edges.

Using the previous theorem we can construct geometric graphs containing no convex pentagons on n points as follows: Since $p(5) = 9$, take eight points such that no five of them are in convex position. Then substitute each point by m or $m + 1$ points, $m \geq 1$, and finally join points that are in different subsets. The graph obtained contains $\lfloor \frac{7}{16}n^2 \rfloor$. See Figure 1. Similar constructions an be used to obtain geometric graphs without convex quadrilaterals.

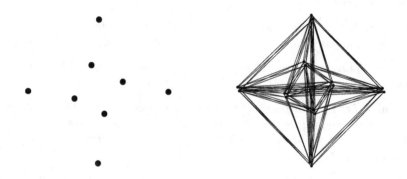

Fig. 1. Constructing a geometric graph with sixteen vertices, no convex pentagons, and $\lfloor \frac{7}{16}n^2 \rfloor$ edges, n=16.

3   Geometric Graphs with Predetermined Vertex Sets

To conclude our paper, we study the following problem: Let P_n be a point set. What is the largest number of edges that a geometric graph may have such that the vertex set of G is P_n, and it contains no convex r-gon?

We start by proving:

Theorem 5 *Let P_n be a point set. Then there is a geometric graph with vertex set P_n containing no convex pentagons with $\lfloor \frac{3}{8}n^2 \rfloor$. Our bound is tight.*

Proof. Assume w.l.o.g. that P_n contains $4m$ points. It is well known that given any point set in general position with $4m$ points on the plane, there exist two

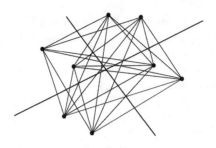

Fig. 2.

intersecting lines such that there are exactly m points in the interior of each of the four wedges into which they divide the plane. Our geometric graph now contains the edges obtained by joining all pairs of points with elements in different wedges. See Figure 2. That the bound is tight follows from the fact that if the elements of P_n are in convex position, by Theorem 2 the graph cannot have more than $\lfloor \frac{3}{8}n^2 \rfloor$ edges. □

We finish by proving a similar result for geometric graphs containing no convex heptagons.

Theorem 6 *Let P_n be a point set. Then there is a geometric graph containing no convex heptagons with vertex set P_n with $\lfloor \frac{5n}{12} \rfloor$ edges. This bound is tight.*

To prove Theorem 6, we will use the following result which is interesting on its own:

Theorem 7 *Let P_n be a point set with n points in general position. Then there exist three concurrent lines, i.e. that intersect at a common point, such that the interior of each of the six wedges determined by them contains exactly $\lfloor \frac{n}{6} \rfloor$ or $\lceil \frac{n}{6} \rceil$ elements of P_n.*

Proof. We prove our result for $n = 6m$. Similar arguments can be applied for the remaining cases. Choose a horizontal line $\mathcal{L}_1$ that leaves $3m$ points in the interior of the semi-plane below it. Find a second line $\mathcal{L}_2$ with positive slope such that the four wedges $W_1, \ldots, W_4$ determined by them contain m, $2m$, m and $2m$ points respectively; see Figure 3(a). Let q be the point of intersection of our lines. Draw two rays emanating from q; the first one r_1, splitting W_2 into two wedges, each containing m points in their interiors. The second ray, r_2, splits W_4 in a similar way; see Figure 3(b).If r_1 and r_2 are collinear, we are done. Suppose w.l.o.g. that the angle Θ_1 between r_1 and r_2 in the clockwise direction (the size of the angle we have to rotate r_1 to make it coincide with r_2) is greater than π.

Rotate $\mathcal{L}_1$ continuously keeping $3m$ points on each of the semi-planes determined by $\mathcal{L}_1$. Simultaneously update the second line, and r_1, and r_2. After a careful rotation of $\mathcal{L}_1$ 180 degrees, we can make $\mathcal{L}_1$ and $\mathcal{L}_2$ coincide with themselves (but with different orientations), and r_1 (resp. r_2) fall on r_2 (resp. r_1). Along the way, Θ changed from its original size to $2\pi - \Theta$, and thus at some point it took the value π. At this point r_1 and r_2 became collinear. Our result follows. □

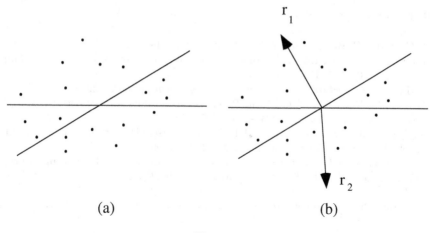

(a)   (b)

Fig. 3.

To prove Theorem 6, choose three lines as in the previous Theorem. As we did in Theorem 5, join all pars of points with elements in different wedges. Clearly the resulting geometric graph contains no convex heptagons. See Figure 4. To prove that our bound is tight, choose P_n in convex position.

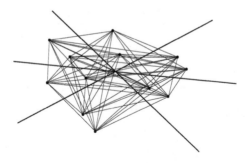

Fig. 4.

At this point we were unable to prove a result similar Theorem 6 for convex quadrilaterals, hexagons. We ask the following question:

Problem 1 *Is it true that given a point set P_n of n points in general position, there is always a geometric graph containing no convex quadrilaterals (resp. hexagons) whose vertex set is P_n with $t_3(n)$ (resp. $t_5(n)$) edges?*

A similar question for convex r-gons $r \geq 8$ is open. However to solve this problems new techniques will be required, as Theorem 7 does not generalize to more than three lines. It is straightforward to find examples of point sets with $8m$ elements for which there are no four concurrent lines that split the point set into equal size subsets.

References

1. R.C., and E.F. Buck, *Math. Mag.*, **22**, 195-198 (1948-49).
2. P.Erdös and G.Szekeres, A Combinatorial Problem in Geometry, *Composito Mathematica* **2**, 463-470 (1935).
3. J.Matouusek, *Lectures on Discrete Geometry*, Springer-Verlag,2002.
4. G. Karoly, J. Pach, and G. Toth, Ramsey-Type results for geometric graphs, I *Discrete Comput. Geom.* **18**, 247-255 (1997).
5. G. Karoly, J. Pach, G. Toth, and P. Valtr, Ramsey-Type results for geometric graphs II, *Discrete Comput. Geom.* **20**, 375-388 (1998).
6. J. Pach, Geometric graph theory. Surveys in combinatorics, 1999 (Canterbury), 167–200, London Math. Soc. Lecture Note Ser., 267, Cambridge Univ. Press, Cambridge, 1999.
7. P. Turán, On an extremal problem in Graph Theory, *Mat. Fiz. Lapok* **48**, 436-452 (1941).

Relaxing Planarity for Topological Graphs⋆

János Pach[1], Radoš Radoičić[2], and Géza Tóth[3]

[1] City College, CUNY and Courant Institute of Mathematical Sciences
New York University, New York, NY 10012, USA
pach@cims.nyu.edu
[2] Department of Mathematics
Massachusetts Institute of Technology, Cambridge, MA 02139, USA
rados@math.mit.edu
[3] Rényi Institute of the Hungarian Academy of Sciences
H-1364 Budapest, P.O.B. 127, Hungary
geza@renyi.hu

Abstract. According to Euler's formula, every planar graph with n vertices has at most $O(n)$ edges. How much can we relax the condition of planarity without violating the conclusion? After surveying some classical and recent results of this kind, we prove that every graph of n vertices, which can be drawn in the plane without three pairwise crossing edges, has at most $O(n)$ edges. For straight-line drawings, this statement has been established by Agarwal et al., using a more complicated argument, but for the general case previously no bound better than $O(n^{3/2})$ was known.

1 Introduction

A *geometric graph* is a graph drawn in the plane so that its vertices are represented by points in general position (i.e., no three are collinear) and its edges by straight-line segments connecting the corresponding points. *Topological graphs* are defined similarly, except that now each edge can be represented by any simple (non-selfintersecting) Jordan arc passing through no vertices other than its endpoints. Throughout this paper, we assume that if two edges of a topological graph G share an interior point, then at this point they properly cross. We also assume, for simplicity, that no three edges cross at the same point and that any two edges cross only a finite number of times. If any two edges of G have at most one point in common (including their endpoints), then G is said to be a *simple* topological graph. Clearly, every geometric graph is simple. Let $V(G)$ and $E(G)$ denote the vertex set and edge set of G, respectively. We will make no notational distinction between the vertices (edges) of the underlying abstract graph, and the points (arcs) representing them in the plane.

⋆ János Pach has been supported by NSF Grant CCR-00-98245, by PSC-CUNY Research Award 63352-0036, and by OTKA T-032458. Géza Tóth has been supported by OTKA-T-038397 and by an award from the New York University Research Challenge Fund.

J. Akiyama and M. Kano (Eds.): JCDCG 2002, LNCS 2866, pp. 221–232, 2003.
© Springer-Verlag Berlin Heidelberg 2003

It follows from Euler's Polyhedral Formula that every simple planar graph with n vertices has at most $3n - 6$ edges. Equivalently, every topological graph with n vertices and more than $3n - 6$ edges has a pair of crossing edges. What happens if, instead of a crossing pair of edges, we want to guarantee the existence of some larger configurations involving several crossings? What kind of *unavoidable* substructures must occur in every geometric (or topological) graph G having n vertices and more than Cn edges, for an appropriate large constant $C > 0$?

In the next four sections, we approach this question from four different directions, each leading to different answers. In the last section, we prove that any topological graph with n vertices and no three pairwise crossing edges has at most $O(n)$ edges. For *simple* topological graphs, this result was first established by Agarwal-Aronov-Pach-Pollack-Sharir [AAPPS97], using a more complicated argument.

2   Ordinary and Topological Minors

A graph H is said to be a *minor* of another graph G if H can be obtained from a subgraph of G by a series of edge contractions (and deletions). If a subgraph of G can be obtained from H by replacing its edges with independent paths between their endpoints, then H is called a *topological minor* of G. Clearly, a topological minor of G is also its (ordinary) minor.

If a graph G with n vertices has no minor isomorphic to K_5 or to $K_{3,3}$, then by Kuratowski's theorem it is planar and its number of edges cannot exceed $3n - 6$. It follows from an old result of Wagner that the same conclusion holds under the weaker assumption that G has no K_5 minor. A few years ago Mader [M98] proved the following famous conjecture of Dirac:

Theorem 2.1. (Mader) *Every graph of n vertices with no topological K_5 minor has at most $3n - 6$ edges.*

If we only assume that G has no topological K_r minor for some $r > 5$, we can still conclude that G is *sparse*, i.e., its number of edges is at most linear in n.

Theorem 2.2. (Komlós-Szemerédi [KSz96], Bollobás-Thomason [BT98]) *For any positive integer r, every graph of n vertices with no topological K_r minor has at most cr^2n edges.*

Moreover, Komlós and Szemerédi showed that the above statement is true with any positive constant $c > 1/4$, provided that r is large enough. Apart from the value of the constant, this theorem is sharp, as is shown by the union of pairwise disjoint copies of a complete bipartite graph of size roughly r^2.

We have a better bound on the number of edges, under the stronger assumption that G has no K_r minor.

Theorem 2.3. (Kostochka [K84], Thomason [T84]) *For any positive integer r, every graph of n vertices with no K_r minor has at most $cr\sqrt{\log r}\,n$ edges.*

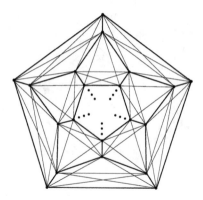

Fig. 1.

The best value of the constant c for which the theorem holds was asymptotically determined in [T01]. The theorem is sharp up to the constant. (Warning! The letters c and C used in several statements will denote *unrelated* positive constants.)

Reversing Theorem 2.3, we obtain that every graph with n vertices and more than $cr\sqrt{\log r}\,n$ edges has a K_r minor. This immediately implies that if the chromatic number $\chi(G)$ of G is at least $2cr\sqrt{\log r}+1$, then G has a K_r minor. According to Hadwiger's notorious conjecture, for the same conclusion it is enough to assume that $\chi(G) \geq r$. This is known to be true for $r \leq 6$ (see [RST93]).

3   Quasi-Planar Graphs

A graph is planar if and only if it can be drawn as a topological graph with no crossing edges. What happens if we relax this condition and we allow r crossings per edge, for some fixed $r \geq 0$?

Theorem 3.1. [PT97] *Let r be a natural number and let G be a simple topological graph of n vertices, in which every edge crosses at most r others. Then, for any $r \leq 4$, we have $|E(G)| \leq (r + 3)(n - 2)$.*

The case $r = 0$ is Euler's theorem, which is sharp. In the case $r = 1$, studied in [PT97] and independently by Gärtner, Thiele, and Ziegler (personal communication), the above bound can be attained for all $n \geq 12$. The result is also sharp for $r = 2$, provided that $n \equiv 5 \pmod{15}$ is sufficiently large (see Figure 1).

However, for $r = 3$, we have recently proved that $|E(G)| \leq 5.5(n - 2)$, and this bound is best possible up to an additive constant [PRTT02]. For very large values of r, a much better upper bound can be deduced from the following theorem of Ajtai-Chvátal-Newborn-Szemerédi [ACNS82] and Leighton [L84]: any topological graph with n vertices and $e > 4n$ edges has at least constant times e^3/n^2 crossings.

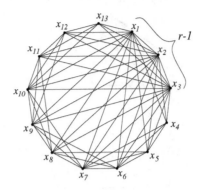

Fig. 2. Construction showing that Theorem 3.4 is sharp ($n = 13$, $r = 4$).

Corollary 3.2. [PRTT02] *Any topological graph with n vertices, whose each edge crosses at most r others, has at most $4\sqrt{r}n$ edges.*

One can also obtain a linear upper bound for the number of edges of a topological graph under the weaker assumption that no edge can cross more than r other edges *incident to the same vertex*. This can be further generalized, as follows.

Theorem 3.3. [PPST02] *Let G be a topological graph with n vertices which contains no $r + s$ edges such that the first r are incident to the same vertex and each of them crosses the other s edges. Then we have $|E(G)| \leq C_s rn$, where C_s is a constant depending only on s.*

In particular, it follows that if a topological graph contains no large *gridlike* crossing pattern (two large sets of edges such that every element of the first set crosses all elements of the second), its number of edges is at most linear in n. It is a challenging open problem to decide whether the same assertion remains true for all topological graphs containing no large *complete* crossing pattern.

For any positive integer r, we call a topological graph r-*quasi-planar* if it has no r pairwise crossing edges. A topological graph is x-*monotone* if all of its edges are x-monotone curves, i.e., every vertical line crosses them at most once. Clearly, every geometric graph is x-monotone, because its edges are straight-line segments (that are assumed to be non-vertical). If the vertices of a geometric graph are in convex position, then it is said to be a *convex* geometric graph.

Theorem 3.4. [CP92] *The maximum number of edges of any r-quasi-planar convex geometric graph with $n \geq 2r$ edges is*

$$2(r - 1)n - \binom{2r - 1}{2}.$$

Theorem 3.5. (Valtr [V98]) *Every r-quasi-planar x-monotone topological graph with n vertices has at most $C_r n \log n$ edges, for a suitable constant C_r depending on r.*

Theorem 3.6. [PSS96] *For any $r \geq 4$, every r-quasi-planar simple topological graph G with n vertices has at most $C_r n (\log n)^{2(r-3)}$ edges, for a suitable constant C_r depending only on r.*

In Section 6, we will point out that Theorem 3.6 remains true even if we drop the assumption that G is simple, i.e., two edges may cross more than once.

For 3-quasi-planar topological graphs we have a linear upper bound.

Theorem 3.7. [AAPPS97] *Every 3-quasi-planar simple topological graph G with n vertices has at most Cn edges, for a suitable constant C.*

In Section 7, we give a short new proof of the last theorem, showing that here, too, one can drop the assumption that no two edges cross more than once (i.e., that G is simple). In this case, previously no bound better than $O(n^{3/2})$ was known. Theorem 3.7 can also be extended in another direction: it remains true for every topological graph G with no $r+2$ edges such that each of the first r edges crosses the last two and the last two edges cross each other. Of course, the constant C in the theorem now depends on r [PRT02].

All theorems in this section provide (usually linear) upper bounds on the number of edges of topological graphs satisfying certain conditions. In each case, one may ask whether a stronger statement is true. Is it possible that the graphs in question can be decomposed into a small number planar graphs? For instance, the following stronger form of Theorem 3.7 may hold:

Conjecture 3.8. *There is a constant k such that the edges of every 3-quasi-planar topological graph G can be colored by k colors so that no two edges of the same color cross each other.*

McGuinness [Mc00] proved that Conjecture 3.8 is true for simple topological graphs, provided that there is a closed Jordan curve crossing every edge of G precisely once. The statement is also true for r-quasi-planar convex geometric graphs, for any fixed r (see [K88], [KK97]).

4   Generalized Thrackles and Their Relatives

Two edges are said to be *adjacent* if they share an endpoint. We say that a graph drawn in the plane is a *generalized thrackle* if any two edges meet an odd number of times, counting their common endpoints, if they have any. That is, a graph is a generalized thrackle if and only if it has no two adjacent edges that cross an odd number of times and no two non-adjacent edges that cross an even number of times. In particular, a generalized thrackle cannot have two non-adjacent edges that are disjoint. Although at first glance this property may appear to be the exact opposite of planarity, surprisingly, the two notions are not that different. In particular, for bipartite graphs, they are equivalent.

Theorem 4.1. [LPS97] *A bipartite graph can be drawn in the plane as a generalized thrackle if and only if it is planar.*

Fig. 3. A generalized thrackle with n vertices and $2n - 2$ edges

Using the fact that every graph G has a bipartite subgraph with at least $|E(G)|/2$ edges, we obtain that if a graph G of n vertices can be drawn as a generalized thrackle, then $|E(G)| = O(n)$.

Theorem 4.2. (Cairns-Nikolayevsky [CN00]) *Every generalized thrackle with n vertices has at most $2n - 2$ edges. This bound is sharp.*

A geometric graph G is a generalized thrackle if and only if it has no two disjoint edges. (The edges are supposed to be *closed* sets, so that two disjoint edges are necessarily non-adjacent.) One can relax this condition by assuming that G has no r pairwise disjoint edges, for some fixed $r \geq 2$. For $r = 2$, it was proved by Hopf-Pannwitz [HP34] that every graph satisfying this property has at most n edges, and that this bound is sharp. For $r = 3$, the first linear bound on the number of edges of such graphs was established by Alon-Erdős [AE89], which was later improved to $3n$ by Goddard-Katchalski-Kleitman [GKK96]. For general r, the first linear bound was established in [PT94]. The best currently known estimate is the following:

Theorem 4.3. (Tóth [T00]) *Every geometric graph with n vertices and no r pairwise disjoint edges has at most $2^9(r - 1)^2 n$ edges.*

It is likely that the dependence of this bound on r can be further improved to linear. If we want to prove the analogue of Theorem 4.3 for topological graphs, we have to make some additional assumptions on the structure of G, otherwise it is possible that any two edges of G cross each other.

Conjecture 4.4. (Conway's Thrackle Conjecture) *Let G be a simple topological graph of n vertices. If G has no two disjoint edges, then $|E(G)| \leq n$.*

For many related results, consult [LPS97], [CN00], [W71]. The next interesting open question is to decide whether the maximum number of edges of a simple topological graph with n vertices and no three pairwise disjoint edges is $O(n)$.

5   Locally Planar Graphs

For any $r \geq 3$, a topological graph G is called r-*locally planar* if G has no selfintersecting path of length at most r. Roughly speaking, this means that the embedding of the graph is planar in a neighborhood of radius $r/2$ around any vertex. In [PPTT02], we showed that there exist 3-locally planar geometric graphs with n vertices and with at least constant times $n \log n$ edges. Somewhat surprisingly (to us), Tardos [T02] managed to extend this result to any fixed $r \geq 3$. He constructed a sequence of r-locally planar geometric graphs with n vertices and a superlinear number of edges (approximately n times the $\lfloor r/2 \rfloor$ times iterated logarithm of n). Moreover, these graphs are bipartite and all of their edges can be stabbed by the same line.

The following positive result is probably very far from being sharp.

Theorem 5.1. [PPTT02] *The maximum number of edges of a 3-locally planar topological graph with n vertices is $O(n^{3/2})$.*

For geometric graphs, much stronger results are known.

Theorem 5.2. [PPTT02] *The maximum number of edges of a 3-locally planar x-monotone topological graph with n vertices is $O(n \log n)$. This bound is asymptotically sharp.*

For 5-locally planar x-monotone topological graphs, we have a slightly better upper bound on the number of edges: $O(n \log n / \log \log n)$. This bound can be further improved under the additional assumption that all edges of the graph cross the y-axis.

Theorem 5.3. [PPTT02] *Let G be an x-monotone r-locally planar topological graph of n vertices all of whose edges cross the y-axis. Then, we have $|E(G)| \leq cn(\log n)^{1/\lfloor r/2 \rfloor}$ for a suitable constant c.*

6   Strengthening Theorem 3.6

In this section, we outline the proof of

Theorem 6.1. *Every r-quasi-planar topological graph with n vertices has at most*

$$f_r(n) := C_r n (\log n)^{4(r-3)}$$

edges, where $r \geq 2$ and C_r is a suitable positive constant depending on r.

Let G be a graph with vertex set $V(G)$ and edge set $E(G)$. The *bisection width* $b(G)$ of G is defined as the minimum number of edges, whose removal splits the graph into two roughly equal subgraphs. More precisely, $b(G)$ is the minimum number of edges running between V_1 and V_2, over all partitions of the vertex set of G into two disjoint parts $V_1 \cup V_2$ such that $|V_1|, |V_2| \geq |V(G)|/3$. The *pair-crossing number* PAIR-CR(G) of a graph G is the minimum number of crossing pairs of edges in any drawing of G.

We need a recent result of Kolman and Matoušek [KM03], whose analogue for ordinary crossing numbers was proved in [PSS96] and [SV94].

Lemma 6.2. [KM03] *Let G be a graph of n vertices with degrees $d_1, d_2, \ldots, d_n$. Then we have*

$$b^2(G) \leq c(\log n)^2 \left(\text{PAIR-CR}(G) + \sum_{i=1}^{n} d_i^2 \right),$$

where c is a suitable constant.

We follow the idea of the original proof of Theorem 3.6. We establish Theorem 6.1 by double induction on r and n. By Theorem 7.1 (in the next section), the statement is true for $r = 3$ and for all n. It is also true for any $r > 2$ and $n \leq n_r$, provided that C_r is sufficiently large in terms of n_r, because then the stated bound exceeds $\binom{n}{2}$. (The integers n_r can be specified later so as to satify certain simple technical conditions.)

Assume that we have already proved Theorem 6.1 for some $r \geq 3$ and all n. Let $n \geq n_{r+1}$, and suppose that the theorem holds for $r+1$ and for all topological graphs having fewer than n vertices.

Let G be an $(r+1)$-quasi-planar topological graph of n vertices. For simplicity, we use the same letter G to denote the underlying abstract graph. For any edge $e \in E(G)$, let $G_e \subset G$ denote the topological graph consisting of all edges of G that cross e. Clearly, G_e is r-quasi-planar. Thus, by the induction hypothesis, we have

$$\text{PAIR-CR}(G) \leq \frac{1}{2} \sum_{e \in E(G)} |E(G_e)| \leq \frac{1}{2}|E(G)|f_r(n).$$

Using the fact that $\sum_{i=1}^{n} d_i^2 \leq 2|E(G)|n$ holds for every graph G with degrees $d_1, d_2, \ldots, d_n$, Lemma 6.2 implies that

$$b(G) \leq \left(c(\log n)^2 |E(G)| f_r(n) \right)^{1/2}.$$

Consider a partition of $V(G)$ into two parts of sizes $n_1, n_2 \leq 2n/3$ such that the number of edges running between them is $b(G)$. Obviously, both subgraphs induced by these parts are $(r+1)$-quasi-planar. Thus, we can apply the induction hypothesis to obtain

$$|E(G)| \leq f_{r+1}(n_1) + f_{r+1}(n_2) + b(G).$$

Comparing the last two inequalities, the result follows by some routine calculation.

7   Strengthening Theorem 3.7

The aim of this section is to prove the following stronger version of Theorem 3.7.

Theorem 7.1. *Every 3-quasi-planar topological graph with n vertices has at most Cn edges, for a suitable constant C.*

Let G be a 3-quasi-planar topological graph with n vertices. Redraw G, if necessary, without creating 3 pairwise crossing edges so that the number of crossings in the resulting topological graph $\tilde{G}$ is as small as possible. Obviously, no edge of $\tilde{G}$ crosses itself, otherwise we could reduce the number of crossings by removing the loop. Suppose that $\tilde{G}$ has two distinct edges that cross at least twice. A region enclosed by two pieces of the participating edges is called a *lens*. Suppose there is a lens ℓ that contains no vertex of $\tilde{G}$. Consider a *minimal* lens $\ell' \subseteq \ell$, by containment. Notice that by swapping the two sides of ℓ', we could reduce the number of crossings without creating any new pair of crossing edges. In particular, $\tilde{G}$ remains 3-quasi-planar. Therefore, we can conclude that

Claim 1. Each lens of $\tilde{G}$ contains a vertex.

We may assume without loss of generality that the underlying abstract graph of G is connected, because otherwise we can prove Theorem 7.1 by induction on the number of vertices. Let $e_1, e_2, \ldots, e_{n-1} \in E(G)$ be a sequence of edges such that $e_1, e_2, \ldots, e_i$ form a tree $T_i \subseteq G$ for every $1 \leq i \leq n - 1$. In particular, $e_1, e_2, \ldots, e_{n-1}$ form a spanning tree of G.

First, we construct a sequence of crossing-free topological graphs (trees), $\tilde{T}_1, \tilde{T}_2, \ldots, \tilde{T}_{n-1}$. Let $\tilde{T}_1$ be defined as a topological graph of two vertices, consisting of the single edge e_1 (as was drawn in $\tilde{G}$). Suppose that $\tilde{T}_i$ has already been defined for some $i \geq 1$, and let v denote the endpoint of e_{i+1} that does not belong to T_i. Now add to $\tilde{T}_i$ the piece of e_{i+1} between v and its first crossing with $\tilde{T}_i$. More precisely, follow the edge e_{i+1} from v up to the point v' where it hits $\tilde{T}_i$ for the first time, and denote this piece of e_{i+1} by $\tilde{e}_{i+1}$. If v' is a vertex of $\tilde{T}_i$, then add v and $\tilde{e}_{i+1}$ to $\tilde{T}_i$ and let $\tilde{T}_{i+1}$ be the resulting topological graph. If v' is in the interior of an edge e of $\tilde{T}_i$, then introduce a new vertex at v'. It divides e into two edges, e' and e''. Add both of them to $\tilde{T}_i$, and delete e. Also add v and $\tilde{e}_{i+1}$, and let $\tilde{T}_{i+1}$ be the resulting topological graph.

After $n - 2$ steps, we obtain a topological tree $\tilde{T} := \tilde{T}_{n-1}$, which (1) is crossing-free, (2) has fewer than $2n$ vertices, (3) contains each vertex of $\tilde{G}$, and (4) has the property that each of its edges is either a *full edge*, or a *piece of an edge* of $\tilde{G}$.

Let D denote the open region obtained by removing from the plane every point belonging to $\tilde{T}$. Define a *convex* geometric graph H, as follows. Traveling around the boundary of D in clockwise direction, we encounter two kinds of different "features": vertices and edges of $\tilde{T}$. Represent each such feature by a different vertex x_i of H, in clockwise order in convex position. Note that the same feature will be represented by several x_i's: every edge will be represented twice, because we visit both of its sides, and every vertex will be represented as many times as its degree in $\tilde{T}$. It is not hard to see that the number of vertices $x_i \in V(H)$ does not exceed $8n$.

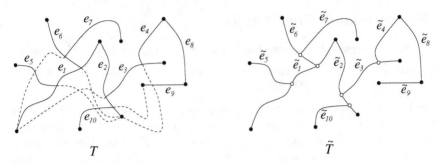

Fig. 4. Constructing $\tilde{T}$ from T

Next, we define the edges of H. Let E be the set of edges of $\tilde{G}\setminus T$. Every edge $e \in E$ may cross $\tilde{T}$ at several points. These crossing points divide e into several pieces, called *segments*. Let S denote the set of all segments of all edges $e \in E$. With the exception of its endpoints, every segment $s \in S$ runs in the region D. The endpoints of s belong to two features along the boundary of D, represented by two vertices x_i and x_j of H. Connect x_i and x_j by a straight-line edge of H. Notice that H has no loops, because if $x_i = x_j$, then, using the fact that $\tilde{T}$ is connected, one can easily conclude that the lens enclosed by s and by the edge of $\tilde{T}$ corresponding to x_i has no vertex of G in its interior. This contradicts Claim 1.

Of course, several different segments may give rise to the same edge $x_i x_j \in E(H)$. Two such segments are said to be of the *same type*. Observe that two segments of the same type cannot cross. Indeed, as no edge intersects itself, the two crossing segments would belong to distinct edges $e_1, e_2 \in E$. Since any two vertices of G are connected by at most one edge, at least one of x_i and x_j corresponds to an edge (and not to a vertex) of $\tilde{T}$, which together with e_1 and e_2 would form a pairwise intersecting triple of edges, contradicting our assumption that G is 3-quasi-planar.

Claim 2. H is a 3-quasi-planar convex geometric graph.

To establish this claim, it is sufficient to observe that if two edges of H cross each other, then the "features" of $\tilde{T}$ corresponding to their endpoints alternate in the clockwise order around the boundary of D. Therefore, any three pairwise crossing edges of H would correspond to three pairwise crossing segments, which is a contradiction.

A *segment* s is said to be *shielded* if there are two other segments, s_1 and s_2, of the same type, one on each side of s. Otherwise, s is called *exposed*. An *edge* $e \in E$ is said to be *exposed* if at least one of its segments is exposed. Otherwise, e is called a *shielded edge*.

In view of Claim 2, we can apply Theorem 3.4 [CP92] to H. We obtain that $|E(H)| \le 4|V(H)| - 10 < 32n$, that is, there are fewer than $32n$ different *types* of segments. There are at most two exposed segments of the same type, so the total number of exposed segments is smaller than $64n$, and this is also an upper bound on the number of exposed edges in E.

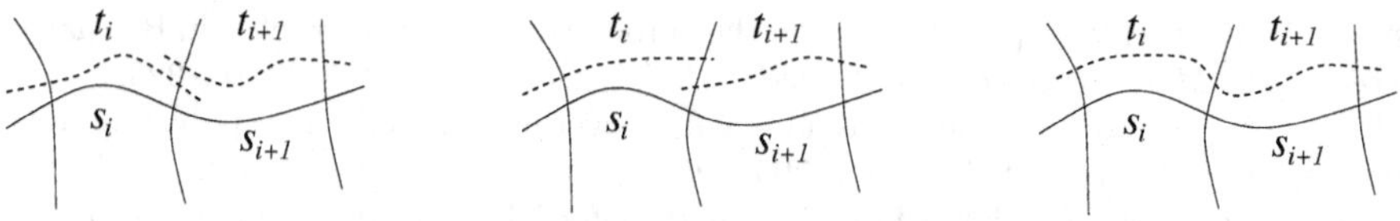

Fig. 5. t_i and t_{i+1} belong to the same edge

It remains to bound the number of shielded edges in E.

Claim 3. There are no shielded edges.

Suppose, in order to obtain a contradiction, that there is a shielded edge $e \in E$. Orient e arbitrarily, and denote its segments by $s_1, s_2, \ldots, s_m \in S$, listed according to this orientation. For any $1 \leq i \leq m$, let $t_i \in S$ be the (unique) segment of the same type as s_i, running closest to s_i on its left side.

Since there is no self-intersecting edge and empty lens in $\tilde{G}$, the segments t_i and t_{i+1} belong to the same edge $f \in E$, for every $i < m$ (see Fig. 5). However, this means that both endpoints of e and f coincide, which is impossible.

We can conclude that E has fewer than $64n$ elements, all of which are exposed. Thus, taking into account the $n - 1$ edges of the spanning tree T, the total number of edges of $\tilde{G}$ is smaller than $65n$.

References

AAPPS97. P. K. Agarwal, B. Aronov, J. Pach, R. Pollack, and M. Sharir, Quasi-planar graphs have a linear number of edges, *Combinatorica* **17** (1997), 1–9.

ACNS82. M. Ajtai, V. Chvátal, M. Newborn, and E. Szemerédi, Crossing-free subgraphs, in: *Theory and Practice of Combinatorics, North-Holland Math. Stud.* **60**, North-Holland, Amsterdam-New York, 1982, 9–12.

AE89. N. Alon and P. Erdős, Disjoint edges in geometric graphs, *Discrete Comput. Geom.* **4** (1989), 287–290.

BT98. Bollobás and A. Thomason, Proof of a conjecture of Mader, Erdős and Hajnal on topological complete subgraphs, *European J. Combin.* **19** (1998), 883–887.

BKV02. P. Braß, G. Károlyi, and P. Valtr, A Turán-type extremal theory for convex geometric graphs, in: *Discrete and Computational Geometry – The Goodman-Pollack Festschrift* (B. Aronov et al., eds.), Springer Verlag, Berlin, 2003, to appear.

CN00. G. Cairns and Y. Nikolayevsky, Bounds for generalized thrackles, *Discrete Comput. Geom.* **23** (2000), 191–206.

CP92. V. Capoyleas and J. Pach, A Turán-type theorem on chords of a convex polygon, *Journal of Combinatorial Theory, Series B* **56** (1992), 9–15.

GKK96. W. Goddard, M. Katchalski, and D. J. Kleitman, Forcing disjoint segments in the plane, *European J. Combin.* **17** (1996), 391–395.

HP34. H. Hopf and E. Pannwitz, Aufgabe Nr. 167, *Jahresbericht der deutschen Mathematiker-Vereinigung* **43** (1934), 114.

KM03. P. Kolman and J. Matoušek, Crossing number, pair-crossing number, and expansion, manuscript.

KSz96. J. Komlós and E. Szemerédi, Topological cliques in graphs II, *Combin. Probab. Comput.* **5** (1996), 79–90.

K84. A. V. Kostochka, Lower bound of the Hadwiger number of graphs by their average degree, *Combinatorica* **4** (1984), 307–316.

K88. A. V. Kostochka, Upper bounds on the chromatic number of graphs (in Russian), *Trudy Inst. Mat. (Novosibirsk), Modeli i Metody Optim.*, **10** (1988), 204–226.

KK97. A. V. Kostochka and J. Kratochvíl, Covering and coloring polygon-circle graphs, *Discrete Math.* **163** (1997), 299–305.

L84. F. T. Leighton, New lower bound techniques for VLSI, *Math. Systems Theory* **17** (1984), 47–70.

LPS97. L. Lovász, J. Pach, and M. Szegedy, On Conway's thrackle conjecture, *Discrete and Computational Geometry*, **18** (1997), 369–376.

M98. W. Mader, $3n - 5$ edges do force a subdivision of K_5, *Combinatorica* **18** (1998), 569–595.

Mc00. S. McGuinness, Colouring arcwise connected sets in the plane I, *Graphs & Combin.* **16** (2000), 429–439.

P99. J. Pach, Geometric graph theory, in: *Surveys in Combinatorics, 1999 (J. D. Lamb and D. A. Preece, eds.)*, London Mathematical Society Lecture Notes **267**, Cambridge University Press, Cambridge, 1999, 167–200.

PPST02. J. Pach, R. Pinchasi, M. Sharir, and G. Tóth, Topological graphs with no large grids, to appear.

PPTT02. J. Pach, R. Pinchasi, G. Tardos, and G. Tóth, Geometric graphs with no self-intersecting path of length three, in: *Graph Drawing* (M. T. Goodrich, S. G. Kobourov, eds.), *Lecture Notes in Computer Science*, **2528**, Springer-Verlag, Berlin, 2002, 295–311.

PRT02. J. Pach, R. Radoičić, and G. Tóth, On quasi-planar graphs, in preparation.

PRTT02. J. Pach, R. Radoičić, G. Tardos, and G. Tóth, Graphs drawn with at most 3 crossings per edge, to appear.

PSS96. J. Pach, F. Shahrokhi, and M. Szegedy, Applications of the crossing number, *Algorithmica* **16** (1996), 111–117.

PT97. J. Pach and G. Tóth, Graphs drawn with few crossings per edge, *Combinatorica* **17** (1997), 427–439.

PT94. J. Pach and J. Törőcsik, Some geometric applications of Dilworth's theorem, *Discrete Comput. Geom.* **12** (1994), 1–7.

PR02. R. Pinchasi and R. Radoičić, On the number of edges in geometric graphs with no self-intersecting cycle of length 4, to appear.

RST93. N. Robertson, P. Seymour, and R. Thomas, Hadwiger's conjecture for K_6-free graphs, *Combinatorica* **13** (1993), 279–361.

SV94. O. Sýkora and I. Vrťo, On VLSI layouts of the star graph and related networks, *Integration, The VLSI Journal* **17** (1994), 83-93.

T02. G. Tardos, On the number of edges in a geometric graph with no short self-intersecting paths, in preparation.

T84. A. Thomason, An extremal function for contractions of graphs, *Math. Proc. Cambridge Philos. Soc.* **95** (1984), 261–265.

T00. G. Tóth, Note on geometric graphs, *J. Combin. Theory, Ser. A* **89** (2000), 126–132.

T01. A. Thomason, The extremal function for complete minors, *J. Combin. Theory Ser. B* **81** (2001), 318–338.

V98. P. Valtr, On geometric graphs with no k pairwise parallel edges, *Discrete and Computational Geometry* **19** (1998), 461–469.

W71. D. R. Woodall, Thrackles and deadlock, in: *Combinatorial Mathematics and its Applications (Proc. Conf., Oxford, 1969)*, Academic Press, London, 1971, 335–347.

On the Size of a Radial Set

Rom Pinchasi

Massachusetts Institute of Technology
77 Massachusetts Avenue, Cambridge, MA 02139-4307
room@math.mit.edu

Abstract. Let G be a finite set in the plane. A point $x \notin G$ is called a *radial point* of G, if every line through x and a point from G includes at least two points of G. In this paper we show that for any line l not passing through the convex hull of G there are at most $(\frac{9}{10} + o(1))|G|$ radial points separated from G by l. As a consequence we prove two nice geometric applications in the plane.

1 Introduction

Definition 1 *Let G be a finite set of points in the real affine plane. We say that a point $x \notin G$ is a* radial point *of G, if every line through x and a point of G includes at least two points of G. A set of points R is called a* radial set *of G if every $x \in R$ is a radial point of G.*

The main result in this paper is the following theorem.

Theorem 2 *Let G be a set of n points in the plane, not contained in a line. Let R be a radial set of G. If R and G can be separated by a line then $|R| < (\frac{9}{10} + o(1))n$.*

It is very significant the the constant multiplier in front of n in the formulation of Theorem 2 is strictly less than 1 (at least when n is large enough). The importance of this will soon become clear when we discuss possible applications of this theorem. In a sense, bringing the constant multiplier to be less than 1 is the principle contribution of this paper and the reason for the cumbersome proof.

A line l is said to be *determined* by a set of points, if it contains at least two points from this set. For every set of non-collinear points in the plane, an old and celebrated theorem conjectured by Sylvester ([S93]) and proved by Gallai ([G44]), guarantees the existence of a line passing through exactly two points of G. In [F96], K. Fukuda conjectured that Gallai-Sylvester theorem can be generalized in the following nice way (this conjecture appears also in [DSF98]).

Conjecture 3 (Da-Silva, Fukuda) *Let G be a set of n green points and m red points in the plane. Assume that the green points can be separated by a line from the red points. If $|m - n| \leq 1$, then G determines a line which includes exactly one green point and exactly one red point.*

J. Akiyama and M. Kano (Eds.): JCDCG 2002, LNCS 2866, pp. 233–245, 2003.
© Springer-Verlag Berlin Heidelberg 2003

It turned out that this conjecture is in fact false, at least for small values of m, n. L. Finschi and K. Fukuda found a counterexample for $n = 4, m = 5$.

A weaker form of Conjecture 3 was proved by Pach and Pinchasi in [PP00]. It is shown there that any set of n red points and n green points, which is not contained in a line, determines a bichromatic line passing through at most two red points and at most two green points.

Definition 4 *Let G be a set of red and green points in the plane. An almost red line is a line, determined by G, which includes exactly one green point and at least one red point. Similarly we define an almost green line.*

In terms of this definition, Conjecture 3 is equivalent to finding a line which is almost green and almost red at the same time.

One corollary of Theorem 2 is that if G satisfies the conditions in Conjecture 3 and $|G|$ is large enough, then G determines an almost green line and also an almost red line (but those two lines may be different).

Corollary 5 *Let R be a set of m red points and B a set of n green points in the plane. Assume that B is not contained in a line and that R and B can be separated by a line. If $n \geq n(\epsilon)$ is large enough and $m \geq (\frac{9}{10} + \epsilon)n$, where $\epsilon > 0$ is arbitrary, then $G = R \cup B$ determines an almost red line.*

Proof: If G does not determine an almost red line, then R is a radial set for B. Therefore, by Theorem 2, $m < (\frac{9}{10} + o(1))n$. ∎

Corollary 6 *Let R be a set of m red points and B a set of n green points, such that none of these two sets is contained in a line. Assume that R and B can be separated by a line and that $|m - n| \leq 1$. Then for large enough n there must exist an almost red line and also an almost green line.* ∎

As for constructions, it is easy to come up with a construction of a set G and a radial set R for G so that R and G can be separated by a line and $|R| = \frac{1}{2}|G|$. For this, take G to be the set of vertices of a regular $2n$-gon, and let R be the n points at infinity that correspond to the directions of the edges of this regular $2n$-gon.

We should also mention here a closely related result by Pach and Sharir. In [PS99], Pach and Sharir show that without any restrictions, if R is a radial set for G, then $|R| = O(|G|)$. The (small) example, shown in Figure (1), in which the white points form a radial set for the black points, shows that in general $|R|$ can be larger than $|G|$, if we don't require that R and G can be separated by a line.

The proof of Theorem 2 is based on the careful analysis of the *flip array* (also called *allowable sequences*) associated with the set $R \cup G$. The method of flip arrays, invented by Goodman and Pollack, is described in Section 2. For further discussion of this method consult [GP93].

The proof of Theorem 2 has two parts. In the first part, described in Section 3, we show that $|R| \leq \frac{9}{5}n$. In the second part, described in Section 4, we further improve this bound and obtain $|R| < (\frac{9}{10} + o(1))n$. But first let us recall the notion of a flip array of a finite planar set.

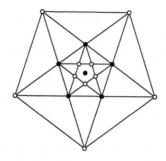

Fig. 1. a 6-point set with a 10-point radial set

2   Flip Arrays – Notations

Let G be a set of n points in the plane. We assume that the points of G have pairwise different x-coordinates, and we number them from 1 to n according to the order of their x coordinates. A *flip array* of G, usually denoted by S_G, is a sequence of permutations in S_n. Each permutation is obtained from G by projecting its points on a directed line l (which is in general position with respect to G), thus getting an ordering of the points according to the order of their projections on l. S_G are all possible permutations obtained this way from G ordered by the slope of the line l on which they were obtained. For a detailed description see Section 2 of [P03].

The first permutation is always the identity, while the last one is $(n, n - 1, \ldots, 1)$.

It is important to note that we think of a permutation P as a sequence of n elements namely, $(P(1), \ldots, P(n))$. We then say that the *element* $P(i)$ is at the *place* i in the permutation P. The relative order of two elements i, j depends on whether $P^{-1}(i)$ is greater or less than $P^{-1}(j)$. If $P^{-1}(i) < P^{-1}(j)$ we say that i is *to the left* of j and that j is *to the right* of i.

A *block* in a permutation P is a sequence of consecutive elements in P. We some times refer to the block as a *region* (containing certain *places* in a permutation) and some times we refer to its *content* (the *elements* which are in that region). We say that a block B is *monotone increasing* if the elements in that block form a monotone increasing sequence from left to right. We define a *monotone decreasing* block similarly.

Notation 7 *Let* $1 \le a < b \le n$. *We denote by* $[a, b]$ *the block which consists of the places* $a, a+1, \ldots, b$ *in a general permutation (considered as a sequence of* n *elements).*

Every permutation in S_G is obtained from its predecessor by flipping a monotone increasing block of elements.

Assume $\sigma \in S_G$ is a permutation in the flip array S_G. For two elements $1 \le x < y \le n$, we say that x and y *change order* in σ, if in σ x is to the right of y (that is $\sigma^{-1}(x) > \sigma^{-1}(y)$) and in the permutations which are prior to σ in S_G, x is to the left of y.

If $P_1, P_2 \in S_G$ are two consecutive permutation so that P_2 is obtained from P_1 by a flip F, then we denote $P_F^- = P_1$ and $P_F^+ = P_2$.

We say that two elements $x, y \in \{1, 2, \ldots, n\}$ *change order* in a flip F, if x and y change order in P_F^+.

Let S_G be a flip array of a finite set G. For $P_1, P_2 \in S_G$, we say that P_1 is *previous* to P_2, if P_1 comes before P_2 in S_G. We then say that P_2 is later than P_1.

Similarly, we say that a flip F_1 is *previous* to a flip F_2 if $P_{F_1}^+$ is previous to $P_{F_2}^+$. We then say that F_2 is *later* than F_1. We say that a flip F *occurs between* a flip F_1 and a flip F_2 (where F_1 is previous to F_2), if F is later than F_1 and F_2 is later than F. In this case we sometimes say that F is between F_1 and F_2.

For $P_1, P_2 \in S_G$. We denote by $[P_1, P_2]$ the permutations in S_G which are not previous to P_1 and not later than P_2. For a flip F and $P_1, P_2 \in S_G$, we say that F is *between* P_1 and P_2 if there are two consecutive permutations $\sigma, \sigma' \in [P_1, P_2]$ so that σ' is obtained from σ by the flip F.

We need the following two simple observations.

Observation 8 *Let S_G be a flip array of a set G. Every two elements change order at some point (permutation) in the flip array S_G. From that point on (i.e., in all permutations that come afterwards in S_G) they are always in inverted order.*

Observation 9 *Let S_G be a flip array of a set G of n points in the plane. If a line M determined by G is represented by a flip of the block $[a, b]$, then there are exactly $a - 1$ points of G in one open half plane bounded by M, and $n - b$ points in the other half plane bounded by M.*

3   Part I of the Proof of Theorem 2

Let $S_{R \cup G}$ be a flip array of the set $R \cup G$. We consider the points of R to be red points, and the points of G to be green points. We can therefore consider red elements and green elements when analyzing the flip array $S_{R \cup G}$. Since R and G can be separated by a line, we may assume that in the initial state of the flip array all the red elements are to the left of the green elements. In other words, the red elements are $\{1, 2, \ldots, |R|\}$ and the green elements are $\{|R| + 1, |R| + 2, \ldots, |R| + |G|\}$. Therefore, if T is a flip whose block B includes red elements as well as green elements, then in P_T^- the red elements in B are to the left of the green elements in B.

Observe that the fact that R is a radial set of G translates, in terms of the flip array $S_{R \cup G}$, to that there is no flip whose block includes exactly one green element and at least one red element.

Definition 10 *A flip T is called* interesting *if the block of T includes the leftmost green element in P_T^- and at least one red element. If T is an interesting flip, then the* red size of T, *which is denoted by $rs(T)$, is the number of red elements in the block of T.*

Claim 11 *Every interesting flip T represents a line which includes an edge of* conv G.

Proof: Let l be the line which is represented by T. As T is an interesting flip, the block of T includes a green element and a red element. Hence, it includes at least two green elements, which means that l is determined by G. The green elements in the block of T are the leftmost green elements in P_T^-. This means that if we project G on a line M which is perpendicular to l, then the images, under this projection, of the points of $G \setminus l$ are on one side of the image of l on M. This shows that l supports conv G. ∎

Claim 12 *Every red element r takes part in exactly one interesting flip T. T is the first flip whose block includes r and a green element.*

Proof: Let r be any red element. Let T be the first flip whose block B includes r and some green element. Such a flip must exist since r changes order with all green elements during the flip array. In P_T^-, all the green elements are to the right of r and therefore there are no green elements to the left of the block B. It follows that B includes the leftmost green element in P_T^-. r is included in B and hence T is an interesting flip.

Assume that r takes part in an interesting flip T' which is later than T. Let b denote the leftmost green element included in B in the permutation P_T^-. Let B' denote the block of T'. Clearly, b is not included in B', for b and r change order already in T. It follows that in P_T^-, b is to the left of the block B' (for it is to the left of r). Therefore, B' does not include the leftmost green element in P_T^-. This is a contradiction to the assumption that T is an interesting flip. ∎

Assignment of Green Elements to Every Interesting Flip

Fix T, an interesting flip. In this section we describe how to assign to T a list of size $rs(T)$ of distinct green elements. Let $B = [a, b]$ denote the block of T. Denote by

$$r_{rs(T)}, r_{rs(T)-1}, \ldots, r_1$$

the red elements in B, from left to right, as they appear in P_T^+.

In P_T^+, B includes the leftmost green elements. B does not include all of the green elements, for otherwise G contained in a line (which is represented by the flip T). Therefore, $r_{rs(T)}$ and some green element must have their order changed at a flip which occurs after T. In particular, there must be a flip, later than T, whose block includes elements from B.

Claim 13 *Let F be the first flip later than T whose block includes $r_{rs(T)}$. If T' is an interesting flip which is later than T. Then T' is also later than F.*

Proof: Assume to the contrary that T' is not later than F. Let $b_1, \ldots, b_k$ (note that $k \geq 2$) denote the green elements in the block of T. In P_T^+, $b_1, \ldots, b_k$ are the leftmost green elements. In every $P \in [P_T^+, P_F^-]$, $r_{rs(T)}$ remains untouched and does not change order with any other element. It follows that in every

$P \in [P_T^+, P_F^-]$ $b_1, \ldots, b_k$ are the leftmost green elements. In particular, this is true for $P_{T'}^-$. T' is an interesting flip and hence its block includes the two leftmost green elements in $P_{T'}^-$. Those two elements must be from $\{b_1, \ldots, b_k\}$. This is a contradiction for every two elements from $\{b_1, \ldots, b_k\}$ change order already in T. ∎

Claim 14 *The first element of B which takes part in a flip which is later than T is r_1 alone.*

Proof: We recall that $B = [a, b]$ and r_1 is at the place b (which is the rightmost place in B) in P_T^+. Let T' be the first flip later than T whose block $B' = [a', b']$ includes an element from B. Clearly, B and B' share exactly one place in common, for otherwise there would be two elements which change order both in T and in T'. Therefore, either $a' = b$ and $b' > b$ in which case r_1 is included in B', or $b' = a$ and $a' < a$. Assume to the contrary that the latter case happens. In P_T^+, the element at the place a (the leftmost element in B) is the leftmost green element. This is true also in $P_{T'}^-$, because T' is the first flip later than T which includes elements from B. Therefore B' includes one green element (which is the leftmost green element) and some nonzero number of red elements, contradicting our assumption that R is a radial set. ∎

Claim 15 *The element which is next to the right of r_1 in P_T^+ is a green element.*

Proof: Assume to the contrary that the element next to the right of r_1 in P_T^+ is a red element which we denote by r. Let $b_1, \ldots, b_k$ denote the green elements in B, from left to right, as they appear in P_T^-. In P_T^-, r is next to the right of b_k. It follows that r already changed order with each one of $b_1, \ldots, b_k$. However, $b_1, \ldots, b_k$ are the leftmost green elements in P_T^- and hence those are the only green elements with which r changed order until right before T. Since R is a radial set of G, at every flip in which r changes order with some green element it changes order with at least two green elements. It follows that there are at least two green elements in B which change order already before T, which is a contradiction. ∎

Let x_0^T denote the green element next to the right of r_1 in P_T^+. In P_T^+, x_0^T is the leftmost green element which is not in the block of T. Let T_1 be the first flip later than T, whose block B_1 includes r_1. We show that in $P_{T_1}^-$ the element x, next to the right of r_1, is a green element.

Indeed, if $x = x_0^T$, then we are done. Assume $x \neq x_0^T$. In P_T^+, x is to the right of x_0^T for it is to the right of r_1 and x_0^T is the the green element next to the right of r_1 in P_T^+. In $P_{T_1}^-$, x_0^T is to the right of x, for otherwise, x_0^T is to the left of r_1 which means that r_1 and x_0^T changed order between T and T_1, contradicting the assumption that T_1 is the first flip later than T whose block includes r_1. It follows now that x and x_0^T change order before T_1 and that $x > x_0^T$. Since x_0^T is green and $x > x_0^T$, we conclude that that x is also green.

Since in $P_{T_1}^-$, the element next to the right of r_1 is green, it follows that all the elements in B_1 except for r_1 are green. Let x_1^T denote the rightmost element of B_1 in $P_{T_1}^-$. Observe that in $P_{T_1}^+$, x_1^T is at the place b and it is the leftmost green element which is not in the block of T.

We now inductively define $x_2^T, \ldots, x_{rs(T)}^T$ and $T_2, \ldots, T_{rs(T)}$. Let $k \geq 1$ and assume we already defined $x_1^T, \ldots, x_k^T$ and $T_1, \ldots, T_k$. Assume also that x_i^T is at the place $b - k + 1$ in $P_{T_k}^+$ and it is the leftmost green element which is not in the block of T. Moreover, assume that in $P_{T_k}^+$, the content of the block $B' = [a, b-k]$ is the same as right after T.

The element at the place $b - k$ in $P_{T_k}^+$ is r_{k+1}. Similar to Claim 14, the first element of B' which takes part in a flip which is later than T_k is r_{k+1} alone. Let T_{k+1} be the first flip, which is later than T_k, whose block includes r_{k+1}. Let B_{k+1} denote the block of T_{k+1}. Then $B_{k+1} = [b-k+1, c]$ for some $c > b-k+1$. Again we can show (since in $P_{T_k}^+$ the element next to the right of r_{k+1} is a green element, namely x_k^T) that the rightmost element of B_{k+1} in $P_{T_{k+1}}^-$ is green. We denote that element by x_{k+1}^T. In $P_{T_{k+1}}^+$ x_{k+1}^T is at the place $b - k + 1$ and it is the leftmost green element which is not in the block of T. The content of the block $[a, b - k]$ in $P_{T_{k+1}}^+$ is the same as in P_T^+. Thus, the inductive construction is completed.

Next, we define x_{-1}^T to be the leftmost green element (in B) in P_T^+. Note that by Claim 13, $T_1, \ldots, T_{rs(T)}$ occur before the first interesting flip which is later than T. Moreover, for every $0 \leq k \leq rs(T)$, x_k^T is the leftmost green element in $P_{T_k}^+$ which is not in the block of T.

Claim 16 $x_0^T < x_1^T < x_2^T < \ldots < x_{rs(T)}^T$, and every two of them change order in one of the permutations in $[P_T^+, P_{T_{rs(T)}}^+]$.

Proof: It is enough to show that for every $0 \leq i < rs(T)$, $x_i^T < x_{i+1}^T$ and that they change order in some $\sigma \in [P_T^+, P_{T_{rs(T)}}^+]$. Denote for convenience $T_0 = T$.

We first show that $x_i^T \neq x_{i+1}^T$. Assume to the contrary that $x_i^T = x_{i+1}^T$. Let $B_{i+1} = [c, d]$ be the block of T_{i+1}. In $P_{T_{i+1}}^-$, r_{i+1} is at the place c and $x_i^T = x_{i+1}^T$ is at the place d. $d > c + 1$, for otherwise B_{i+1} includes only two elements one green and one red, contradicting our assumption that R is a radial set of G. Let x denote the green element at the place $c+1$ in $P_{T_{i+1}}^-$. In $P_{T_i}^+$, x_i^T is the leftmost green element not in B, and therefore is to the left of x. In $P_{T_{i+1}}^-$, x is to the left of x_i^T (which is then at the place d). It follows that x and x_i^T change order before T_{i+1}. This is a contradiction for their order changes in T_{i+1}. Hence $x_i^T \neq x_{i+1}^T$.

In $P_{T_i}^+$, x_i^T is the leftmost green element which is not in B. In particular, it is to the left of x_{i+1}^T. In $P_{T_{i+1}}^+$, x_{i+1}^T is the leftmost green element which is not in B. In particular it is to the left of x_i^T. This shows that $x_i^T < x_{i+1}^T$ and that x_i^T and x_{i+1}^T change order in some $\sigma \in [P_{T_i}^+, P_{T_{i+1}}^+]$. ∎

We are now ready to define the list of $rs(T)$ green elements which we assign to T. We take this list to be

$$x_{-1}^T, x_1^T, x_2^T, \ldots, x_{rs(T)-1}^T.$$

Definition 17 *Define* $Green(r_j) = x_j^T$ *for* $1 \leq j \leq rs(T)-1$ *and* $Green(r_{rs(T)}) = x_{-1}^T$.

Remark: $Green(r_j)$ is well defined, for every red element takes part in exactly one interesting flip (Claim 12).

Counting the Number
of Assigned Green Elements

Claim 18 *Let T be an interesting flip. Then x_{-1}^T is not a member of any list which is assigned to any interesting flip T' which is later than T.*

Proof: Let T' be an interesting flip, later than T. In P_T^+, x_{-1}^T is the leftmost green element. Denote $b = x_{-1}^T$. Let F be any flip, later than T, whose block B' includes b. Since R is a radial set of G, B' includes at least one more green element b'. In P_T^+, b' is to the right of b. Since b and b' change order at F, it follows that $b < b'$ and that in P_F^+, b' is to the left of b.

In particular, it follows that in $P_{T'}^+$, b is not the leftmost green element, and hence $b \neq x_{-1}^{T'}$. Assume $b = x_j^{T'}$ for some $1 \leq j \leq rs(T') - 1$. By Claim 16, $x_{j-1}^{T'} < b$ and they change order at a flip F' which is later than T'. This is a contradiction for, as we have seen in the previous paragraph, we should have $b < b'$, for every green element b' in the block of F'. ∎

Claim 19 *Let T be an interesting flip. If $rs(T) \geq 3$ then none of $x_1^T, \ldots, x_{rs(T)-2}^T$ is a member a list which is assigned to any interesting flip T', which later than T.*

Proof: Let T' be an interesting flip which occurs later than T. Let $1 \leq i \leq rs(T) - 2$ and assume that $x_i^T = x_j^{T'}$ where $j = -1$ or $1 \leq j \leq rs(T') - 1$. By Claim 16, x_{i+1}^T and x_{i+2}^T change order with x_i^T and move to its left already before T'. We consider two possible cases.

Case 1. $j \neq -1$. In some $\sigma \in S_{R \cup G}$ which is later than or equal to $P_{T'}^+$, $x_j^{T'}$ is the leftmost green element which is not in the block of T (recall the construction of $x_0^{T'}, \ldots, x_{rs(T')}^{T'}$). Hence x_{i+1}^T and x_{i+2}^T must both be in the block of T' which is impossible since they change order already before T'.

Case 2. $j = -1$. Then $x_j^{T'}$ is leftmost green element in $P_{T'}^+$. This is a contradiction for x_{i+1}^T and x_{i+2}^T are both to the left of $x_j^{T'}$ in $P_{T'}^+$. ∎

Claim 20 *Let T be an interesting flip with $rs(T) \geq 2$. $x_{rs(T)-1}^T$ may be a member of at most two more lists that are assigned to interesting flips which are later than T.*

Proof: Assume to the contrary that $x_{rs(T)-1}^T$ in included in the lists assigned to S_1, S_2, and S_3, which are interesting flips that occur in that order later than T.

Denote $x = x_{rs(T)-1}^T$ and $b = x_{rs(T)}^T$. By Claim 16, $x < b$, and they change order already before S_1. Therefore, in $P_{S_1}^-$, b is to the left of x and this is true also for every permutation which is later than $P_{S_1}^-$.

Assume by contradiction that $x = x_i^{S_1} = x_j^{S_2} = x_k^{S_3}$.

First note that none of i, j, k equals -1. This is because x cannot be the leftmost green element once b moves to its left (which happens before S_1).

Since $i \neq -1$ then $x = x_i^{S_1}$ is, in some $\sigma \in S_{R \cup G}$ which is later than S_1, the leftmost green element which is not in the block of S_1. It follows that b is included in the block of S_1. Similarly, b is included also in the blocks of S_2 and S_3. By Claim 11, S_1, S_2 and S_3 represent three lines which include three edges of conv G. This is a contradiction, for there is no point which belongs three different edges of conv G. ∎

Claim 21 *Let T be an interesting flip with $rs(T) \geq 2$. If $x_{rs(T)-1}^T = x_j^S$ for some interesting flip S which is later than T, then $j = 1$.*

Proof: In the proof of Claim 20 we saw that $j \neq -1$. Assume that $j > 1$. We first show that x_1^S cannot be in the block B of T. For assume it is, then by Claim 16, $x_0^S < x_1^S$ and they change order after S. Therefore, in P_T^-, x_0^S is to the left of x_1^S. Since T is an interesting flip, B includes the leftmost green elements in P_T^-. It follows that $x_0^S \in B$ as well. This is a contradiction for x_0^S and x_1^S cannot change order in T, as they change order in some permutation which is later than P_S^+.

Let $\sigma = P_{T_{rs(T)-1}}^+$ (recall the definition of $T_1, \ldots, T_{rs(T)}$). σ is previous to the permutation P_S^- and in σ $x_{rs(T)-1}^T$ is the leftmost green element which is not in the block of T. In particular, x_1^S is to the right of $x_{rs(T)-1}^T$ in σ. By Claim 16, $x_{rs(T)-1}^T = x_j^S > x_1^S$. It follows that x_1^S and x_j^S change order already before S, contradicting Claim 16. ∎

Definition 22 *A green element x is called* labeled *if it is included in a list assigned to some interesting flip T. A labeled element is called* A-labeled *if it is x_{-1}^T for some T, and it is called* B-labeled *otherwise.*

Claim 23 *The number of labeled elements is at least $\frac{5}{9}|R|$.*

Proof: Let T be an interesting flip. From Claims 18, 19, and 21, it follows that except for x_1^T and $x_{rs(T)-1}^T$, every other green element which is included in the list assigned to T is not a member of any list assigned to any other interesting flip $S \neq T$. x_1^T and $x_{rs(T)-1}^T$ may be included in at most three different lists each.

Denote by $T^1, T^2, \ldots, T^l$ the interesting flips in the flip array $S_{R \cup G}$. By Claim 12, every red element takes part in exactly one interesting flip. Therefore,

$$\sum_{i=1}^{l} rs(T^i) = |R|.$$

Define

$$l_1 = \#\{1 \leq i \leq l \,|\, rs(T^i) = 1\},$$
$$l_2 = \#\{1 \leq i \leq l \,|\, rs(T^i) = 2\}.$$

If T is an interesting flip and $rs(T) \geq 3$ then $x_{-1}^T, x_2^T, \ldots x_{rs(T)-2}^T$ appear only in the list assigned to T. Regardless of $rs(T)$, x_1^T and $x_{rs(T)-1}^T$ may be members of at most two more lists assigned to other interesting flips. It follows that the number of labeled elements is at least

$$l_1 + 1\frac{1}{3}l_2 + \sum_{rs(T)\geq 3} (rs(T) - 2 + \frac{2}{3}),$$

which is greater than or equal to

$$\frac{1\frac{2}{3}}{3}(l_1 + 2l_2 + \sum_{rs(T)\geq 3} rs(T)) = \frac{5}{9}\sum_{i=1}^{l} rs(T^i) = \frac{5}{9}|R|.$$

∎

4   Part II of the Proof of Theorem 2

In Section 3, we saw that the set of labeled elements is of size at least $\frac{5}{9}|R|$. Roughly speaking, we are going to show that at most $\frac{1}{2} + o(1)$ of the green elements are labeled and that will clearly imply Theorem 2.

Definition 24 *If x is B-labeled and T is the first interesting flip so that $x = x_j^T$ for some $1 \leq j \leq rs(T) - 1$, then we say that x is labeled at $P_{T_j}^+$ (recall the definition of $T_1, \ldots, T_{rs(T)}$).*
If x is A-labeled, say $x = x_{-1}^T$ for some interesting flip T, then we say that x is labeled at P_T^+.

Claim 25 *If x is B-labeled, then x cannot belong to any block of an interesting flip.*

Proof: Suppose that $x = x_j^T$ ($j \neq -1$) and that x belongs to the block of an interesting flip S. Clearly $S \neq T$. We split into two cases.
Case 1. S occurs before T. By Claim 16, $x_0^T < x_j^T$ and they change order after T. It follows that in P_S^-, x_0^T is to the left of x_j^T and therefore also belongs to the block of S. This is a contradiction, for x_0^T and x_j^T cannot change order in S, as they change order after T.
Case 2. S occurs after T. By Claim 16, $x_j^T < x_{rs(T)}^T$ and they change order before S. It follows that in P_S^-, $x_{rs(T)}^T$ is to the left of x_j^T and therefore also belongs to the block of S. This is a contradiction, for $x_{rs(T)}^T$ and x_j^T cannot change order in S, as they change order before S. ∎

Claim 26 *Assume that x is A-labeled, $x = x_{-1}^T$ for an interesting flip T. Let F be any flip which occurs before T, then x is the rightmost element of the block of F in P_F^-.*

Proof: Assume not. Let b be the rightmost element of the block of F in P_F^-. Since x is a green element, so is b. In P_F^+, b moves to the left of x. This is a contradiction for $x = x_{-1}^T$ is the leftmost green element in P_T^+. ∎

Claim 27 *Suppose that x is B-labeled, say $x = x_j^T$ where $j \neq -1$ and T is an interesting flip. Let F be a flip which occurs before T and whose block B_F includes x. Then in P_F^-, x is in one of the two rightmost places in B_F. Moreover, let w be the rightmost element of B_F in P_F^-. If $w \neq x$ then w cannot be B-labeled in any permutation which is later than P_F^+.*

Proof: Assume to the contrary that x is not in one of the two rightmost places in B_F. Then the green elements y, z in the two rightmost places of B_F, move to the left of x by the flip F. Since $x = x_j^T$, then for some σ which is later than T, x is the leftmost green element which is not in the block of T. Therefore y and z must be in the block of T. This is a contradiction for y and z change order already in F.

To see the second assertion of the claim, assume that x is not the leftmost element of B_F in P_F^- and that w is B-labeled at a permutation which is later than P_F^+. w moves to the left of x by the flip F. Let σ be the permutation at which x is labeled. We know that σ is later than F. In σ, x is the leftmost green element which is not in the block of T. Therefore w must belong to the block of T, contradicting Claim 25. ∎

Claim 28 *Suppose that x is B-labeled, say $x = x_j^T$ where $j \neq -1$ and T is an interesting flip. Then there are at most two A-labeled elements that move to the left of x before it is labeled.*

Proof: Assume to the contrary that y_1, y_2, y_3 are three A-labeled elements which move to the left of x before it is labeled. Let σ be the permutation at which x is labeled. In σ, x is the leftmost green element which is not in the block of T. Therefore y_1, y_2, y_3 belong to the block of T. Without loss of generality assume that in P_T^-, y_1, y_2, y_3 are in that order from left to right in the block of T. y_2 cannot be A-labeled before T, because it moves to the left of y_1 by the flip T. y_2 cannot be A-labeled after T, because y_3 moves to the left of y_2 by the flip T. Clearly, y_2 is not A-labeled in T, for it is not the leftmost green element in P_T^+. This contradicts the assumption that y_2 is A-labeled. ∎

Fix k, a positive integer. Let $r_1, \ldots r_k$ be the first k red elements that take part in an interesting flip (if $r < r'$ are two red elements which take part in the same interesting flip T, we regard r as a red element which takes part in T before r' does).

Define a graph $\mathcal{H}$ whose vertices are the green elements. Connect two green elements, x and y, by an edge if there is a flip F whose block B_F includes at least one of $r_1, \ldots, r_k$ and in P_F^-, x and y are the two rightmost elements of B_F. Denote by A the set of A-labeled vertices, by B the set of B-labeled vertices, and by Z the set of vertices which are not labeled.

We recall the definition of $Green(r)$ (Definition 17) where r is a red element.

Claim 29 *Let x be a green element which is labeled at a permutation σ. Let $1 \leq i \leq k$ and suppose that x and r_i change order in a permutation which is later than σ. Then for some $j < i$, $Green(r_j) = x$.*

Proof: Let r be the red element such that $Green(r) = x$. It is enough to show that r takes part in an interesting flip before r_i does. In σ, r is to the right of x but r_i is to the left of x, because r_i and x change order in a permutation which is later than σ. We recall (Claim 12) that the first flip in which r gets flipped together with a green element is an interesting flip.

Case 1. x is A-labeled. In σ, x is the leftmost green element. It follows now that r takes part in an interesting flip before r_i does.

Case 2. x is B-labeled, and say $x = x_j^T$ where $j \neq -1$. In σ, x is the leftmost green element which is not in the block of T. Since in σ r_i is to the left of x and r is to the right of x, then r takes part in an interesting flip which is previous or equal to T and r_i takes part in an interesting flip which is later than or equal to T. If r and r_i both take part in T, then clearly $r < r_i$ (as r is to the right of r_i in σ). In either case it follows that r takes part in an interesting flip before r_i does. ∎

Corollary 30 (and a definition) *There are at most k green elements which get flipped together with some r_i after they are labeled. Denote the set of those elements by C.*

Notation 31 *Let $A' = A \setminus C$ and $B' = B \setminus C$.*

Corollary 32 *Assume F is a flip whose block includes some r_i where $1 \leq i \leq k$ and an element $x \in A' \cup B'$. Let σ be the permutation at which x is labeled. Then σ is later than P_F^+.*

By Claim 26, there are no edges between vertices of A'. By Claim 27, there are no edges between vertices of B'. By Claim 28, every vertex of B' is connected to at most two vertices of A'. It follows that there are at most $2|A' \cup B'|$ edges connecting two vertices from $A' \cup B'$.

By Claims 26 and 27, every flip whose block includes a vertex $x \in A' \cup B'$ and some r_i ($1 \leq i \leq k$), contributes 1 to the degree of that vertex and a total of at most 2 to the sum of the degrees in $\mathcal{H}$. There are at most $\binom{k}{2}$ flips the block of whom includes more than one of $\{r_1, \ldots, r_k\}$. Hence, the sum of the degrees of all vertices in $A' \cup B'$ is at least $k(|A' \cup B'|) - k^2$. It follows that the number E of edges between vertices of Z and $A' \cup B'$ is at least $k(|A' \cup B'|) - k^2 - 4|A' \cup B'| - E_C$, where E_C is the number of edges between a vertex from $A' \cup B'$ and a vertex from C. This is because there are at most $2|A' \cup B'|$ edges connecting two vertices from $A' \cup B'$.

For every two green elements which are connected by and edge in $\mathcal{H}$, we say that the edge (xy) is associated with r_i ($1 \leq i \leq k$), if there is a flip F in which x, y, r_i take part, and in P_F^-, the two right most elements of the block of F are x

and y. By the definition of $\mathcal{H}$, for every edge is associated with some r_i (at least one). For every $1 \leq i \leq k$, there are at most $\frac{n}{2}$ edges in $\mathcal{H}$ associated with r_i. Indeed, at every flip F in which r_i takes part, it contributes at most one edge to $\mathcal{H}$. At every such flip r_i changes order with at least two green elements. It follows that the number of those flips is at most $\frac{n}{2}$.

We conclude that $E \leq \frac{kn}{2} - E_C$. Combining the lower and upper bounds on E, we get

$$\frac{k-4}{k}|A' \cup B'| < k + \frac{n}{2}.$$

We conclude that

$$|A \cup B| = |A' \cup B'| + |C| \leq |A' \cup B'| + k \leq k + \frac{k}{k-4}\left(k + \frac{n}{2}\right).$$

Since this is true for every $k \leq |R|$, this shows that $|A \cup B| < (\frac{1}{2} + o(1))n$. In view of Section 3, in which we proved that $|A \cup B| \geq \frac{5}{9}|R|$, this completes the proof of Theorem 2. ∎

References

DSF98. P.F. Da Silva and K. Fukuda, Isolating points by lines in the plane, *Journal of Geometry* **62** (1998), 48–65.

F96. K. Fukuda, Question raised at the Problem Session of the AMS-IMS-SIAM Joint Summer Research Conference on Discrete and Computational Geometry: Ten Years Later, Mount Holyoke College, South Hadley, Massachusetts, 1996.

G44. T. Gallai, Solution of problem 4065, *American Mathematical Monthly* **51** (1944), 169–171.

GP93. J. E. Goodman and R. Pollack, Allowable sequences and order types in discrete and computational geometry, Chapter V in: *New Trends in Discrete and Computational Geometry (J. Pach, ed.)*, Springer-Verlag, Berlin, 1993, 103–134.

PP00. J. Pach and R. Pinchasi, Bichromatic lines with few points, *Journal of Combinatorial Theory* A **90** (2000), 326–335.

PS99. J. Pach and M. Sharir, Radial points in the plane, *European J. Combin.* **22** (2001), no. 6, 855–863.

P03. R. Pinchasi, Lines with many points on both sides, *Discrete and Computational Geometry,* to appear.

S93. J.J. Sylvester, Mathematical question 11851, *Educational Times* **59** (1893), 98–99.

Tight Bounds for Visibility Matching of f-Equal Width Objects

David Rappaport

School of Computing, Queen's University
Kingston, Ontario, K7L 3N6, Canada

Abstract. Let s denote a compact convex object in $\mathbb{R}^2$. The f-width of s is the perpendicular distance between two distinct parallel lines of support of s with direction f. A set of disjoint convex compact objects in $\mathbb{R}^2$ is of equal f-width if there exists a direction f such that every pair of objects have equal f-width. A visibility matching, for a set of equal f-width objects is a matching using non-crossing lines of site in the visibility graph of the set. In this note we establish tight bounds on the size of a maximal visibility matching for a set of f-equal width objects by showing that $\lfloor \frac{2n}{3} \rfloor \leq h(n) \leq \frac{2n}{3}$.

1 Introduction

Let S denote a set of non-intersecting compact convex objects in $\mathbb{R}^2$. We say that two objects a and b from the set S *see each other* if there exists a straight line segment l with one point in a and one point in b such that l lies in the complement of S - $\{a, b\}$. We call such a line segment a *line of sight*. The *visibility graph* of S, denoted by $\text{Vis}(S)$, associates a vertex to each object of S, and an edge between two vertices if and only if the associated objects see each other.

The combinatorial structure of the visibility graph for sets in $\mathbb{R}^2$ has been studied extensively. Some results on the combinatorial structure of the visibility graph of line segments can be found in [1,2,4,6,8].

In [4] the notion of set of f-equal width objects is introduced. Let s denote a compact object in $\mathbb{R}^2$. The f-width of s is the perpendicular distance between two distinct parallel lines of support of s with direction f. A set of compact objects in $\mathbb{R}^2$ is of f-equal width if there exists a direction f such that every pair of objects have f-equal width. A *visibility matching*, for a set of f-equal width objects is a matching using non-crossing lines of sight in the visibility graph of the set. Hosono [3] has shown that size of a maximal visibility matching of $2n$ f-equal width objects $h(n)$ satisfies the inequalities $\lfloor \frac{2n}{3} \rfloor \leq h(n) \leq \lfloor \frac{4n}{5} \rfloor$. In this note we establish tight bounds on the size of a maximal visibility matching for a set of f-equal width objects by showing that $\lfloor \frac{2n}{3} \rfloor \leq h(n) \leq \frac{2n}{3}$.

Some related results regarding the visibility graph of disjoint convex objects appear in a paper by Hosono, Meijer, and Rappaport [5] where it is shown that a set of translates of disjoint convex bodies admit a Hamilton path. In [7] it shown that the visibility graph of a set of disjoint congruent discs is Hamiltonian.

J. Akiyama and M. Kano (Eds.): JCDCG 2002, LNCS 2866, pp. 246–250, 2003.

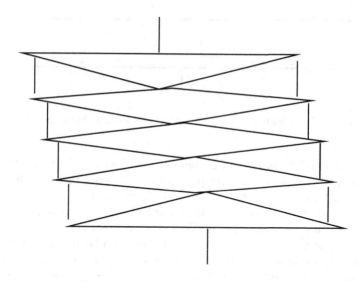

Fig. 1. This example leads to a maximal matching of at most $\frac{2n}{3}$.

2   Lower Bound

Throughout we assume that S is a set of f-equal width objects and f is horizontal.

Let Γ denote an arbitrary compact subset of $\mathbb{R}^2$. We say that a line l *supports* Γ if $l \cap \Gamma \neq \emptyset$, and a closed halfplane bounded by l contains Γ. If a point $\gamma \in \Gamma$ is contained in l then we say that γ is *extreme*.

For an element $s \in S$, let NORTH(s) be used to denote a horizontal line of support passing through an extreme point in s with maximum y coordinate, and let SOUTH(s) be used to denote a horizontal line of support passing through an extreme point in s with minimum y coordinate. Let $s, t \in S$, then $s < t$ if NORTH(s) is above NORTH(t), or when s and t are supported by the same horizontal line, then s is to the left of t. We use $<_{min}(S)$ to denote the least element of S using the $<$ ordering.

Consider a set S of f-equal width objects. We can partition the set S based on the sweep of a horizontal line . Let S_0 denote the subset of S intersected SOUTH$(<_{min}(S))$. Let $\Sigma_i = S - S_0 \cup \ldots \cup S_{i-1}$. We give a recursive definition for S_i the subset of S intersected by SOUTH$(<_{min}(\Sigma_i))$.

Observe that if the cardinality of the smallest S_i is 2 or more then we obtain our lower bound by simply matching objects within the same subset of our partition. It remains to settle the issue when there are subsets of cardinality 1.

Let $S_{j-1}, j \leq n-1$ be a subset of cardinality 1, whose only element is called s. If S_j is of cardinality 1 then there is a line of sight between the element in S_j and s and they can be matched. Furthermore, if the cardinality of S_j is 2 or 4 or more then the matched objects in S_j overcome the unmatched s and the bound is achieved. Thus consider the case where S_j is of cardinality 3, and let a left

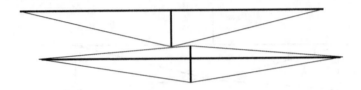

Fig. 2. The anatomy of a gem with its skeleton in bold lines.

to right labelling of the objects of S_j be a, b, c. If s sees a or c we match all of the objects in the two sets. For example s sees a and since there is a horizontal line intersecting a, b, c and no other element of Σ_j then b must see c. Suppose on the other hand that s does not see either a or c. There is a horizontal line that intersects s and no other elements of Σ_{j-1} so s must see at least one object in S_j namely b. However, this implies that b is higher than both a or c (otherwise s could see a or c) so there is a horizontal line of sight between a and c. Therefore we match all of the objects in the union of S_{j-1} and S_j. Finally we have the case where S_{n-1} is of cardinality 1. In this event the best lower bound we can guarantee is $2n/3 - 1$.

3   Upper Bound

We give a construction that achieves the stated upper bound. Consider $3N$ f-equal width objects, as shown in Figure 1. The objects are a collection of $2N$ vertical unit length line segments called *bars* and N *gems*. Each gem is built upon a skeleton made up of a unit length vertical line segment and a horizontal line segment. See Figure 2.

Let $n = N - 1$ We can number the gems $g_0, \ldots, g_n$ from top to bottom. The horizontal segment of g_i is located a distance of i/n from the top of the vertical segment. Let c be a positive constant. Each horizontal segment of the gems skeleton is of length $2cn$ and is bisected by the vertical. The gems are placed one below the other, such that g_i is slightly to the right of g_{i-1}, and pushed up so that the gems are almost touching. That is there exists small positive values ϵ and $\epsilon_0, \epsilon_1, \ldots, \epsilon_n$ so that the top of g_i is positioned at coordinate $(i\epsilon, -i + \epsilon_i)$. We put a bar above the top gem and a bar below the bottom gem, and respectively call them top and bot. We can now fit the remaining $2n$ bars between the gems on the left and on the right. We refer to these bars as $l_1 \ldots l_n$ and $r_1 \ldots r_n$. There exists a positive value δ such that l_i can be placed directly above the left end of g_i so that the vertical gap between the segment and the boundaries of g_{i-1} and g_i is at least δ. On the right we place r_i directly below the right extreme of g_{i-1} again maintaining a gap of at least δ between the bar and its neighbouring gems. We prove that this construction is possible without overlapping any of the objects.

Lemma 1. *The vertical distance between the left extreme of g_i and the boundary of g_{i-1} is at least $1 + 2\delta$*

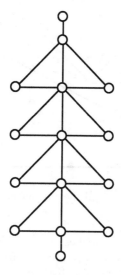

Fig. 3. The visibility graph for the objects.

Proof: Consider the distance between the two consecutive horizontal segments, say between g_{i-1} and g_i. This distance is at least $1 + 1/n - \epsilon_i$. Using similar triangles observe that the vertical distance of the boundary of g_{i-1} from the horizontal line segment is at most

$$\frac{\epsilon(1 - \frac{i}{n})}{cn} \leq \frac{\epsilon}{cn}.$$

Also observe that

$$\epsilon_i < \frac{\epsilon}{cn}, i = 0, \dots n.$$

Therefore the vertical distance between the boundary of g_{i-1} and the left extreme of g_i is at least $1 + \frac{1}{n} - \frac{2\epsilon}{cn}$. If we set c to 4ϵ we have the desired result, then δ becomes $1/4n$. $\qquad\square$

We claim that the visibility graph of such a set of objects is isomorphic to a graph G with vertex set $\{g_0, g_1, \dots, g_n, l_1, l_2, \dots, l_n, r_1, r_2, \dots, r_n, \text{top}, \text{bot}, \}$, and edges $\{(l_i, g_i), (l_i, g_{i-1}), (r_i, g_i), (r_i, g_{i-1}), (g_i, g_{i-1}), (\text{top}, g_0), (\text{bot}, g_n)\}$, for $i = 1 \dots n$. See Figure 3.

Theorem 1. $\mathrm{Vis}(S)$ *is isomorphic to* G.

Proof: The inclusion of the prescribed edges is not in question and can easily be verified. We show that no other edges exist in the visibility graph. It is clear that top only sees g_0 and bot only sees g_n. Observe that the right and left extremes of the gems are on a straight line (they could also be put on the arc of a parabola, to avoid using degeneracies in our argument) it is clear the g_i sees g_{i-1} and g_{i+1} and no other gems. Similarly the bars r_i see no other r_j and no l_j, for $i \neq j$. Furthermore we claim that l_i cannot see r_i. The gap between g_i and

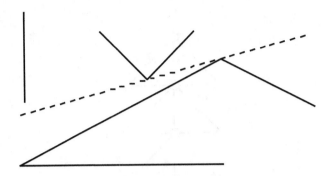

Fig. 4. Detail of a pair of gems and the amount of sight leak between them.

g_{i+1} can be made as small as one wishes. Thus the only lines of sight that leak through the space between g_i and g_{i+1} are arbitrarily close to the boundaries of the respective gems. To be specific, the distance between gems effects the angle between the sight leak and the boundary of g_i. Thus we can make this angle as small as we like. On the other hand we also have control over the size of δ which is guaranteed to be some positive value. See Figure 4.

Since the bars are of distance at least δ from the gems there is no line of sight between any r_i and l_i. Thus we have shown that Vis(S) is isomorphic to G. □

It is now a routine matter to verify that a maximal matching in the derived visibility graph saturates $2N$ of the $3N$ objects.

References

1. P. Bose, G. Toussaint: Growing a tree from its branches, Journal of Algorithms, 19 (1995) 86–103.
2. M. Hoffmann and C. S. T'oth, Segment endpoint visibility graphs are Hamiltonian, In Proc. 13th Canadian Conference on Computational Geometry (CCCG '01), Waterloo(ON), Canada, (2001)109–112,To appear in Computational Geometry: Theory and Applications, 2003.
3. K. Hosono On an estimate of the size of the maximum matching for a disjoint family of compact convex sets in the plane, Discrete Appl. Math. 113 (2001) 291–298.
4. K. Hosono and K. Matsuda, On the perfect matching of disjoint convex sets by line segments, Discrete Appl. Math. 118 (2002) 223–238.
5. K. Hosono, H. Meijer, D. Rappaport, On the visibility graph of convex translates, Discrete Appl. Math. 113 (2001) 195–210.
6. A. Mirzaian, Hamiltonian triangulations and circumscribing polygons of disjoint line segments, Comput. Geom. Theory Appl., 2, 1 (1992) 15–30.
7. D. Rappaport The visibility graph of congruent discs is Hamiltonian, to appear in Computational Geometry: Theory and Applications, (accepted Aug. 8, 2002).
8. D. Rappaport Computing simple circuits from a set of line segments is NP-complete, SIAM J. Comput., 18, 6, (1989) 1128–1139.

Long Paths through Specified Vertices in 3-Connected Graphs

Toshinori Sakai

Research Institute of Educational Development, Tokai University
2-28-4 Tomigaya, Shibuya-ku, Tokyo 151-8677, Japan
tsakai@ried.tokai.ac.jp

Abstract. Let $h \geq 6$ be an integer, let G be a 3-connected graph with $|V(G)| \geq h - 1$, and let x and z be distinct vertices of G. We show that if for any nonadjacent distinct vertices u and v in $V(G) - \{x, z\}$, the sum of the degrees of u and v in G is greater than or equal to h, then for any subset Y of $V(G) - \{x, z\}$ with $|Y| \leq 2$, G contains a path which has x and z as its endvertices, passes through all vertices in Y, and has length at least $h - 2$. We also state a similar result for cycles in 2-connected graphs.

1 Introduction

All graphs considered in this paper are finite simple undirected graphs with no loops and no multiple edges. For two vertices x and z, a path having x as its initial vertex and z as its terminal vertex is called an (x, z)-path. An (x, z)-path of length at least m is referred to as an $(x, z; m)$-path. Furthermore, for a vertex set Y (we allow the possibility that $Y \cap \{x, z\} \neq \phi$), an (x, z)-path passing through all vertices in Y is referred to as an (x, Y, z)-path, and an (x, Y, z)-path of length at least m is referred to as an $(x, Y, z; m)$-path. If Y consists of a single vertex, say y, then an (x, Y, z)-path and an $(x, Y, z; m)$-path are also referred to as an (x, y, z)-path and an $(x, y, z; m)$-path, respectively. The following Theorem A appears in [1] together with a similar result for cycles.

Theorem A. *Let k and d be integers with $d \geq k \geq 3$, and let G be a k-connected graph with $|V(G)| \geq 2d - 1$ and $\delta(G) \geq d$. Let x and z be distinct vertices of G, and let Y be a subset of $V(G) - \{x, z\}$ with cardinality $k - 1$. Then G contains an $(x, Y, z; 2d - 2)$-path.*

In [4], it is shown that for $k \geq 4$ in Theorem A, the same conclusion holds even if we replace the condition imposed on the minimum degree by the following weaker condition:

$$\max\{d_G(u), d_G(v)\} \geq d \quad \text{for any nonadjacent}$$
$$\text{distinct vertices } u \text{ and } v \text{ in } V(G) - \{x, z\}.$$

For $k = 3$ and $k = 2$, we obtained the following results, respectively.

J. Akiyama and M. Kano (Eds.): JCDCG 2002, LNCS 2866, pp. 251–260, 2003.
© Springer-Verlag Berlin Heidelberg 2003

Theorem 1. *Let h be an integer with $h \geq 6$, and let G be a 3-connected graph with $|V(G)| \geq h - 1$. Let x and z be distinct vertices of G, and let Y be a subset of $V(G) - \{x, z\}$ with cardinality at most 2. Suppose that*

$$
\left.
\begin{array}{l}
d_G(u) + d_G(v) \geq h \quad \text{for any nonadjacent} \\
\text{distinct vertices } u \text{ and } v \text{ in } V(G) - \{x, z\}.
\end{array}
\right\} \tag{1}
$$

Then G contains an $(x, Y, z; h - 2)$-path.

Theorem 2. *Let h be an integer with $h \geq 4$. Let G be a 2-connected graph with $|V(G)| \geq h$, and let Y be a subset of $V(G)$ with cardinality at most 2. Suppose that $d_G(u) + d_G(v) \geq h$ for any nonadjacent distinct vertices u and v of G. Then G contains a cycle which passes through all vertices in Y and has length at least h.*

Since the proofs of Theorems 1 and 2 follow essentially the same line of argument, we prove only Theorem 1 in this paper. Our notation is standard with the possible exception of the following:

Let G be a graph and let A be a subset of $V(G)$. For a subgraph H of G, we let $N_H(A) = \cup_{v \in A}(N_G(v) \cap V(H))$, $n_H(A) = |N_H(A)|$. A vertex a is often identified with the set $\{a\}$. For example, we write $N_H(a)$ and $n_H(a)$ for $N_H(\{a\})$ and $n_H(\{a\})$, respectively. For a vertex a and a subset B of $V(G)$, a path is called an (a, B)-path if its initial vertex is a and its terminal vertex belongs to B, and no other vertex on it belongs to B. Let $P = u_0 u_1 \cdots u_m$ be a path. We denote the length m of P by $l(P)$. For two vertices u_i, u_j on P with $i \leq j$, we let $P[u_i, u_j]$, $P[u_i, u_j)$, $P(u_i, u_j]$, and $P(u_i, u_j)$ denote the "subpaths" $u_i u_{i+1} u_{i+2} \cdots u_{j-1} u_j$, $u_i u_{i+1} u_{i+2} \cdots u_{j-1}$, $u_{i+1} u_{i+2} \cdots u_{j-1} u_j$, and $u_{i+1} u_{i+2} \cdots u_{j-1}$ of P, respectively. The path obtained by "tracing P backward" is denoted by P^{-1}; that is to say, $P^{-1} = u_m \cdots u_1 u_0$.

We conclude this section with propositions which we need in our proof of Theorem 1. Proposition B appears in [3] as Theorem 2; Proposition C appears in [2] as Lemma 2.1; and Proposition D appears in [5] as Corollary.

Proposition B. *Let $h \geq 6$ be an integer, and let G be a 3-connected graph with $|V(G)| \geq h - 1$. Let x and z be distinct vertices of G, and suppose that $d_G(u) + d_G(v) \geq h$ for any nonadjacent distinct vertices u and v in $V(G) - \{x, z\}$. Then G contains an $(x, z; h - 2)$-path passing through all vertices whose degree in G is strictly less than $\frac{h}{2}$.*

Proposition C. *Let $d \geq 1$ be an integer, let G be a nonseparable graph, and let x, y, z and w be vertices of G with $x \neq z$. Suppose that $d_G(u) \geq d$ for all vertices u in $V(G) - \{x, z, w\}$ and, in the case where $y \in \{x, z, w\}$, suppose further that $d_G(y) \geq \min\{d, 3\}$. Then G contains an $(x, y, z; d)$-path.*

Proposition D. *Let $d \geq 1$ be an integer, let G be a nonseparable graph, and let x, z and y be vertices of G with $x \neq z$. Suppose that $\max\{d_G(u), d_G(v)\} \geq d$ for any nonadjacent distinct vertices u and v in $V(G) - \{x, z\}$, and $d_G(y) \geq d$. Then G contains an $(x, y, z; d)$-path.*

2   Proof of Theorem 3

Throughout this section, let h, G, x, z, Y be as in Theorem 1. We proceed by induction on $|Y|(\leq 2)$. Let S denote the set of vertices in $V(G) - \{x, z\}$ whose degree in G is strictly less than $\frac{h}{2}$. By (1), any two distinct vertices in S are adjacent to each other. If $|V(G)| = h - 1$ or $Y \subseteq S$, then the result follows from Proposition B. Also if $|V(G)| \geq h$ and $S = \phi$, the result follows from Theorem A by letting $d = \lceil \frac{h}{2} \rceil$ and $k = 3$. Thus we may assume that $|V(G)| \geq h$, $Y \not\subseteq S$ (so $Y \neq \phi$) and $S \neq \phi$. By the induction hypothesis, G contains an $(x, Y', z; h - 2)$-path for any proper subset Y' of Y. Let P be a longest (x, z)-path such that $|Y - V(P)| \leq 1$ and $S \cap Y \subset V(P)$. Then P is an $(x, z; h - 2)$-path. Hence if $Y \subseteq V(P)$, then there is nothing to be proved. Thus we may assume $Y \not\subseteq V(P)$. By our choice of P, $Y - V(P)$ consists of a single vertex whose degree is greater than or equal to $\frac{h}{2}$, and $|V(P) \cap Y| \leq 1$. Write $Y - V(P) = \{y\}$. We have $d_G(y) \geq \frac{h}{2}$. On the other hand, it follows from the definition of S and the 3-connectedness of G that $\frac{h}{2} > d_G(v) \geq 3$ for all $v \in S$. Since $S \neq \phi$, we have $\frac{h}{2} > 3$, and hence

$$d_G(y) \geq \tfrac{h}{2} > 3, \; h \geq 7 \text{ and } |V(P)| \geq h - 1 \geq 6. \tag{2}$$

Let $\mathcal{H}$ denote the set of connected components of $G - V(P)$. For each $H \in \mathcal{H}$, define $\mathcal{B}_H$ as follows: if H is separable, then let $\mathcal{B}_H$ be the set of endblocks of H; if H is nonseparable, then let $\mathcal{B}_H = \{H\}$. Set $\mathcal{B} = \bigcup_{H \in \mathcal{H}} \mathcal{B}_H$. For each $B \in \mathcal{B}$, denote by H_B the unique element of $\mathcal{H}$ such that $B \in \mathcal{B}_{H_B}$, and set $U_B = V(H_B)$. Let C denote the set of cutvertices of $G - V(P)$. For each $B \in \mathcal{B}$, set $A_B = V(B) - C$ and define δ_B as folows: if H_B is separable, or if H_B is nonseparable and $A_B \subseteq S$, then let $\delta_B = \min_{v \in A_B} d_G(v)$; if H_B is nonseparable and $A_B \not\subseteq S$ then let $\delta_B = \min_{v \in A_B - S} d_G(v)$. For each $H \in \mathcal{H}$ and subsets X_1 and X_2 of $V(H)$, let $\mathcal{V}(X_1, X_2)$ denote the set of those pairs (u, v) of vertices u and v in $N_P(V(H))$ such that u occurs before v on P, and such that there exist $a \in N_H(u)$ and $b \in N_H(v)$ such that $a \neq b$; and either $a \in X_1$ and $b \in X_2$, or $a \in X_2$ and $b \in X_1$. We first show the following lemma concerning $X_1 = A_B$, $X_2 = U_B$ and vertices $a, b \in U_B$:

Lemma 1. Let $B \in \mathcal{B}$. Let a and b be vertices in U_B such that $a \neq b$ and $\{a, b\} \cap A_B \neq \phi$. Then there exists an $(a, b; \delta_B - n_P(A_B))$-path in H_B. Furthermore, if at least one of the following conditions (i), (ii) or (iii) holds, then there exists an $(a, y, b; \delta_B - n_P(A_B))$-path in H_B:

   (i) $y \in A_B$;
   (ii) $\{y\} = V(B) \cap C$ and $d_B(y) \geq 3$; or
   (iii) $|\{a, b\} \cap A_B| = 1$ and there is an (a, y, b)-path in H_B.

Proof. We may assume $a \in A_B$. Define b' as follows. First assume $b \in V(B)$. In this case, we let $b' = b$. Next assume $b \notin V(B)$. Then H_B is separable. In this case, we let b' be the unique vertex in $V(B) \cap C$. Further define y^* as follows: if (i) or (ii) holds, let $y^* = y$; otherwise, let y^* be a vertex in A_B such that $d_B(y^*) \geq \delta_B$. We apply Proposition C or D to B according to whether H_B is separable or nonseparable, respectively. Then we see that there exists an $(a, y^*, b'; \delta_B - n_P(A_B))$-path R in B. In the case where (i) or (ii) holds, R is an $(a, y, b'; \delta_B - n_P(A_B))$-path. This completes the proof for the case where $b \in A_B$. Thus assume that $b \notin A_B$. Let R' be a (b', b)-path in H_B. In the case where (iii) holds, let Q be an (a, y, b)-path in H_B and let $R' = Q[b', b]$. Then RR' is a path with the desired properties. $\qquad\square$

For each $H \in \mathcal{H}$ and subsets $X_1, X_2 \subseteq V(H)$, set $\mathcal{V}_i(X_1, X_2) = \{(u, v) \in \mathcal{V}(X_1, X_2) \mid |V(P(u, v)) \cap Y| = i\}$, $i = 0, 1$. For each $B \in \mathcal{B}$, let $\mathcal{W}(B) = \mathcal{V}(A_B, U_B)$ and $\mathcal{W}_i(B) = \mathcal{V}_i(A_B, U_B)$, $i = 0, 1$,

$$\mathcal{W}_{1,+}(B) = \{(u, v) \in \mathcal{W}_1(B) \mid V(P(u, v)) \cap Y \not\subseteq S\},$$
$$\mathcal{W}_{1,-}(B) = \{(u, v) \in \mathcal{W}_1(B) \mid V(P(u, v)) \cap Y \subseteq S\}$$

and define the subset $\mathcal{W}^*(B)$ of $\mathcal{W}(B)$ as follows: if the condition (i) or (ii) in Lemma 1 is satisfied, then let $\mathcal{W}^*(B) = \mathcal{W}(B)$; otherwise let $\mathcal{W}^*(B)$ denote the set of those pairs (u, v) in $\mathcal{W}(B)$ such that there exist distinct vertices $a \in N_{H_B}(u)$ and $b \in N_{H_B}(v)$ satisfying the condition (iii) in Lemma 1 (so $\mathcal{W}^*(B) = \phi$ when $y \notin U_B$). For each $B \in \mathcal{B}$, set $\mathcal{W}_0^*(B) = \mathcal{W}_0(B) \cap \mathcal{W}^*(B)$, $\mathcal{W}_{1,+}^*(B) = \mathcal{W}_{1,+}(B) \cap \mathcal{W}^*(B)$ and $\mathcal{W}_{1,-}^*(B) = \mathcal{W}_{1,-}(B) \cap \mathcal{W}^*(B)$. The following Lemma 2 follows from the definition of $\mathcal{W}(B)$, $\mathcal{W}^*(B)$ and from Lemma 1.

Lemma 2. *Let $B \in \mathcal{B}$ and let $(u, v) \in \mathcal{W}(B)$. Then there exists a $(u, v; \delta_B - n_P(A_B)+2)$-path whose inner vertices lie in U_B. Furthermore, if $(u, v) \in \mathcal{W}^*(B)$, then there exists a $(u, y, v; \delta_B - n_P(A_B)+2)$-path whose inner vertices lie in U_B.*

By the maximality of $l(P)$ and the definition of $\mathcal{W}_0(B)$ and $\mathcal{W}_{1,+}^*(B)$, we obtain:

Lemma 3. *Let $B \in \mathcal{B}$. Then the following hold.*

(i) $l(P[u, v]) \geq 2$ for any $u, v \in N_P(U_B)$ such that u occurs before v on P.
(ii) $l(P[u, v]) \geq \delta_B - n_P(A_B) + 2$ for any u, v such that $(u, v) \in \mathcal{W}_0(B) \cup \mathcal{W}_{1,+}^*(B)$.

Let H be the connected component in $\mathcal{H}$ such that $y \in V(H)$ and set $U = V(H)$.

Case 1. *H is separable:*
In this case, we often consider subsets $\mathcal{V}(A_{B_1}, Z_2)$ of $\mathcal{W}^*(B_1)$ and $\mathcal{V}(A_{B_2}, Z_1)$ of $\mathcal{W}^*(B_2)$, where B_1 and B_2 are elements of $\mathcal{B}_H$ and Z_1 and Z_2 are subsets of U such that $y \in Z_1 \cap Z_2$ and $Z_1 \cup Z_2 = U$.

For each $B \in \mathcal{B}_H$, let c_B denote the unique vertex in $V(B) \cap C$. We choose distinct blocks $B_1, B_2 \in \mathcal{B}_H$ and define subgraphs K_1, K_2 of H as follows. First assume that $y \in V(B)$ for some $B \in \mathcal{B}_H$. In this case, we take $B_1, B_2 \in \mathcal{B}_H$ so that $y \in V(B_1) \cup V(B_2)$ and define K_1, K_2 as follows: if $y \in A_{B_1} \cup A_{B_2}$, then for the index $i \in \{1, 2\}$ with $y \in A_{B_i}$ let $K_i = B_i - c_{B_i}$, and for the other index j let $K_j = H$; if $y = c_{B_1} = c_{B_2}$, then let $K_1 = H - A_{B_2}$ and $K_2 = H - A_{B_1}$; otherwise, for indices $i, j \in \{1, 2\}$ with $i \neq j$ and $y = c_{B_i}$, let b_0 be the (unique) vertex in C with $b_0 \neq y$ such that b_0 and y are in the same block of H and such that every (y, A_{B_j})-path passes through b_0, let K_i be the connnected component of $H - b_0$ which contains y, and let $K_j = H - A_{B_i}$. Next assume that $y \notin V(B)$ for any $B \in \mathcal{B}_H$. In this case, we first take distinct vertices b_1 and b_2 in C with $b_1 \neq y$ and $b_2 \neq y$ such that b_1 and y are in the same block of H, and b_2 and y are in the same block of H (so b_1 and b_2 are in the same block of H when $y \notin C$). Let K_1 be the connected component of $H - b_2$ which contains y and let K_2 be the connected component of $H - b_1$ which contains y. We take subgraphs $B_1 \in \mathcal{B}_H$ of $H - V(K_2)$ and $B_2 \in \mathcal{B}_H$ of $H - V(K_1)$.

For each $i = 1, 2$, set $Z_i = V(K_i)$, write $A_{B_i} = A_i$, $c_{B_i} = c_i$ and $\delta_{B_i} = \delta_i$ for simplicity, and let $A^* = A_1 \cup A_2$. Then $\mathcal{V}(A_i, Z_j) \subseteq \mathcal{W}^*(B_i)$ for each $(i, j) \in \{(1, 2), (2, 1)\}$, $y \in Z_1 \cap Z_2$ and $Z_1 \cup Z_2 = U$. Since $G - c_1$ and $G - c_2$ are both 2-connected, $n_P(A_1) \geq 2$ and $n_P(A_2) \geq 2$, and hence $\mathcal{V}(A_1, A_2) \neq \phi$.

Case 1.1. $\mathcal{V}_0(A_1, A_2) \neq \phi$:

Choose $(u_0, v_0) \in \mathcal{V}_0(A_1, A_2)$ so that $P[u_0, v_0]$ is minimal. By symmetry, we may assume $u_0 \in N_P(A_1)$, $v_0 \in N_P(A_2)$ and $y \notin A_2$. By the minimality of $P[u_0, v_0]$, $N_{P(u_0, v_0)}(A^*) = \phi$. Take $a \in N_H(u_0) \cap A_1$ and $b \in N_H(v_0) \cap A_2$. Since there is an $(a, y, c_2; \delta_1 - n_P(A_1))$-path in $H - A_2$ and there is a $(c_2, b; \delta_2 - n_P(A_2))$-path in B_2, there is an (a, y, b)-path Q in H of length at least $\delta_1 + \delta_2 - (n_P(A_1) + n_P(A_2)) \geq h - (n_P(A_1) + n_P(A_2))$. Consider the path $P_1 = P[x, u_0] u_0 a Q b v_0 P[v_0, z]$. Then P_1 is an (x, Y, z)-path, and $l(P_1[u_0, v_0]) = l(Q) + 2 \geq h - (n_P(A_1) + n_P(A_2)) + 2$. On the other hand, since $N_{P(u_0, v_0)}(A^*) = \phi$, $l(P[x, u_0]) + l(P[v_0, z]) \geq 2(n_P(A^*) - 2)$ by Lemma 3(i). Consequently, we obtain $l(P_1) \geq h - 2 + (n_P(A^*) - n_P(A_1)) + (n_P(A^*) - n_P(A_2)) \geq h - 2$, as desired.

Case 1.2. $\mathcal{V}_0(A_1, A_2) = \phi$:

Recall $\mathcal{V}(A_1, A_2) \neq \phi$ and take $(u_0, v_0) \in \mathcal{V}_1(A_1, A_2)$. We may assume $u_0 \in N_P(A_1)$ and $v_0 \in N_P(A_2)$. Let y' be the unique vertex in $V(P) \cap Y$.

Case 1.2.1. $n_{P[x, y']}(A_1) \geq 2$:

Recall $n_P(A_1) \geq 2$ and $n_P(A_2) \geq 2$. Then in view of the assumption of Case 1.2, we obtain $N_{P[x, y']}(A_2) = \phi$, $n_{P[y', z]}(A_2) \geq 2$ and $N_{P[y', z]}(A_1) = \phi$.

We first consider the case where all of the following conditions hold: $y = c_1 = c_2$, $N_G(y) \subseteq A^*$, $d_{B_1}(y) \leq 2$ and $d_{B_2}(y) \leq 2$. Since $d_G(y) \geq 4$ by (2), $d_{B_1}(y) = d_{B_2}(y) = 2$ by assumption. Then $h \leq 8$ again by (2). Also since $|V(B_1)| \geq d_{B_1}(y) + 1 = 3$ and $N_P(A_1) \subseteq V(P[x, y'])$, it follows from the 3-connectedness of G that there exists $(u, v) \in \mathcal{V}_0(A_1, A_1)$. Let R be a (u, y, v)-path whose inner vertices are in $V(B_1)$ and consider the path $P' = P[x, u] R P[v, z]$. Then P' is an (x, Y, z)-path. Since $l(P'[u, v]) \geq 4$ and $l(P'[v, z]) = l(P'[v, y']) + l(P'[y', z]) \geq 2$, we obtain $l(P') \geq 6 \geq h - 2$. Therefore we may assume that

$$\max\{d_{B_1}(y), d_{B_2}(y)\} \geq 3 \text{ when } y = c_1 = c_2 \text{ and } N_G(y) \subseteq A^* \text{ hold.} \quad (3)$$

Lemma 4. *There exist $u_1, v_1 \in N_P(U)$ and a (u_1, y, v_1)-path R_1 whose inner vertices lie in U such that u_1 occurs before v_1 on P, $N_{P(u_1,v_1)}(A^*) = \phi$ and such that either $u_1, v_1 \in V(P[x, y'])$ and $l(R_1) \geq \delta_1 - n_P(A_1) + 1$, or else $u_1, v_1 \in V(P[y', z])$ and $l(R_1) \geq \delta_2 - n_P(A_2) + 1$.*

Proof. First consider the case where $y \in A_1$. If $|V(B_1)| \geq 3$, then by the 3-connectedness of G and the assumption that $N_P(A_1) \subset V(P[x, y'])$, $\mathcal{V}_0(A_1, A_1) \subseteq \mathcal{W}_0^*(B_1) \neq \phi$. Choose $(u_1, v_1) \in \mathcal{W}_0^*(B_1)$ so that $P[u_1, v_1]$ is minimal. Then $N_{P(u_1,v_1)}(A^*) = \phi$ and the desired conclusion follows from Lemma 2. Next assume $|V(B_1)| = 2$. In this case, $A_1 = \{y\}$ and $n_P(y) = d_G(y) - d_{B_1}(y) \geq 3$ by (2). Let u_1 be the vertex in $N_P(y)$ which is closest to x on P and let v_1 be the vertex in $N_P(y)$ which is closest to u_1 on P. Then $N_{P(u_1,v_1)}(A^*) = \phi$ and the path $R_1 = u_1 y v_1$ has length $2 = d_{B_1}(y) + 1 = d_G(y) - n_P(y) + 1 = \delta_1 - n_P(A_1) + 1$.

Next consider the case where $y = c_1$. If $d_{B_1}(y) \geq 3$, then $|V(B_1)| \geq 4$, and we can argue as in the case where $y \in A_1$ and $|V(B_1)| \geq 3$. Assume next that there exists $u \in N_P(Z_1 \cap Z_2)$. Then $\mathcal{W}_0^*(B_1) \supseteq \mathcal{V}_0(A_1, Z_2) \neq \phi$ or $\mathcal{W}_0^*(B_2) \supseteq \mathcal{V}_0(A_2, Z_1) \neq \phi$ according to whether $u \in V(P[x, y'])$ or $u \in V(P((y', z]))$. Then by choosing $(u_1, v_1) \in \mathcal{W}_0^*(B_1) \cup \mathcal{W}_0^*(B_2)$ so that $P[u_1, v_1]$ is minimal, we can obtain the desired conclusion. Thus assume that $d_{B_1}(y) \leq 2$ and $N_P(Z_1 \cap Z_2) = \phi$. Then we see $y = c_2$ and $\mathcal{B} = \{B_1, B_2\}$ by the assumption that $N_P(Z_1 \cap Z_2) = \phi$ and by the 3-connectedness of G (if $y \notin V(B_2)$ and the block containing y and b_0 consists only of y and b_0, where b_0 is as in the second paragraph of Case 1, we also make use of the fact that $d_G(y) \geq 4$). Since we also have $N_P(y) = \phi$ (recall $y \in Z_1 \cap Z_2$), $d_{B_2}(y) \geq 3$ by (3), and we can argue as in the case where $d_{B_1}(y) \geq 3$.

We can argue similarly in the case where $y \in A_2$ or $y = c_2$. Now consider the case where $y \notin V(B_1) \cup V(B_2)$. Let b_1 and b_2 be as defined in the definition of K_1 and K_2. Since $G - \{b_1, b_2\}$ is connected, there exists $u \in N_P(Z_1 \cap Z_2)$. Hence $\mathcal{W}_0^*(B_1) \supseteq \mathcal{V}_0(A_1, Z_2) \neq \phi$ or $\mathcal{W}_0^*(B_2) \supseteq \mathcal{V}_0(A_2, Z_1) \neq \phi$ according to whether $u \in V(P[x, y'])$ or $u \in V(P(y', z])$, and we can obtain the desired conclusion in this case as well. $\square$

Let u_1, v_1 and R_1 be as in Lemma 4. By symmetry, we may assume that $u_1, v_1 \in V(P[x, y'])$. Consider the path $P_1 = P[x, u_1] R_1 P[v_1, z]$. Then P_1 is an (x, Y, z)-path. By the choice of R_1, $l(P_1[u_1, v_1]) = l(R_1) \geq \delta_1 - n_P(A_1) + 1$. On the other hand, we have $l(P_1[x, u_1]) + l(P_1[v_1, y']) = l(P[x, u_1]) + l(P[v_1, y']) \geq (n_{P[x,u_1]}(A_1) - 1) + (n_{P[v_1,y']}(A_1) - 1) \geq n_P(A_1) - 2$. Hence, we obtain $l(P_1[x, y']) \geq \delta_1 - 1$. Now it suffices to show that $l(P_1[y', z]) \geq \delta_2 - 1$ because $\delta_1 + \delta_2 \geq h$. If $|V(B_2)| = 2$, then using Lemma 3(i), we obtain $l(P_1[y', z]) = l(P[y', z]) \geq 2(n_{P[y',z]}(A_2) - 1) = 2(n_P(A_2) - 1) \geq (n_P(A_2) - 1) + 1 = n_P(A_2) = \delta_2 - 1$. Thus consider the case where $|V(B_2)| \geq 3$. Then we can choose $(u_2, v_2) \in \mathcal{V}_0(A_2, A_2) \subseteq \mathcal{W}_0(B_2)$ such that $N_{P(u_2,v_2)}(A_2) = \phi$. Then $l(P[u_2, v_2]) \geq \delta_2 - n_P(A_2) + 2$ by Lemma 3(ii), and $l(P[y', u_2]) + l(P[v_2, z]) \geq (n_{P[y',u_2]}(A_2) - 1) + (n_{P[v_2,z]}(A_2) - 1) = n_P(A_2) - 2$. Consequently, we now obtain $l(P_1[y', z]) = l(P[y', z]) \geq \delta_2$, as desired.

Case 1.2.2. $N_{P[x,y']}(A_1) = \{u_0\}$:

Recall $n_P(A_1) \geq 2$ and $n_P(A_2) \geq 2$ again. Then $N_{P[y',z]}(A_1) \neq \phi$, and in view of the assumption of Case 1.2, we obtain $N_{P[y',z]}(A_1) = \{v_0\}$ and $N_{P(y',z]}(A_2) = \{v_0\}$, and furthermore $N_{P[x,y']}(A_2) = \{u_0\}$.

Since G is 3-connected, $N_P(U) - \{u_0, v_0\} \neq \phi$. By symmetry, we may assume that there exists $t \in N_{P[x,y']}(U) - \{u_0\}$. In the case where $y' \in N_P(U)$, we let $t = y'$. We have $t \in N_P(Z_1 - A^*)$ or $t \in N_P(Z_2 - A^*)$ because $Z_1 \cup Z_2 = U$. By symmetry, we may assume $t \in N_P(Z_2 - A^*)$. Write $\{q_1, q_2\} = \{u_0, t\}$ so that q_1 occurs before q_2 on P, and let $q_3 = v_0$. We have $y' \in V(P[q_2, q_3))$. Since $u_0 \in N_P(A_1)$ and $t \in N_P(Z_2 - A_1)$, we get $(q_1, q_2) \in \mathcal{W}_0^*(B_1)$. Also since $u_0 \in N_P(A_1)$, $t \in N_P(Z_2 - A_2)$ and $v_0 \in N_P(A_2)$, we get $(q_2, q_3) \in \mathcal{W}(B_2)$ and $(q_1, q_3) \in \mathcal{W}(B_2)$. Let R_1, R_2 and R_3 be a longest (q_1, y, q_2)-path, a longest (q_2, q_3)-path and a longest (q_1, q_3)-path, respectively, such that their inner vertices lie in U. Since $n_P(A_1) = n_P(A_2) = 2$ in this case, it follows from Lemma 2 and the maximality of the R_i that $l(R_1) \geq \delta_1$, $l(R_2) \geq \delta_2$ and $l(R_3) \geq \delta_2$. For reference in Case 2.2.1, we use the following estimate of each $l(R_i)$:

$$l(R_1) \geq \delta_1 - 1, \quad l(R_2) \geq \delta_2 - 1 \quad \text{and} \quad l(R_3) \geq \delta_2 - 1. \tag{4}$$

Consider the path $P_1 = P[x, q_1]R_1P[q_2, z]$. Then P_1 is an (x, Y, z)-path and

$$l(P_1[q_1, q_2]) = l(R_1) \geq \delta_1 - 1. \tag{5}$$

We show $l(P_1) \geq h - 2$ in most of the following subcases.

Subcase 1. $q_2 = y'$:

Since $(q_2, q_3) = (y', v_0) \in \mathcal{W}_0(B_2)$ in this subcase, and since $n_P(A_2) = 2$, it follows from Lemma 3(ii) that $l(P_1[q_2, q_3]) = l(P[q_2, q_3]) \geq \delta_2$. Adding (5) and this, we obtain $l(P_1) \geq l(P_1[q_1, q_3]) \geq \delta_1 + \delta_2 - 1 \geq h - 1$.

Subcase 2. $q_2 \neq y'$:

In this subcase, $t \neq y'$, and hence $y' \notin N_P(U)$ by the choice of t.

Subcase 2.1. $N_{P(q_1, q_2)}(y') \neq \phi$:

Let u be a vertex in $N_{P(q_1, q_2)}(y')$ and let $R = uy'$. Consider the path $P_2 = P[x, q_1]R_1 P^{-1}[q_2, u]RP[y', z]$. Then P_2 is an (x, Y, z)-path. We show $l(P_2) \geq h - 2$ in this subcase. We have $l(P_2[q_1, q_2]) + l(P_2[u, y']) = l(R_1) + l(R) \geq \delta_1$ by (4). Now it suffices to show that $l(P_2[q_2, u]) + l(P_2[y', q_3]) \geq \delta_2 - 2$. For this purpose, consider the path $P_3 = P[x, u]RP^{-1}[y', q_2]R_2P[q_3, z]$. Since P_3 is an (x, y', z)-path, the maximality of $l(P)$ implies $l(P[u, q_3]) \geq l(P_3[u, q_3]) = l(R) + l(P[q_2, y']) + l(R_2)$, and hence $l(P_2[q_2, u]) + l(P_2[y', q_3]) = l(P[u, q_2]) + l(P[y', q_3]) \geq l(R) + l(R_2) \geq \delta_2$.

Recall that $P_1 = P[x, q_1]R_1P[q_2, z]$. We show $l(P_1) \geq h - 2$ in the remaining subcases.

Subcase 2.2. $N_{P[x,q_1)}(y') \neq \phi$:

Let u be a vertex in $N_{P[x,q_1)}(y')$ and let $R = uy'$. Consider the path $P_2 = P[x,u]RP^{-1}[y',q_1]R_3P[q_3,z]$. Since P_2 is an (x,y',z)-path, the maximality of $l(P)$ implies $l(P[u,q_3]) \geq l(P_2[u,q_3]) = l(R) + l(P[q_1,y']) + l(R_3)$, and hence $l(P_1[u,q_1]) + l(P_1[y',q_3]) = l(P[u,q_1]) + l(P[y',q_3]) \geq l(R) + l(R_3) \geq \delta_2$. By this and (5), we get $l(P_1) > l(P_1[u,q_1]) + l(P_1[q_1,q_2]) + l(P_1[y',q_3]) \geq \delta_1 + \delta_2 - 1 \geq h - 1$.

Subcase 2.3. $N_{P(q_3,z]}(y') \neq \phi$:

Let u be a vertex in $N_{P(q_3,z]}(y')$ and let $R = uy'$. Then we can argue as in the proof for Subcase 2.2 with P_2 being replaced by $P_2' = P[x,q_2]R_2P^{-1}[q_3,y']R^{-1}P[u,z]$ to obtain $l(P_1[q_2,y']) + l(P_1[q_3,u]) \geq \delta_2$. From this together with (5), we obtain $l(P_1) \geq \delta_1 + \delta_2 - 1 \geq h - 1$.

Subcase 2.4. $N_P(y') \subseteq \{q_1\} \cup V(P[q_2,q_3])$:

Subcase 2.4.1. $N_G(y') \subseteq \{q_1\} \cup V(P[q_2,q_3])$:

In this subcase, $l(P_1[q_2,q_3]) = l(P[q_2,q_3]) \geq |(N_G(y') \cup \{y'\}) - \{q_1\}| - 1 \geq d_G(y') - 1$. By this and (5), $l(P_1) \geq \delta_1 + d_G(y') - 2 \geq h - 2$ since $y' \notin N_P(A_1) \cup \{x,z\}$.

Subcase 2.4.2. $N_G(y') \not\subseteq \{q_1\} \cup V(P[q_2,q_3])$:

Take $w \in N_G(y') - V(P)$. Let H' be the element of $\mathcal{H}$ such that $w \in V(H')$ and let $U' = V(H')$. We have $H' \neq H$. In the case where there exists $u \in N_P(U') - (\{q_1\} \cup V(P[q_2,q_3]))$, we can argue as in Subcases 2.1, 2.2 or 2.3 by letting R be a (u,y')-path whose inner vertices lie in U'. Thus assume $N_P(U') \subseteq \{q_1\} \cup V(P[q_2,q_3])$. Take $B_3 \in \mathcal{B}_{H'}$. Write $A_3 = A_{B_3}$ and $\delta_3 = \delta_{B_3}$ for simplicity. Since $\delta_1 + \delta_3 \geq h$, it suffices to show that $l(P_1[q_2,q_3]) \geq \delta_3 - 1$.

  Assume first that there exists $(u,v) \in \mathcal{W}(B_3)$ such that $u,v \in V(P[q_2,q_3])$. Then we can choose $(u',v') \in \mathcal{W}(B_3)$ such that $u',v' \in V(P[u,v])$ and $N_{P(u',v')}(U') = \phi$. Since $y' \in N_P(U')$ by the choice of H', $y' \notin V(P(u',v'))$. Therefore $(u',v') \in \mathcal{W}_0(B_3)$, and hence $l(P[u',v']) \geq \delta_3 - n_P(A_3) + 2$ by Lemma 3(ii). On the other hand, since $N_P(A_3) \subseteq N_P(U') \subseteq \{q_1\} \cup V(P[q_2,u']) \cup V(P[v',q_3])$, $l(P[q_2,u']) + l(P[v',q_3]) \geq n_P(A_3) - 3$. Hence we obtain $l(P_1[q_2,q_3]) = l(P[q_2,q_3]) \geq \delta_3 - 1$.

  Assume next that no two vertices $u,v \in V(P[q_2,q_3])$ satisfy $(u,v) \in \mathcal{W}(B_3)$. Since $N_P(U') \subseteq \{q_1\} \cup V(P[q_2,q_3])$ and G is 3-connected, it follows that $n_{P[q_2,q_3]}(A_3) \geq 1$ and that if $|U'| \geq 2$ then there exist two independent edges joining U' and $V(P[q_2,q_3])$. Then the assumption implies that $U' = A_3$ and $|U'| = 1$, and hence $l(P_1[q_2,q_3]) = l(P[q_2,q_3]) \geq 2(n_P(U') - 2) \geq (n_P(U') - 2) + 1 = \delta_3 - 1$ by Lemma 3(i) and the 3-connectedness of G.

Case 2. H is nonseparable:

We have $\mathcal{W}^*(H) = \mathcal{W}(H) = \mathcal{V}(U,U)$ in this case. Since $y \in U \not\subseteq S$, $\delta_H \geq \frac{h}{2}$ by the definition of δ_H.

Case 2.1. There exist $(u_0, v_0), (u_0', v_0') \in \mathcal{W}_0(H) \cup \mathcal{W}_{1,+}(H)$ such that $E(P[u_0, v_0]) \cap E(P[u_0', v_0']) = \phi$:

We have $\{(u_0, v_0), (u_0', v_0')\} \cap \mathcal{W}_0(H) \neq \phi$. By symmetry, we may assume that u_0 occurs before u_0' on P and that $(u_0, v_0) \in \mathcal{W}_0(H)$. Then there exists $(u_1, v_1) \in \mathcal{W}_0(H)$ with $u_1, v_1 \in V(P[u_0, v_0])$ such that $N_{P(u_1, v_1)}(U) = \phi$. Hence by Lemma 2, there exists a $(u_1, y, v_1; \delta_H - n_P(U) + 2)$-path R whose inner vertices lie in U. Furthermore, since $(u_0', v_0') \in \mathcal{W}_0(H) \cup \mathcal{W}_{1,+}(H)$, we can take $(u_2, v_2) \in \mathcal{W}_0(H) \cup \mathcal{W}_{1,+}(H)$ with $u_2, v_2 \in V(P[u_0', v_0'])$ such that $N_{P(u_2, v_2)}(U) = \phi$. Consider the path $P_1 = P[x, u_1]RP[v_1, z]$. Then P_1 is an (x, Y, z)-path and $l(P_1[u_1, v_1]) = l(R) \geq \delta_H - n_P(U) + 2$. On the other hand, since $(u_2, v_2) \in \mathcal{W}_0(H) \cup \mathcal{W}_{1,+}(H) = \mathcal{W}_0(H) \cup \mathcal{W}_{1,+}^*(H)$, it follows from Lemma 3(ii) that $l(P_1[u_2, v_2]) = l(P[u_2, v_2]) \geq \delta_H - n_P(U) + 2$. Furthermore, $l(P_1[x, u_1]) + l(P_1[v_1, u_2]) + l(P_1[v_2, z]) \geq 2(n_P(U) - 3)$ by Lemma 3(i). Consequently, we obtain $l(P_1) \geq 2\delta_H - 2 \geq h - 2$.

Case 2.2. Otherwise:

Case 2.2.1. $|U| \geq 3$:
By the 3-connectedness of G, there exist pairwise disjoint edges $a_i q_i$ with $a_i \in U$ and $q_i \in V(P)$, $1 \leq i \leq 3$. We may assume that q_1, q_2, q_3 occur on P in this order. Then $(q_1, q_2), (q_2, q_3), (q_1, q_3) \in \mathcal{W}(H)$. Hence by the assumption of Case 2.2, $|\{(q_1, q_2), (q_2, q_3)\} \cap \mathcal{W}_0(H)| = |\{(q_1, q_2), (q_2, q_3)\} \cap \mathcal{W}_{1,-}(H)| = 1$ and $V(P) \cap Y$ consists of one vertex in S. By symmetry, we may assume $(q_1, q_2) \in \mathcal{W}_0(H)$ and $(q_2, q_3) \in \mathcal{W}_{1,-}(H)$. Write $V(P(q_2, q_3)) \cap Y = \{y'\}$. By the assumption of Case 2.2, we obtain

$$N_{P[x, q_1]}(U - \{a_1\}) = N_{P(q_1, q_2)}(U - \{a_1, a_2\}) = N_{P(q_2, y')}(U - \{a_2\})$$
$$= N_{P[y', q_3)}(U - \{a_3\}) = N_{P(q_3, z]}(U - \{a_3\}) = \phi. \tag{6}$$

Since (6) in particular implies that $y' \notin N_P(U)$ and since $y' \in S$, $U \cap S = \phi$ by (1) and $\delta_H > h - \frac{h}{2} = \frac{h}{2}$.

First consider the case where $n_P(U) \geq \frac{h}{2} - 1$. Since $(q_1, q_2) \in \mathcal{W}_0(H)$, there exists $(u_1, v_1) \in \mathcal{W}_0(H)$ with $u_1, v_1 \in V(P[q_1, q_2])$ such that $N_{P(u_1, v_1)}(U) = \phi$. Let R be a longest (u_1, y, v_1)-path whose inner vertices lie in U and consider the path $P_1 = P[x, u_1]RP[v_1, z]$. Then P_1 is an (x, Y, z)-path. Since $l(R) \geq 4$ by the assumption of Case 2.2.1 and the maximality of $l(R)$, $l(P_1) = l(P[x, u_1]) + l(R) + l(P[v_1, z]) > 2(n_{P[x, u_1]}(U) - 1) + 4 + 2(n_{P[v_1, z]}(U) - 1) = 2n_P(U) \geq h - 2$. Thus we may assume that

$$n_P(U) < \frac{h}{2} - 1. \tag{7}$$

Lemma 5.

  (I) *At least one of the following holds:*
    (i) *there exists an $(a_1, y, a_2; \delta_H - 3)$-path Q_1 in H; or*
    (ii) *$y \neq a_1$, $y \in N_H(q_2)$ and there exists an $(a_1, y; \delta_H - 3)$-path Q_1' in H.*
  (II) *In H there exist an $(a_2, a_3; \delta_H - 3)$-path Q_2 and an $(a_1, a_3; \delta_H - 3)$-path Q_3.*

Proof. First note that it follows from (6) that $d_H(v) \geq \delta_H - 3$ for all $v \in U - \{a_1, a_2, a_3\}$ (recall $U \cap S = \phi$). If $y \notin \{a_1, a_2, a_3\}$ or $d_H(y) \geq \min\{\delta_H - 3, 3\}$, then (I)(i) and (II) follow from Proposition C. Thus consider the case where $y \in \{a_1, a_2, a_3\}$ and $d_H(y) = 2 < \delta_H - 3$ (so $\delta_H \geq 6$). Since $n_P(y) = d_G(y) - d_H(y) > \frac{h}{2} - 2$, $N_P(U) = N_P(y)$ by (7). Then we must have $n_P(w) \leq 3$ for all $w \in U - \{y\}$ by the assumption of Case 2.2. Therefore $\delta_H - (|U| - 1) \leq 3$, and hence $|U| \geq 4$ because $\delta_H \geq 6$. Take $w' \in U - \{a_1, a_2, a_3\}$. Then $d_H(w') \geq \delta_H - 3$ by (6), and hence (II) follows from Proposition C. Furthermore, from Proposition C, (I)(i) follows when $y = a_1$, and (I)(ii) follows when $y \neq a_1$. □

Let Q_1, Q_1', Q_2, Q_3 be as in Lemma 5. If (I)(i) of Lemma 5 holds, then let $R_1 = q_1 a_1 Q_1 a_2 q_2$; otherwise, let $R_1 = q_1 a_1 Q_1' y q_2$. Furthermore define R_2 and R_3 by $R_2 = q_2 a_2 Q_2 a_3 q_3$ and $R_3 = q_1 a_1 Q_3 a_3 q_3$. We have $l(R_i) \geq \delta_H - 1$ for $i = 1, 2, 3$. Now we can argue as in Subcase 2 of Case 1.2.2 by letting $\delta_1 = \delta_2 = \delta_H$.

Case 2.2.2. $|U| \leq 2$:
In this case, we have $n_P(y) \geq \delta_H - 1 \geq \frac{h}{2} - 1$, and hence $n_P(y) \geq 3$ by (2). Let $r = n_P(U)$ and label the vertices in $N_P(U)$ along P as $u_1, u_2, \cdots, u_r$.

First consider the case where $r \geq \frac{h}{2}$. In this case, $r \geq 4$ by (2). Hence we can take $u_i \in N_P(y)$ with $u_i \neq u_r$ and $V(P(u_i, u_{i+1})) \cap Y = \phi$. Thus we can take a (u_i, y, u_{i+1})-path R whose inner vertices lie in U. Then the path $P_1 = P[x, u_i]RP[u_{i+1}, z]$ is an (x, Y, z)-path and $l(P_1) \geq 2(n_{P[x, u_i]}(U) - 1) + 2 + 2(n_{P[u_{i+1}, z]}(U) - 1) \geq 2r - 2 \geq h - 2$. Now consider the case where $r < \frac{h}{2}$. In this case, $r < d_G(y)$ by (2). Hence we can write $U = \{y, w\}$, where $w \neq y$, and we have $N_P(U) = N_P(y)$. Then it follows from the assumption of Case 2.2 that $n_P(w) \leq 3$. On the other hand, since $n_P(w) = d_G(w) - 1 \geq 2$ by the 3-connectedness of G and since $r = n_P(y) \geq 3$, it again follows from the assumption of Case 2.2 that there exist exactly two indices $1 \leq i < j \leq r - 1$ such that $|\{(u_i, u_{i+1}), (u_j, u_{j+1})\} \cap \mathcal{W}_0(H)| = |\{(u_i, u_{i+1}), (u_j, u_{j+1})\} \cap \mathcal{W}_{1,-}(H)| = 1$, and $V(P) \cap Y$ consists of one vertex in S. Write $V(P) \cap Y = \{y'\}$. By symmetry, we may assume $y' \in V(P(u_j, u_{j+1}))$. Since $y' \in S$ and $y' \notin N_P(w)$, $w \notin S$ and $d_G(w) > \frac{h}{2}$ by (1). Thus $\frac{h}{2} < d_G(w) = n_P(w) + 1 \leq 4$, and hence $h \leq 7$. Let R be a (u_i, U, u_{i+1})-path and consider the path $P_1 = P[x, u_i]RP[u_{i+1}, z]$. Then P_1 is an (x, Y, z)-path and $l(P_1) \geq l(R) + l(P[u_j, u_{j+1}]) \geq 3 + 2 = 5 \geq h - 2$.

This completes the proof of Theorem 1.

References

1. Egawa, Y., Glas, R., and Locke, S.C.: Cycles and paths through specified vertices in k-connected graphs, *J. Combinat. Theory* Ser.B 52 (1991), 20-29
2. Locke, S.C.: A generalization of Dirac's theorem, *Combinatorica* 5 (1985), 149-159
3. Sakai, T.: Long cycles and paths through vertices with small degree, submitted
4. Sakai, T.: Long paths and cycles through specified vertices in k-connected graphs, *Ars Combin.* **58** (2001), 33–65
5. Sakai, T.: Long paths through a specified vertex in a 2-connected graph, *Tokoha Gakuen University Research Review, Faculty of Education* 15 (1994), 59-64

On the Number of Intersections
of Three Monochromatic Trees in the Plane

Kazuhiro Suzuki

Department of Computer Science and Communication Engineering
Kogakuin University Nishi-Shinjuku, Shinjuku-ku, Tokyo 163-8677 Japan

Abstract. Let R, B, and G be three disjoint sets of points in the plane such that the points of $X = R \cup B \cup G$ are in general position. In this paper, we prove that we can draw three spanning geometric trees on R, on B, and on G such that every edge intersects at most three segments of each other tree. Then the number of intersections of the trees is at most $3|X| - 9$. A similar problem had been previously considered for two point sets.

1 Introduction

Let G be a finite graph without loops or multiple edges. We denote by $V(G)$ and $E(G)$ the set of vertices and the set of edges of G, respectively. For a set X, we denote by $|X|$ the cardinality of X. A *geometric graph* $G = (V(G), E(G))$ is a graph drawn in the plane (two-dimensional Euclidean space) such that $V(G)$ is a set of points in the plane, the points of which are in general position (i.e., no three lie on the same line), and $E(G)$ is a set of (possibly crossing) straight-line segments whose endpoints belong to $V(G)$. If a geometric graph G is a tree then we call G a *geometric tree*.

Let R and B be two disjoint sets of red points and blue points in the plane, respectively, such that the points of $R \cup B$ are in general position. For a set X of points in the plane, we denote by $\mathrm{conv}(X)$ the convex hull of X, which is the smallest convex set containing X. In 1996, Tokunaga [1] showed that the minimum number of crossings of two spanning geometric trees without crossings on R and on B is determined by the number of edges in the boundary of $\mathrm{conv}(R \cup B)$ joining points of R to those of B.

We consider a similar problem for three disjoint sets R, B, and G of red points, blue points, and green points in the plane, respectively, and have the following theorem.

Theorem 1. *Let R, B, and G be three disjoint sets of points in the plane such that the points of $X = R \cup B \cup G$ are in general position. Then we can draw three spanning geometric trees without crossings on R, on B, and on G such that the number of intersections of the trees is at most $3|X| - 9$.*

This follows from the following theorem that is the main theorem of this paper.

J. Akiyama and M. Kano (Eds.): JCDCG 2002, LNCS 2866, pp. 261–272, 2003.
© Springer-Verlag Berlin Heidelberg 2003

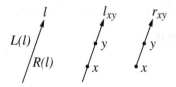

Fig. 1. Definitions of $R(l)$, $L(l)$, l_{xy}, and r_{xy}.

Theorem 2.   *Let R, B, and G be three disjoint sets of points in the plane such that the points of $X = R \cup B \cup G$ are in general position. Then we can draw three spanning geometric trees on R, on B, and on G that satisfy the following four conditions:*

(i) each geometric tree is without crossings;
(ii) every edge intersects at most three edges of each other tree;
(iii) one specified point $x \in X$ on the boundary of conv(X) is of degree one; and
(iv) the edge incident to x intersects at most two edges of each other tree.

This theorem is proved in Section 3. Let us see now that Theorem 2 implies Theorem 1.

Let T_r, T_b, and T_g be three geometric trees as in Theorem 2, and let F_x be a forest $(V(T_r) \cup V(T_b) \cup V(T_g), E(T_r) \cup E(T_b) \cup E(T_g)) = (X, E(T_r) \cup E(T_b) \cup E(T_g))$. By Theorem 2, each edge of F_x intersects at most 6 edges and $|E(F_x)| = |E(T_r)| + |E(T_b)| + |E(T_g)| = (|V(T_r)| - 1) + (|V(T_b)| - 1) + (|V(T_g)| - 1) = |X| - 3$. On the other hand, one intersection corresponds to two edges. Thus the number of intersections of the trees is at most $6(|X| - 3)/2 = 3|X| - 9$.

We conclude this section with the following conjecture.

Conjecture 1.   Let R, B, and G be three disjoint sets of points in the plane such that the points of $X = R \cup B \cup G$ are in general position. Then we can draw three spanning geometric trees without crossings on R, on B, and on G such that every edge intersects at most two edges of each other tree.

2   Preliminaries

In this paper, we deal only with *directed lines* in order to define the right side of a line and the left side of it. Thus a *line* means a directed line. A line l dissects the plane into three pieces: l and two open half-planes $R(l)$ and $L(l)$, where $R(l)$ and $L(l)$ denote the *open half-planes* which are on the right side and on the left side of l, respectively. For two points x and y, we denote by l_{xy} and r_{xy} a line in the direction from x to y and a ray emanating from x to y, respectively. (See Fig. 1.)

In order to prove Theorem 2, we need the following lemma.

Lemma 1.   *Let R and B be two disjoint sets of points in the plane such that the points of $X = R \cup B$ are in general position. Let l be a line passing through*

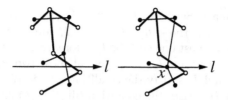

Fig. 2. Examples of Lemma 1.

at most one point in X. Then we can draw two spanning geometric trees T_r on R and T_b on B that satisfy the following six conditions:
(i) each geometric tree is without crossings;
(ii) each segment intersects at most three segments;
(iii) if l passes through no point in X then l intersects at most one edge of each tree;
(iv) if l passes through one point $x \in R$ then l and T_r intersect only at x, and l intersects at most one edge of T_b;
(v) if l passes through one point $x \in B$ then l and T_b intersect only at x, and l intersects at most one edge of T_r; and
(vi) each segment crossing l intersects at most two segments.
(See Fig. 2.)

Proof. We prove the lemma by mathematical induction on $|X|$. If $|X| \le 2$ then the lemma follows immediately. Thus we may assume $|X| \ge 3$, so $|R| \ge 2$ or $|B| \ge 2$, say $|R| \ge 2$.

Claim 1   *We may assume that $L(l) \cap B \ne \emptyset$, $L(l) \cap R \ne \emptyset$, $R(l) \cap B \ne \emptyset$, and $R(l) \cap R \ne \emptyset$.*

Proof. Suppose $|B| = 0$. There exist at least three vertices of R on the boundary of $conv(X) = conv(R)$. Two vertices of these vertices are in $L(l) \cup l$ or in $R(l) \cup l$, say in $L(l) \cup l$. Let x and y be two vertices in $X \cap (L(l) \cup l)$ such that x is adjacent to y on the boundary of $conv(X) = conv(R)$. Without loss of generality, we may assume that x is not on l. By applying the inductive hypothesis to $X - x$ and l, we can obtain one spanning geometric tree T_r' on $R - x$ that satisfies the six conditions of lemma 1. Since x is a vertex on the boundary of $conv(X) = conv(R)$, the segment xy intersects no other edges of T_r'. Therefore by setting $T_r = T_r' \cup \{x\} \cup \{xy\}$, we can obtain the desired trees.

Suppose $|B| = 1$. Let x be a vertex in B. By applying the inductive hypothesis to $X - x$ and l, we can obtain one spanning geometric tree T_r' on R that satisfies the six conditions of lemma 1. Therefore by setting $T_r = T_r'$ and $T_b = (\{x\}, \emptyset)$, we can obtain the desired trees.

Thus we may assume $|R| \ge 2$ and $|B| \ge 2$.

Suppose that $L(l) \cap X = \emptyset$ or $R(l) \cap X = \emptyset$, say $L(l) \cap X = \emptyset$. Let x and y be two vertices in R. By applying the inductive hypothesis to $X - x$ and l_{xy}, we can obtain two spanning geometric trees T_r' on $R - x$ and T_b' on B that satisfy the six conditions of lemma 1. Set $T_r = T'_r \cup \{x\} \cup \{xy\}$ and $T_b = T'_b$. Then by the condition (iv) of Lemma 1, the segment xy and T'_r intersect only at y, and the segment xy intersects at most one edge of T'_b. Moreover, by the condition (vi) of Lemma 1, the segment crossing l_{xy} intersects at most two edges of T'_r and one segment xy. Thus T_r and T_b satisfy the condition (i) and (ii) of Lemma 1. Since $L(l) \cap X = \emptyset$, T_r and T_b satisfy the condition (iii),(iv),(v) and (vi) of lemma 1. Therefore we can obtain the desired trees.

Thus we may assume that $L(l) \cap X \neq \emptyset$ and $R(l) \cap X \neq \emptyset$.

Suppose that $L(l) \cap B = \emptyset$ or $L(l) \cap R = \emptyset$ or $R(l) \cap B = \emptyset$ or $R(l) \cap R = \emptyset$, say $L(l) \cap B = \emptyset$. If $R(l) \cap R = \emptyset$ then l separates X into R and B. By applying the inductive hypothesis to R and l, we can obtain one spanning geometric tree T'_r on R that satisfies the six conditions of lemma 1. By applying the inductive hypothesis to B and l, we can obtain one spanning geometric tree T'_b on B that satisfies the six conditions of lemma 1. Therefore by setting $T_r = T'_r$ and $T_b = T'_b$, we can obtain the desired trees. Thus we may assume $R(l) \cap R \neq \emptyset$. If $l \cap R \neq \emptyset$ then let x be a vertex in $l \cap R$. If $l \cap R = \emptyset$ then let x be a vertex in $L(l) \cap R$. Let y be a vertex in $R(l) \cap R$. By applying the inductive hypothesis to $R \cap (L(l) \cup l)$ and l_{xy}, we can obtain one spanning geometric tree T'_r on $R \cap (L(l) \cup l)$ that satisfies the six conditions of lemma 1. By applying the inductive hypothesis to $X \cap R(l)$ and l_{xy}, we can obtain two spanning geometric trees T^r_r on $R \cap R(l)$ and T^b_r on $B \cap R(l)$ that satisfy the six conditions of lemma 1. Therefore by setting $T_r = T^l_r \cup T^r_r \cup \{xy\}$ and $T_b = T^r_b$, we can obtain the desired tree.

Thus we may assume $L(l) \cap B \neq \emptyset$, $L(l) \cap R \neq \emptyset$, $R(l) \cap B \neq \emptyset$, and $R(l) \cap R \neq \emptyset$. ∎

We consider two cases whether or not $l \cap X$ is empty.

Case 1 $l \cap X = \emptyset$.

Let l' be a line parallel to l such that $l' \cap X \neq \emptyset$, $X \cap L(l) \cap R(l') = \emptyset$, and l' has the same direction as l. Let x be a vertex in $l' \cap X$. Without loss of generality, we may assume that x is a red point. Let r_x and r^*_x be two rays lying on the line l' and having the same starting point x such that r_x has the same direction as l' and r^*_x has the opposite direction of l'. By a suitable clockwise rotation of r_x and a suitable counterclockwise rotation of r^*_x around x, we can find two distinct vertices y and z and two lines l_{xy} and l_{xz} such that $X \cap R(l') \cap L(l_{xy}) = \emptyset$ and $X \cap R(l') \cap R(l_{xz}) = \emptyset$. Since $X \cap L(l) \cap R(l') = \emptyset$, $y, z \in R(l) \cap X$. (See Fig. 3.)

Claim 2   *We may assume that y or z is a vertex in B.*

Proof.   Suppose that y and z are vertices in R. Now $R(l) \cap B \neq \emptyset$ and $L(l) \cap B \neq \emptyset$. Let w and w' be two vertices in $R(l) \cap B$ and in $L(l) \cap B$, respectively. At least one segment in $\{xy, xz\}$ does not intersect ww', without loss of generality, we may assume that xy does not intersect ww'. (See Fig. 4.)

By applying the inductive hypothesis to $L(l) \cap X$ and $l_{ww'}$, we can obtain two spanning geometric trees T^l_r on $L(l) \cap R$ and T^l_b on $L(l) \cap B$ that satisfy the six conditions of Lemma 1. By applying the inductive hypothesis to $R(l) \cap X$ and $l_{ww'}$, we can obtain two spanning geometric trees T^r_r on $R(l) \cap R$ and T^r_b on

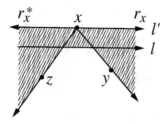

Fig. 3. The starting situation of the first case.

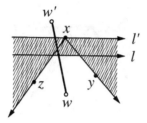

Fig. 4. We may assume that xy does not intersect ww'.

$R(l) \cap B$ that satisfy the six conditions of Lemma 1. Set $T_r = T_r^l \cup T_r^r \cup \{xy\}$ and $T_b = T_b^l \cup T_b^r \cup \{ww'\}$. Then by the condition (iv) of Lemma 1, the segment ww' and T_b^l intersect only at w', and the segment ww' intersects at most one edge of T_b^l. Similarly, the segment ww' and T_b^r intersect only at w, and the segment ww' intersects at most one edge of T_r^r. Moreover, by the condition (vi) of Lemma 1, a segment crossing $l_{ww'}$ intersects at most two segments and one segment ww'. Thus T_r and T_b satisfy the six conditions of Lemma 1. Therefore we can obtain the desired trees. Consequently, in this situation, the lemma is proved. ■

By Claim 2, without loss of generality, we may assume y is a vertex in B.

Claim 3   *We may assume that $L(l_{xy}) \cap X = \emptyset$.*

Proof.   Suppose $L(l_{xy}) \cap B \neq \emptyset$. Let y' be a vertex in $L(l_{xy}) \cap B$. Now $R(l) \cap R \neq \emptyset$, so let x' be a vertex in $R(l) \cap R$. (See Fig. 5.)

By applying the inductive hypothesis to $L(l) \cap X$ and $l_{yy'}$, we can obtain two spanning geometric trees T_r^y on $L(l) \cap R$ and T_b^y on $L(l) \cap B$ that satisfy the six conditions of Lemma 1. By applying the inductive hypothesis to $R(l) \cap X$ and $l_{xx'}$, we can obtain two spanning geometric trees T_r^x on $R(l) \cap R$ and T_b^x on $R(l) \cap B$ that satisfy the six conditions of Lemma 1. Set $T_r = T_r^x \cup T_r^y \cup \{xx'\}$ and $T_b = T_b^x \cup T_b^y \cup \{yy'\}$. Then by the condition (iv) of Lemma 1, the segment xx' and T_r^x intersect only at x', and the segment xx' intersects at most one edge of T_b^x. Similarly, the segment yy' and T_b^y intersect only at y', and the segment yy' intersects at most one edge of T_r^y. Moreover, by the condition (vi) of Lemma 1, the segment crossing $l_{xx'}$ intersects at most two edges of T_r^x and one segment

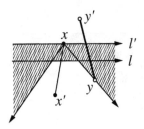

Fig. 5. We can find a red point x' and a blue point y'.

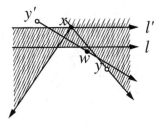

Fig. 6. We can find two segment xw and yy' which do not intersect.

xx', and the segment crossing $l_{yy'}$ intersects at most two edges of T_b^y and one segment yy'. Thus T_r and T_b satisfy the six conditions of Lemma 1. Therefore we can obtain the desired trees.

Hence we can assume $L(l_{xy}) \cap B = \emptyset$. Suppose $L(l_{xy}) \cap R \neq \emptyset$. Note that $L(l_{xy}) \cap X \neq \emptyset$ and $R(l_{xy}) \cap X \neq \emptyset$. By applying the inductive hypothesis to $(R(l_{xy}) \cup l_{xy}) \cap X$ and l, we can obtain two spanning geometric trees T_r^r on $(R(l_{xy}) \cup l_{xy}) \cap R$ and T_b^r on $(R(l_{xy}) \cup l_{xy}) \cap B$ that satisfy the six conditions of Lemma 1. By applying the inductive hypothesis to $(L(l_{xy}) \cup l_{xy}) \cap X$ and l, we can obtain two spanning geometric trees T_r^l on $(L(l_{xy}) \cup l_{xy}) \cap R$ and T_b^l on $(L(l_{xy}) \cup l_{xy}) \cap B$ that satisfy the six conditions of Lemma 1. Set $T_r = T_r^r \cup T_r^l$ and $T_b = T_b^r \cup T_b^l$. Since $L(l_{xy}) \cap B = \emptyset$, $V(T_b^l) = \{y\}$. Then l intersects at most one edge of T_r^r and at most one edge of T_b^r. Thus T_r and T_b satisfy the six conditions of Lemma 1. Therefore we can obtain the desired trees.

Hence we may assume $L(l_{xy}) \cap R = \emptyset$, namely, $L(l_{xy}) \cap X = \emptyset$. ∎

Let y' be a vertex in $L(l) \cap B$. By a suitable clockwise rotation of $r_{y'x}$ around y', we can find a vertex $w \in R(l) \cap X$ such that $R(l) \cap L(l_{y'w}) \cap X = \emptyset$.

Suppose that w is a vertex in R. Then since $w \neq y$, two segments xw and yy' do not intersect. (See Fig. 6.)

By applying the inductive hypothesis to $L(l) \cap X$ and $l_{yy'}$, we can obtain two spanning geometric trees T_r^l on $L(l) \cap R$ and T_b^l on $L(l) \cap B$ that satisfy the six conditions of Lemma 1. By applying the inductive hypothesis to $R(l) \cap X$ and $l_{yy'}$, we can obtain two spanning geometric trees T_r^r on $L(l) \cap R$ and T_b^r on $L(l) \cap B$ that satisfy the six conditions of Lemma 1. Set $T_r = T_r^l \cup T_r^r \cup \{xw\}$ and $T_b = T_b^l \cup T_b^r \cup \{yy'\}$. Then l intersects exactly two segments xw and yy',

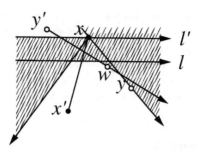

Fig. 7. If w is a blue point then we can use the inductive hypothesis.

and no edges of T_r and T_b intersects xw. By the condition (iv) of Lemma 1, the segment yy' and T_b^l intersect only at y', and the segment yy' intersects at most one edge of T_r^l. Similarly, the segment yy' and T_b^r intersect only at y, and the segment yy' intersects at most one edge of T_r^r. Moreover, by the condition (vi) of Lemma 1, a segment crossing $l_{yy'}$ intersects at most two edges of T_b^l and T_b^r and one segment yy'. Thus T_r and T_b satisfy the six conditions of Lemma 1. Therefore we can obtain the desired trees.

Suppose that w is a vertex in B. Let x' be a vertex in $R(l) \cap R$. Note that no geometric graphs on $X \cap L(l)$ intersect the segment xx', and no geometric graphs on $X \cap R(l)$ intersect the segment wy'. (See Fig. 7.)

By applying the inductive hypothesis to $L(l) \cap X$ and $l_{wy'}$, we can obtain two spanning geometric trees T_r^l on $L(l) \cap R$ and T_b^l on $L(l) \cap B$ that satisfy the six conditions of Lemma 1. By applying the inductive hypothesis to $R(l) \cap X$ and $l_{xx'}$, we can obtain two spanning geometric trees T_r^r on $R(l) \cap R$ and T_b^r on $R(l) \cap B$ that satisfy the six conditions of Lemma 1. Set $T_r = T_r^l \cup T_r^r \cup \{xx'\}$ and $T_b = T_b^l \cup T_b^r \cup \{wy'\}$. Then by the condition (iv) of Lemma 1, the segment xx' and T_r^r intersect only at x', and the segment xx' intersects at most one edge of T_b^r. Similarly, the segment wy' and T_b^l intersect only at y', and the segment wy' intersects at most one edge of T_r^l. Moreover, by the condition (vi) of Lemma 1, the segment crossing $l_{xx'}$ intersects at most two edges of T_r^r and one segment xx', and the segment crossing $l_{wy'}$ intersects at most two edges of T_b^l and one segment wy'. Thus T_r and T_b satisfy the six conditions of Lemma 1. Therefore we can obtain the desired trees. Consequently, in this case, the lemma is proved.

Case 2 $l \cap X \neq \emptyset$.

Let x be a vertex in $l \cap X$. Without loss of generality, $x \in R$. Let y be a vertex in $L(l) \cap B$, and y' be a vertex in $R(l) \cap B$.

By applying the inductive hypothesis to $(L(l) \cup l) \cap X$ and $l_{yy'}$, we can obtain two spanning geometric trees T_r^l on $(L(l) \cup l) \cap R$ and T_b^l on $(L(l) \cup l) \cap B$ that satisfy the six conditions of Lemma 1. By applying the inductive hypothesis to $(R(l) \cup l) \cap X$ and $l_{yy'}$, we can obtain two spanning geometric trees T_r^r on $(R(l) \cup l) \cap R$ and T_b^r on $(R(l) \cup l) \cap B$ that satisfy the six conditions of Lemma 1. Set $T_r = T_r^l \cup T_r^r$ and $T_b = T_b^l \cup T_b^r \cup \{yy'\}$. Then by the condition (iv) of Lemma 1, the segment yy' and T_b^l intersect only at y, and the segment yy' intersects at most one edge of T_r^l. Similarly, the segment yy' and T_r^r intersect only at y', and the segment yy' intersects at most one edge of T_r^r. Moreover, by the condition (vi) of Lemma 1, a segment crossing $l_{yy'}$ intersects at most two edges of T_b^l and T_b^r and one segment yy'. Thus T_r and T_b satisfy the six conditions of Lemma 1. Therefore we can obtain the desired trees. Consequently the proof of the lemma is complete. ∎

3   Proof of Theorem 2

We prove the theorem by mathematical induction on $|X|$. If $|X| = 1$ then the theorem follows immediately. Thus we may assume $|X| \geq 2$. Without loss of generality, we may assume that x is a vertex in R. Let xy be a segment with endpoints x and $y \in X$ at the boundary of $conv(X)$.

Suppose that y is a vertex in R. By applying the inductive hypothesis to $X - x$, we can obtain three spanning geometric trees T_r' on $X - x$, T_b' on B, and T_g' on G that satisfy the four conditions of Theorem 2. Set $T_r = T_r' \cup \{x\} \cup \{xy\}$, $T_b = T_b'$, and $T_g = T_g'$. Since no segments intersects xy, we can obtain the desired trees.

Thus we may assume that y is a vertex in $B \cup G$. Without loss of generality, we may assume that y is a vertex in B. Suppose $|R| = 1$. By applying the inductive hypothesis to $X - x$, we can obtain two spanning geometric trees T_b' on B and T_g' on G that satisfy the four conditions of Theorem 2. Set $T_r = (\{x\}, \emptyset)$, $T_b = T_b'$, and $T_g = T_g'$. We can obtain the desired trees.

Thus we may assume $|R| \geq 2$. Suppose $G = \emptyset$. Let x' be a vertex in R that is not x. By applying Lemma 1 to $X - x$ and $l_{xx'}$, we can obtain two spanning geometric trees T_r' on $X - x$ and T_b' on B that satisfy the six conditions of Lemma 1. Set $T_r = T_r' \cup \{x\} \cup \{xx'\}$ and $T_b = T_b'$. Then by the condition (iv) of Lemma 1, the segment xx' and T_r' intersect only at x', and the segment xx' intersects at most one edge of T_b. Moreover, by the condition (vi) of Lemma 1, a segment crossing $l_{xx'}$ intersects at most two edges of T_r and one segment xx'. Thus T_r and T_b satisfy the four conditions of Theorem 2. Therefore we can obtain the desired trees.

Thus we may assume $G \neq \emptyset$. Without loss of generality, we may assume $(X - \{x, y\}) \subset L(l_{yx})$. By a suitable counterclockwise rotation of r_{yx} around y, we can find a vertex $z \in R$ and a line l_{yz} such that $R(l_{yz}) \cap R = \emptyset$.

Suppose $R(l_{yz}) \cap G = \emptyset$. Since $G \neq \emptyset$, $L(l_{yz}) \cap G \neq \emptyset$. Thus $R(l_{yz}) \cap X \neq \emptyset$ and $L(l_{yz}) \cap X \neq \emptyset$. By applying the inductive hypothesis to $(R(l_{yz}) \cup l_{yz}) \cap X$, we can obtain two spanning geometric trees T_r^r on $(R(l_{yz}) \cup l_{yz}) \cap R$ and T_b^r on $(R(l_{yz}) \cup l_{yz}) \cap B$ that satisfy the four conditions of Theorem 2. By applying the inductive hypothesis to $(L(l_{yz}) \cup l_{yz}) \cap X$, we can obtain three spanning geometric trees T_r^l on $(L(l_{yz}) \cup l_{yz}) \cap R$, T_b^l on $(L(l_{yz}) \cup l_{yz}) \cap B$, and T_g^l on $(L(l_{yz}) \cup l_{yz}) \cap G$ that satisfy the four conditions of Theorem 2. Set $T_r = T_r^r \cup T_r^l$, $T_b = T_b^r \cup T_b^l$, and $T_g = T_g^l$. We can obtain the desired trees.

Thus we may assume $R(l_{yz}) \cap G \neq \emptyset$. By a suitable clockwise rotation of r_{yz} around y, we can find a vertex $w \in G$ and a line l_{yw} such that $R(l_{yz}) \cap L(l_{yw}) \cap$

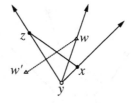

Fig. 8. One of two cases such that xz and ww' intersect.

$G = \emptyset$. By applying the inductive hypothesis to the vertex set $(L(l_{yz}) \cap X) \cup \{y, z, w\}$ that w is specified, we can obtain three spanning geometric trees T'_r on $(L(l_{yz}) \cap R) \cup \{z\}$, T'_b on $(L(l_{yz}) \cap B) \cup \{y\}$, and T'_g on $(L(l_{yz}) \cap G) \cup \{w\}$ that satisfy the four conditions of Theorem 2. Set $T_r = T'_r \cup \{x\} \cup \{xz\}$. Let w' be an endpoint of edge incident to w of T'_g that is not w.

We want to find two spanning geometric trees T''_b on $(R(l_{yz}) \cap B) \cup \{y\}$ and T''_g on $R(l_{yz}) \cap G$ such that $T_b = T'_b \cup T''_b$ and $T_g = T'_g \cup T''_g$. We consider four cases.

Case 1 xz and ww' intersect at a point p, and z is in $R(l_{ww'})$. (See Fig. 8.)

Set

$$A_1 = R(l_{yz}) \cap R(l_{xz}) \cap R(l_{ww'}) \cap L(l_{yw}) \cap X,$$
$$A_2 = R(l_{yz}) \cap L(l_{xz}) \cap R(l_{ww'}) \cap L(l_{yw}) \cap X,$$
$$A_3 = R(l_{yz}) \cap R(l_{xz}) \cap L(l_{ww'}) \cap L(l_{yw}) \cap X,$$
$$A_4 = R(l_{yz}) \cap L(l_{xz}) \cap L(l_{ww'}) \cap L(l_{yw}) \cap X, and$$
$$Y = R(l_{yw}) \cap L(l_{yx}) \cap X.$$

Note that $A_1 \cup A_2 \cup A_3 \cup A_4 \subset B$ and $Y \subset B \cup G$.

Claim 1   *We may assume that $A_2 \neq \emptyset$.*

Proof.   Suppose that $A_2 = \emptyset$. Let S_4 be a geometric tree obtained by joining y to each vertex of A_4. If $A_1 \neq \emptyset$ then let a_1 be a vertex in A_1 and let S_1 be a geometric tree obtained by joining a_1 to each vertex of $A_1 - a_1$. If $A_3 \neq \emptyset$ then let a_3 be a vertex in A_3 and let S_3 be a geometric tree obtained by joining a_3 to each vertex of $A_3 - a_3$. If $A_1 \neq \emptyset$ and $A_3 \neq \emptyset$ then set $T^A_b = S_1 \cup S_3 \cup S_4 \cup \{a_1 a_3, a_3 y\}$. If $A_1 \neq \emptyset$ and $A_3 = \emptyset$ then set $T^A_b = S_1 \cup S_4 \cup \{a_1 y\}$. If $A_1 = \emptyset$ and $A_3 \neq \emptyset$ then set $T^A_b = S_3 \cup S_4 \cup \{a_3 y\}$. If $A_1 = \emptyset$ and $A_3 = \emptyset$ then set $T^A_b = S_4$. By applying Lemma 1 to $Y \cup \{y, w\}$ and l_{xz}, we can obtain two spanning geometric trees T^Y_b on $(Y \cap B) \cup \{y\}$ and T^Y_g on $(Y \cap G) \cup \{w\}$ that satisfy the six conditions of Lemma 1. Set $T''_b = T^A_b \cup T^Y_b$ and $T''_g = T^Y_g$. Then T_r, T_b, and T_g satisfy the four conditions of Theorem 2. ∎

By applying Lemma 1 to $(R(l_{yz}) \cap R(l_{ww'}) \cap X) \cup \{w\}$ and l_{xz}, we can obtain two spanning geometric trees T^r_b on $R(l_{yz}) \cap R(l_{ww'}) \cap B$ and T^r_g on

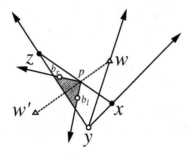

Fig. 9. We can safely join b_r and b_l.

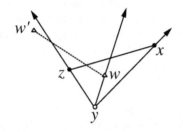

Fig. 10. The other case such that xz and ww' intersect.

$(R(l_{yz}) \cap R(l_{ww'}) \cap G) \cup \{w\}$ that satisfy the six conditions of Lemma 1. By applying Lemma 1 to $(R(l_{yz}) \cap L(l_{ww'}) \cap (X - x)) \cup \{w, y\}$ and l_{xz}, we can obtain two spanning geometric trees T_b^l on $(R(l_{yz}) \cap L(l_{ww'}) \cap B) \cup \{y\}$ and T_g^l on $(R(l_{yz}) \cap L(l_{ww'}) \cap G) \cup \{w\}$ that satisfy the six conditions of Lemma 1. By a suitable clockwise rotation of $r_{pw'}$ and a suitable counterclockwise rotation of $r_{pw'}$ around p, we can find two vertices $b_r \in A_2$ and $b_l \in A_4 \cup \{y\}$ and two lines l_{pb_r} and l_{pb_l} such that $R(l_{yz}) \cap L(l_{pb_r}) \cap R(l_{pb_l}) \cap X = \emptyset$. (See Fig. 9.)

Set $T_b'' = T_b^r \cup T_b^l \cup \{b_r b_l\}$ and $T_g'' = T_g^r \cup T_g^l$. Then T_r, T_b, and T_g satisfy the four conditions of Theorem 2. Consequently, in this case, the theorem is proved.

Case 2   *xz and ww' intersect, and z is in $L(l_{ww'})$. (See Fig. 10.)*

Set

$$A_1 = R(l_{yz}) \cap R(l_{xz}) \cap R(l_{ww'}) \cap L(l_{yw}) \cap X,$$
$$A_2 = R(l_{yz}) \cap R(l_{xz}) \cap L(l_{ww'}) \cap L(l_{yw}) \cap X,$$
$$A_3 = R(l_{yz}) \cap L(l_{xz}) \cap R(l_{ww'}) \cap L(l_{yw}) \cap X,$$
$$A_4 = R(l_{yz}) \cap L(l_{xz}) \cap L(l_{ww'}) \cap L(l_{yw}) \cap X, and$$
$$Y = R(l_{yw}) \cap L(l_{yx}) \cap X.$$

Note that $A_1 \cup A_2 \cup A_3 \cup A_4 \subset B$ and $Y \subset B \cup G$.

Claim 2   *We may assume that $A_2 \neq \emptyset$.*

Proof.   Suppose that $A_2 = \emptyset$. Let S_4 be a geometric tree obtained by joining y to each vertex of A_4. If $A_1 \neq \emptyset$ then let a_1 be a vertex in A_1 and let S_1 be a geometric tree obtained by joining a_1 to each vertex of $A_1 - a_1$. If $A_3 \neq \emptyset$ then let a_3 be a vertex in A_3 and let S_3 be a geometric tree obtained by joining a_3 to each vertex of $A_3 - a_3$. If $A_1 \neq \emptyset$ and $A_3 \neq \emptyset$ then set $T_b^A = S_1 \cup S_3 \cup S_4 \cup \{a_1 a_3, a_3 y\}$. If $A_1 \neq \emptyset$ and $A_3 = \emptyset$ then set $T_b^A = S_1 \cup S_4 \cup \{a_1 y\}$. If $A_1 = \emptyset$ and $A_3 \neq \emptyset$ then set $T_b^A = S_3 \cup S_4 \cup \{a_3 y\}$. If $A_1 = \emptyset$ and $A_3 = \emptyset$ then set $T_b^A = S_4$. By applying Lemma 1 to $Y \cup \{y, w\}$ and l_{xz}, we can obtain two spanning geometric trees T_b^Y on $(Y \cap B) \cup \{y\}$ and T_g^Y on $(Y \cap G) \cup \{w\}$ that satisfy the six conditions of Lemma 1. Set $T_b'' = T_b^A \cup T_b^Y$ and $T_g'' = T_g^Y$. Then T_r, T_b, and T_g satisfy the four conditions of Theorem 2. ∎

Claim 3   *We may assume that $A_1 \cup A_3 \neq \emptyset$.*

Proof.   Suppose that $A_1 = A_3 = \emptyset$. Let S_4 be a geometric tree obtained by joining y to each vertex of A_4. Let a_2 be a vertex in A_2 and S_2 be a geometric tree obtained by joining a_2 to each vertex of $A_2 - a_2$. Then set $T_b^A = S_2 \cup S_4 \cup \{a_2 y\}$. By applying Lemma 1 to $Y \cup \{y, w\}$ and l_{xz}, we can obtain two spanning geometric trees T_b^Y on $(Y \cap B) \cup \{y\}$ and T_g^Y on $(Y \cap G) \cup \{w\}$ that satisfy the six conditions of Lemma 1. Set $T_b'' = T_b^A \cup T_b^Y$ and $T_g'' = T_g^Y$. Then T_r, T_b, and T_g satisfy the four conditions of Theorem 2. ∎

By applying Lemma 1 to $(R(l_{yz}) \cap R(l_{ww'}) \cap (X - x)) \cup \{w\}$ and l_{xz}, we can obtain two spanning geometric trees T_b^r on $R(l_{yz}) \cap R(l_{ww'}) \cap B$ and T_g^r on $(R(l_{yz}) \cap R(l_{ww'}) \cap G) \cup \{w\}$ that satisfy the six conditions of Lemma 1. By applying the inductive hypothesis to $(R(l_{yw}) \cap L(l_{ww'}) \cap X) \cup \{w, y\}$, we can obtain two spanning geometric trees T_b^l on $(R(l_{yw}) \cap L(l_{ww'}) \cap B) \cup \{y\}$ and T_g^l on $(R(l_{yw}) \cap L(l_{ww'}) \cap G) \cup \{w\}$ that satisfy the four conditions of Theorem 2. Let S_4 be a geometric tree obtained by joining y to each vertex of A_4. Let a_2 be a vertex in A_2 and S_2 be a geometric tree obtained by joining a_2 to each vertex of $A_2 - a_2$. By a suitable clockwise rotation of $r_{ww'}$ around w, we can find a vertex b in $A_1 \cup A_3$ and a line l_{wb} such that $R(l_{yz}) \cap R(l_{ww'}) \cap L(l_{wb}) \cap X = \emptyset$. If $b \in A_1$ then set $T_b'' = T_b^r \cup T_b^l \cup S_2 \cup S_4 \cup \{ba_2, a_2 y\}$ and $T_g'' = T_g^r \cup T_g^l$. Then T_r, T_b, and T_g satisfy the four conditions of Theorem 2. If $b \in A_3$ then set $T_b'' = T_b^r \cup T_b^l \cup S_2 \cup S_4 \cup \{by, a_2 y\}$ and $T_g'' = T_g^r \cup T_g^l$. Then T_r, T_b, and T_g satisfy the four conditions of Theorem 2. Consequently, in this case, the theorem is proved. (See Fig. 11.)

We can prove the following cases by a method similar to the proof of case 1 and case 2.

Case 3   *xz and ww' do not intersect, and z is in $R(l_{ww'})$.*

Case 4   *xz and ww' do not intersect, and z is in $L(l_{ww'})$.*

Consequently the theorem is proved.

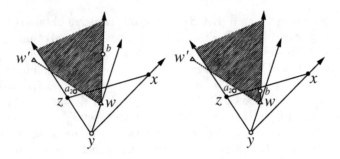

Fig. 11. In each case, we can get a desired blue tree.

References

1. S. Tokunaga, Crossing number of two connected geometric graphs, *Info. Proc. Let.* **59** (1996), 331-333.

Open Problems in Geometric Methods for Instance-Based Learning*

Godfried Toussaint

School of Computer Science
McGill University
Montréal, Québec, Canada
godfried@cs.mcgill.ca

Abstract. In the typical approach to instance-based learning, random data (the training set of patterns) are collected and used to design a decision rule (classifier). One of the most well known such rules is the k-nearest-neighbor decision rule in which an unknown pattern is classified into the majority class among its k nearest neighbors in the training set. In the past fifty years many approaches have been proposed to improve the performance of this rule. More recently geometric methods have been found to be the best. Here we mention a variety of open problems of a computational geometric nature that arize in these methods. To provide some context and motivation for these open problems we briefly describe the methods and list some key references.

1 Introduction and Motivation

In the typical non-parametric classification problem (see Aha [2], Devroye, Györfy and Lugosi [7], Duda, Hart and Stork [9], McLachlan [19], O'Rourke and Toussaint [20]) we have available a set of d measurements or observations (also called a feature vector) taken from each member of a data set of n objects (patterns) denoted by $\{X, Y\} = \{(X_1, Y_1), (X_2, Y_2), ..., (X_n, Y_n)\}$, where X_i and Y_i denote, respectively, the feature vector on the ith object and the class label of that object. One of the most attractive decision procedures, conceived by Fix and Hodges in 1951, is the nearest-neighbor rule (1-*NN*-rule). Let Z be a new pattern (feature vector) to be classified and let X_j be the feature vector in $\{X, Y\} = \{(X_1, Y_1), (X_2, Y_2), ..., (X_n, Y_n)\}$ closest to Z. The nearest neighbor decision rule classifies the unknown pattern Z into class Y_j. Figure 1 depicts the decision boundary of the 1-*NN*-rule. The feature space is partitioned into convex polyhedra (polygons in the plane). This partitioning is called the *Voronoi* diagram. Each pattern (X_i, Y_i) in $\{X, Y\}$ is surrounded by its Voronoi polyhedron consisting of those points in the feature space closer to (X_i, Y_i) than to (X_j, Y_j) for all $j \neq i$. The 1-*NN*-rule classifies a new pattern Z that falls into the Voronoi polyhedron of pattern X_j into class Y_j. Therefore the decision boundary of the 1-*NN*-rule is determined by those portions of the Voronoi diagram that separate

---

* This research was supported by NSERC and FCAR.

J. Akiyama and M. Kano (Eds.): JCDCG 2002, LNCS 2866, pp. 273–283, 2003.
© Springer-Verlag Berlin Heidelberg 2003

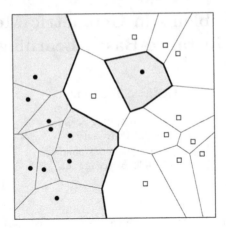

Fig. 1. The nearest neighbor decision boundary is a subset of the Voronoi diagram.

patterns belonging to different classes. In the example depicted in Figure 1 the decision boundary is shown in bold lines and the resulting decision region of one class is shaded.

A key feature of this decision rule (also called *lazy learning* [2], *instance-based learning* [1], [5], and *memory-based reasoning* [27]) is that it performs remarkably well considering that no explicit knowledge of the underlying distributions of the data is used. Consider for example the two class problem and denote the *a priori* probabilities of the two classes by $P(C_1)$ and $P(C_2)$, the *a posteriori* probabilities by $P(C_1|X)$ and $P(C_2|X)$, and the mixture probability density function by

$$p(X) = P(C_1)p(X|C_1) + P(C_2)p(X|C_2),$$

where $p(X|C_i)$ is the class-conditional probability density function given class $C_i, i = 1, 2$. In 1967 Cover and Hart [6] showed, under some continuity assumptions on the underlying distributions, that the asymptotic error rate of the 1-*NN* rule, denoted by $P_e[1\text{-}NN]$, is given by

$$P_e[1 - NN] = 2\mathbf{E}_X[P(C_1|X)P(C_2|X)],$$

where $\mathbf{E}_X$ denotes the expected value with respect to the mixture probability density function $p(X)$. They also showed that $P_e[1\text{-}NN]$ is bounded from above by twice the Bayes error (the error of the best possible rule). More precisely, and for the more general case of M pattern classes the bounds proved by Cover and Hart [6] are given by:

$$P_e \le P_e[1 - NN] \le P_e(2 - MP_e/(M - 1)),$$

where P_e is the optimal Bayes probability of error given by:

$$P_e = 1 - \mathbf{E}_X[\max\{P(C_1|X), P(C_2|X), ..., P(C_M|X)\}]. \tag{1}$$

and

$$P_e[1 - NN] = 1 - \mathbf{E}_X[\sum_{i=1}^{M} P(C_i|X)^2]. \tag{2}$$

Stone [28] and Devroye [8] generalized these results by proving the bounds for all distributions. These bounds imply that the nearest neighbor of Z contains at least half of the total discrimination information contained in an infinite-size training set. Furthermore, a simple generalization of this rule called the k-NN-rule, in which a new pattern Z is classified into the class with the most members present among the k nearest neighbors of Z in $\{X, Y\}$, can be used to obtain good estimates of the Bayes error (Fukunaga and Hostetler [11]) and its probability of error asymptotically approaches the Bayes error (Devroye et al. [7]). The measure $P_e[1\text{-}NN]$ turns up in a surprising variety of related problems sometimes in disguise. For example, it is also the error rate of the *proportional prediction* randomized decision rule considered by Goodman and Kruskal [13] (see also Toussaint [32]).

Since its conception, many pattern recognition practicioners have unfairly criticized the NN-rule, on the grounds of several mistaken assumptions. These mistaken assumptions are: that *all* the data $\{X, Y\}$ must be stored in order to implement such a rule; that to determine the nearest neighbor of a pattern to be classified, distances must be computed between the unknown vector Z and *all* members of $\{X, Y\}$; and that such nearest neighbor rules are not well suited for fast parallel implementation. In fact, all three assumptions are incorrect, and computational geometric progress in the 1980's and 1990's along with faster and cheaper hardware has made the k-NN-rules a practical reality for pattern recognition applications.

In practice the size of the training set $\{X, Y\}$ is not infinite. This raises two fundamental questions of both practical and theoretical interest. How fast does the error rate $P_e[\text{k-}NN]$ approach the Bayes error P_e as n approaches infinity, and what is the finite-sample performance of the k-NN-rule (Psaltis, Snapp and Venkatesh [22], Kulkarni, Lugosi and Venkatesh [17]). These questions have in turn generated a variety of further questions about several aspects of k-NN-rules relevant in practice. Such questions include the following. How can the storage of the training set be reduced without degrading the performance of the decision rule? How should the reduced training set be selected to represent the different classes? How large should k be? How should a value of k be chosen? Should all k neighbors be equally weighted when used to decide the class of an unknown pattern? If not, how should the weights be chosen? Should all the features (attributes) we weighted equally and if not how should the feature weights be chosen? Which distance metric should be used? How can the rule be made robust to overlapping classes or noise present in the training data? How can the rule be made invariant to scaling of the measurements? How can the nearest neighbors of a new point be computed efficiently? What is the smallest neural network that can implement nearest neighbor decision rules? Geometric proximity graphs such as Voronoi diagrams, their duals, and their relatives provide elegant solutions to most of these problems. For a survey of these techniques the reader is referred to [36]. In the following sections open problems concerning some of these geometric methods are proposed.

2   Reducing the Size of the Stored Training Data

2.1   Hart's Condensed Rule and Its Relatives

In 1968 Hart was the first to propose an algorithm for reducing the size of the stored data for the nearest neighbor decision rule [14]. Hart defined a *consistent* subset of the data as one that classified the remaining data correctly with the nearest neighbor rule. He then proposed an algorithm for selecting a consistent subset by heuristically searching for data that were near the decision boundary. The algorithm is simple. Let C denote the desired final consistent subset. Initially C is empty. First a random element from $\{X, Y\}$ is transferred to C. Then C is used as a classifier with the *1-NN* rule to classify all the remaining data in $\{X, Y\}$. During this scan of $\{X, Y\}$ whenever an element is incorrectly classified by C it is transferred from $\{X, Y\}$ to C. Thus $\{X, Y\}$ is shrinking and C is growing. This scan of $\{X, Y\}$ is repeated as long as at least one element is transferred from $\{X, Y\}$ to C during a complete pass of the remaining data in $\{X, Y\}$. The resulting patterns (feature vectors) are also called *support* vectors because they "support" the decision boundary (Vapnik [39]). The motivation for the preceeding heuristic is the intuition that data far from the decision boundary are not needed and that if an element is misclassified it must lie close the the decision boundary. By construction the resulting reduced set C classifies all the training data $\{X, Y\}$ correctly and hence it is referred to here as a *training-set consistent* subset. In the literature Hart's algorithm is called *CNN* and the resulting subset of $\{X, Y\}$ is called a *consistent* subset. Here the longer term *training-set consistent* is used in order to distingish it from another interesting type of subset: one that determines *exactly* the same decision boundary as the entire training set $\{X, Y\}$. The latter kind of subset will be called *decision-boundary consistent*. Clearly decision-boundary consistency implies training-set consistency but the converse is not necessarily true. Empirical results have shown that Hart's *CNN* rule considerably reduces the size of the training set and does not greatly degrade performance on a separate testing (validation) set. It is also easy to see that using a naive brute-force algorithm the complexity of computing the condensed subset of $\{X, Y\}$ is $O(dn^3)$. In practice no efforts have been made to use other than brute-force algorithms. It is an open problem to determine how much this complexity can be reduced.

In an attempt to obtain consistent subsets independent of the order in which the data are processed, and to further reduce the size of the consistent subset Gates [12] proposed what he called the *reduced nearest neighbor rule (RNN)*. The *RNN* rule consists of first performing *CNN* and subsequently a post-processing step in which elements of C are visited and deleted from C if their deletion does not result in misclassifying any elements in $\{X, Y\}$. Experimental results confirmed that *RNN* yields a slightly smaller training-set consistent subset of $\{X, Y\}$ than that obtained with *CNN* [12]. The complexity of computing such a *reduced* consistent subset is an open problem.

2.2 Minimal Size Training-Set Consistent Subsets

The first researchers to deal with computing a *minimal-size* training-set consistent subset were Ritter et al. [23]. They proposed a procedure they called a *selective* nearest neighbor rule *SNN* to obtain a minimal-size training-set consistent subset of $\{X, Y\}$, call it S, with one additional property that Hart's *CNN* does not have. Any training-set consistent subset C obtained by *CNN* has the property that every element of $\{X, Y\}$ is nearer to an element in C of the same class than to any element in C of a different class. On the other hand, the training-set consistent subset S of Ritter et al. [23] has the additional property that every element of $\{X, Y\}$ is nearer to an element in S of the same class than to any element, in the *complete* set, $\{X, Y\}$ of a different class. This additional property of *SNN* tends to keep points closer to the decision boundary than does *CNN*. The additional property allows Ritter et al. [23] to compute the selected subset S without testing all possible subsets of $\{X, Y\}$. Nevertheless, their algorithm still runs in time exponential in n in the worst case (see Wilfong [40]). However, Wilson and Martinez [41] and Wilson [42] claim that the average running time of *SNN* is $O(dn^3)$. No theoretical results along these lines are known.

Wilfong [40] showed in 1991 that the problem of finding the smallest size training-set consistent subset is NP-complete when there are three or more classes. The complexity for the case of *two* classes remains open. Furthermore, Wilfong showed that even for only two classes the problem of finding the smallest size training-set consistent *selective* subset (Ritter et al. [23]) is also NP-complete.

3 Proximity Graph Methods

The most fundamental and natural proximity graph defined on a set of points $\{X, Y\}$ is the *nearest neighbor graph* or *NNG*. Here each point in $\{X, Y\}$ is joined by an edge to its nearest neighbor (Paterson and Yao [21]). Another ubiquitous proximity graph, that contains the nearest neighbor graph as a subgraph, is the minimum spannig tree (*MST*) Zahn [43]. For a problem such as instance-based learning the most useful proximity graphs are adaptive in the sense that the number of edges they contain is a function of how the data are distributed. The minimum spanning tree is not adaptive; for n points it always contains $n - 1$ edges. In 1980 the relative neighborhood graph (*RNG*) was proposed as a tool for extracting the shape of a planar pattern (see Toussaint [35], [33], [30]). An example of the planar *RNG* is shown in Figure 2, but such definitions are readily extended to higher dimensions (see Su and Chang [29]). Proximity graphs have many applications in pattern recognition (see Toussaint [34], [38], [37]). There is a vast literature on proximity graphs and the reader is directed to Jaromczyk and Toussaint [16] for a start. The most well known proximity graphs besides those mentioned above are the Urquhart Graph *UG* [3], the Gabriel graph *GG* and the Delaunay triangulation *DT*. All these are nested together in the following relationship:

$$NNG \subseteq MST \subseteq RNG \subseteq UG \subseteq GG \subseteq DT \tag{3}$$

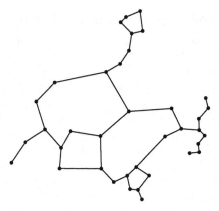

Fig. 2. The *relative-neighborhood-graph* of a planar set of points.

3.1   Decision-Boundary-Consistent Subsets

In 1979 Toussaint and Poulsen [38] used d-dimensional Voronoi diagrams to delete redundant members of $\{X, Y\}$ resulting in a subset of $\{X, Y\}$ that implements *exactly* the same decision boundary as would be obtained using all of $\{X, Y\}$. For this reason the method is called *Voronoi condensing*. The algorithm in [38] is surprisingly simple. Two points in $\{X, Y\}$ are called *Voronoi neighbors* if their corresponding Voronoi polyhedra share a face. First mark each point X_i if all its Voronoi neighbors belong to the same class as X_i. Then discard all marked points. The remaining points form the Voronoi condensed subset $\{X, Y\}$. Voronoi condensing does not change the error rate of the resulting decision rule because the nearest nighbor decision boundary with the reduced set is identical to that obtained by using the entire set. For this reason the Voronoi condensed subset is called *decision-boundary consistent*. Clearly decision-boundary consistency implies training-set consistency but the converse is not necessarily so. The most important consequence of this property is that all the theory developed for the 1-*NN* rule continues to hold true when the rule is preprocessed with Voronoi condensing. While this approach to editing sometimes does not discard a large fraction of the training data (say 90 percent), that information in itself is extremely important to the pattern classifier designer because the fraction of the data discarded is a measure of the resulting reliability of the decision rule. If few points are discarded it means that the feature space is relatively empty because few points are completely "surrounded" by points of the same class. This means more training data are urgently needed to be able to obtain reliable and robust estimates of the future performance of the rule.

3.2   Condensing Prototypes via Proximity Graphs

In 1985 Toussaint, Bhattacharya and Poulsen [37] generalized Voronoi condensing so that it would discard more points in a judicious and organized manner so as not to degrade performance unnecessarily. To better understand the rationale behind their proximity-graph-based methods it is useful to cast the Voronoi condensing algorithm in its dual form. The dual of the Voronoi diagram is the Delaunay triangulation. In this setting Voronoi condensing can be described as follows. Compute the Delaunay triangulation of $\{X, Y\}$. Mark a vertex X_i of the triangulation if all its (graph) neighbors belong to the same class as that of X_i. Finally discard all the marked vertices. The remaining points of $\{X, Y\}$ form the Voronoi condensed set. The methods proposed in [37] substitute the Delaunay triangulation by a subgraph of the triangulation. Since a subgraph has fewer edges, its vertices have lower degree on the average. This means the probability that all the graph neighbors of X_i belong to the same class as that of X_i is higher, which implies more elements of $\{X, Y\}$ will be discarded. By selecting an appropriate subgraph of the Delaunay triangulation one can control the number of elements of $\{X, Y\}$ that are discarded. Furthermore by virtue of the fact that the graph is a subgraph of the Delaunay triangulation and that the latter yields a decision-boundary consistent subset, we are confident in degrading the performance as gracefully as possible. Experimental results obtained in [37] suggested that the Gabriel graph is the best in this respect.

Also in 1985 and independently of Toussaint, Bhattacharya and Poulsen [37], Ichino and Sklansky [15] suggested the same condensing idea but with a different proximity graph that is not necessarily a subgraph of the Delaunay triangulation. They proposed a graph which they call the *rectangular-influence* graph or RIG defined as follows. Two points X_i and X_j in $\{X, Y\}$ are joined by an edge if the smallest orthogonal hyper-rectangle that contains both X_i and X_j contains no other point of $\{X, Y\}$. An orthogonal hyper-rectangle has its edges parallel to the coordinate axes. Figure 3 shows the *rectangle-of-influence* neighbors of a point. Some results on characterizing planar RIG's and recognizing if graphs can be drawn as RIG's in the plane are known [18]. Not surprisingly, condensing the training set with the RIG does not guarantee a decision-boundary consistent subset. On the other hand recall that the RIG has the nice property that it is scale-invariant, which can be useful in some classification problems.

Since the Gabriel graph and the rectangular-influence graph yield good performance for condensing the prototypes in practice, it follows that there is interest in computing these graphs efficiently in high dimensions. Brute force algorithms are simple but run in $O(dn^3)$ time, where d is the dimension. In [37] a heuristic is proposed for computing the Gabriel graph in expected time closer to $O(dn^2)$ than to $O(dn^3)$. Finding algorithms for computing these graphs in $o(dn^3)$ worst-case time is an open problem.

3.3   Proximity-Graph-Neighbor Decision Rules

The classical approaches to k-NN decision rules are rigid in at least two ways: (1) they obtain the k nearest neighbors of the unknown pattern Z based purely on distance information, and (2) the parameter k is fixed. Thus they disregard how the nearest neighbors are distributed around Z and beg the question of what the value of k should be. These problems are solved naturally and efficiently with proximity-graph-neighbor decision rules.

In 1985 Ichino and Sklansky [15] proposed the rectangular-influence graph neighbor decision rule in which Z is classified by a majority vote of its rectangle-

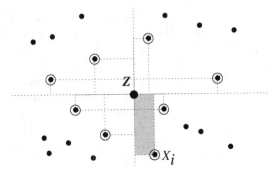

Fig. 3. The *rectangle-of-influence* neighbors of a point are scale-invariant.

of-influence graph neighbors. Devroye et al. [7] call this rule the *layered nearest neighbor rule* and have shown that if there are no ties it is asymptotically Bayes optimal. An additional useful and distinguishing property of this rule is that it is scale-invariant.

Recently new geometric definitions of neighborhoods have been proposed and new nearest neighbor decision rules based on other proximity graphs (Jaromczyk and Toussaint [16]) have been investigated. In 1996 Devroye et al. [7] proposed the *Gabriel neighbor rule* which takes a majority vote among all the Gabriel neighbors of Z among $\{X, Y\}$. Independently, Sánchez, Pla and Ferri [24], [25] proposed similar rules with other graphs as well as the Gabriel and relative neighborhood graphs. Sebban, Nock and Lallich [26] also proposed using the relative neighborhood graph decision rule in the context of selecting prototypes. Thus in these approaches both the value of k and the distance of the neighbors vary locally and adapt naturally to the distribution of the data around Z. Note that these methods also automatically and implicitly assign different "weights" to the nearest geometric neighbors of Z.

A radically different decision rule based on the overall length of proximity graphs (rather than majority votes) was proposed by Ekin, Hammer, Kogan and Winter [10]. These authors proposed computing a discriminant function for each class as follows. In the design stage of the classifier the minimum Steiner tree is computed for the training points in each class. When a query point is to be classified it is inserted in the Steiner tree of each class. Then for each class the ratio between the lengths of the Steiner trees before and after insertion is calculated. Finally the query point is classified to the class having the smallest ratio. Since computing the minimum Steiner tree is NP-hard they propose an approximation for use in practice.

The Gabriel-graph-neighbor rule (Devroye et al. [7], Sánchez et al. [24]) and the rectangular-influence-graph rule (Ichino and Sklansky [15]) open a variety of algorithmic problems in d dimensions. For any particular graph (relative neighbor, Gabriel, etc.) and a training set $\{X, Y\}$, in order to classify an unknown Z we would like to be able to answer quickly the query: which elements of $\{X, Y\}$ are the graph neighbors of Z? For the rectangular-influence-graph some results have been recently obtained by Carmen Cortes, Belen Palop and Mario Lopez.

Another fertile area which has not been investigated concerns the efficient computation of the graph neighbors of a query point, approximately, rather than exactly, with the constraint that the error in classification not be adversely affected.

3.4   Error Estimation with Proximity Graphs

For a given data set $\{X, Y\}$ and decision rule the key question is: what probability of misclassification (error) can we expect in the future [31]. One of the most popular methods for estimating this probability is the *rotation* method in which a small subset of the training set is removed for testing. Then the classifier is designed without this small subset, and finally the classifier is used to classify the test set. The number of erroneous classifications is the error estimate. This is repeated while "rotating" the test subset until all data have been used as a test set. With classifiers such as the proximity-graph neighbor decision rule it is more efficient to first compute the graph for the entire set of data and then delete the test subsets. Hence it would be desirable to have efficient algorithms for inserting and deleting vertices from the most useful proximity graphs such as the Gabriel graph, the relative-neighborhood graph and the rectangular-influence graph. One of the least biased methods for estimating the performance is the *leave-one-out* method. This is a special case of the rotation method in which the size of the test set is a single pattern. Note that in this case no insertions or deletions are required. It suffices to compute the proximity graph for all the data and then classify all the vertices by a majority vote of its neighbors using a simple graph traversal algorithm.

It seems difficult in practice to obtain a condensing algorithm that works well in *all* situations. A comparison of many algorithms led Brighton and Mellish [4] to conclude that algorithms tend to fall in two classes determined by whether the data distrubutions are either homogeneous or not. Algorithms that are good in one situation are not good in the other. Therefore a challenging open problem is to find an algorithm that works well in *all* situations. If such an algorithm exists it is likely to use proximity graphs such as the relative neighborhood graph that are locally highly sensitive to the distribution of the data.

References

1. D. W. Aha, D. Kibler, and M. Albert. Instance-based learning algorithms. In *Machine Learning, 6*, pages 37–66. Kluwer, Boston, Mass., 1991.
2. D. W. Aha, editor. *Lazy Learning.* Kluwer, Norwell, MA, 1997.
3. D. V. Andrade and L. E. de Figueiredo. Good approximations for the relative neighborhood graph. In *Proc. 13th Canadian Conference on Computational Geometry*, University of Waterloo, August 13-15 2001.
4. Henry Brighton and Chris Mellish. On the consistency of information filters for lazy learning algorithms. In J. Zitkow and J. Rauch, editors, *Principles of Data Mining and Knowledge Discovery*. Springer-Verlag, Berlin, 1999.

5. Henry Brighton and Chris Mellish. Advances in instance selection for instance-based learning algorithms. *Data Mining and Knowledge Discovery*, 6:153–172, 2002.
6. Thomas M. Cover and Peter E. Hart. Nearest neighbor pattern classification. *IEEE Transactions on Information Theory*, 13:21–27, 1967.
7. Luc Devroye, László Györfi, and Gábor Lugosi. *A Probabilistic Theory of Pattern Recognition*. Springer-Verlag New York, Inc., 1996.
8. Luc Devroye. On the inequality of Cover and Hart. *IEEE Transactions on Pattern Analysis and Machine Intelligence*, 3:75–78, 1981.
9. Richard O. Duda, Peter E. Hart, and David G. Stork. *Pattern Classification*. John Wiley and Sons, Inc., New York, 2001.
10. Oya Ekin, Peter L. Hammer, Alexander Kogan, and Pawel Winter. Distance-based classification methods. *INFOR*, 37:337–352, 1999.
11. Keinosuke Fukunaga and L. D. Hostetler. K-nearest-neighbor Bayes-risk estimation. *IEEE Transactions on Information Theory*, 21:285–293, 1975.
12. W. Gates. The reduced nearest neighbor rule. *IEEE Transactions on Information Theory*, 18:431–433, 1972.
13. L. A. Goodman and W. H. Kruskal. Measures of association for cross classifications. *J. Amer. Statistical Association*, pages 723–763, 1954.
14. Peter E. Hart. The condensed nearest neighbor rule. *IEEE Transactions on Information Theory*, 14:515–516, 1968.
15. Manabu Ichino and Jack Sklansky. The relative neighborhood graph for mixed feature variables. *Pattern Recognition*, 18:161–167, 1985.
16. J. W. Jaromczyk and Godfried T. Toussaint. Relative neighborhood graphs and their relatives. *Proceedings of the IEEE*, 80(9):1502–1517, 1992.
17. Sanjeev R. Kulkarni, Gábor Lugosi, and Santosh S. Venkatesh. Learning pattern classification - a survey. *IEEE Transactions on Information Theory*, 44:2178–2206, 1998.
18. Giuseppe Liotta, Anna Lubiw, Henk Meijer, and Sue Whitesides. The rectangle of influence drawability problem. *Computational Geometry: Theory and Applications*, 10:1–22, 1998.
19. Geoffrey J. McLachlan. *Discriminant Analysis and Statistical Pattern Recognition*. John Wiley and Sons, Inc., New York, 1992.
20. J. O'Rourke and G. Toussaint. Pattern recognition. In J. E. Goodman and J. O'Rourke, editors, *Handbook of Discrete and Computational Geometry*, chapter 43, pages 797–814. CRC Press LLC, Boca Raton, 1997.
21. M. S. Paterson and F. F. Yao. On nearest-neighbor graphs. In *Automata, Languages and Programming*, volume 623, pages 416–426. Springer, 1992.
22. Demetri Psaltis, Robert R. Snapp, and Santosh S. Venkatesh. On the finite sample performance of the nearest neighbor classifier. *IEEE Transactions on Information Theory*, 40:820–837, 1994.
23. G. L. Ritter, H. B. Woodruff, S. R. Lowry, and T. L. Isenhour. An algorithm for a selective nearest neighbor decision rule. *IEEE Transactions on Information Theory*, 21:665–669, November 1975.
24. J. S. Sánchez, F. Pla, and F. J. Ferri. On the use of neighborhood-based non-parametric classifiers. *Pattern Recognition Letters*, 18:1179–1186, 1997.
25. J. S. Sánchez, F. Pla, and F. J. Ferri. Improving the k-NCN classification rule through heuristic modifications. *Pattern Recognition Letters*, 19:1165–1170, 1998.
26. Marc Sebban, Richard Nock, and Stéphane Lallich. Boosting neighborhood-based classifiers. In *Proceedings of the 18th International Conference on Machine Learning*. Williams College, MA, 2001.

27. C. Stanfill and D. L. Waltz. Toward memory-based reasoning. *Communications of the ACM*, 29:1213–1228, December 1986.

28. C. Stone. Consistent nonparametric regression. *Annals of Statistics*, 8:1348–1360, 1977.

29. T.-H. Su and R.-C. Chang. On constructing the relative neighborhood graph in Euclidean k-dimensional spaces. *Computing*, 46:121–130, 1991.

30. Godfried T. Toussaint and Robert Menard. Fast algorithms for computing the planar relative neighborhood graph. In *Proc. Fifth Symposium on Operations Research*, pages 425–428, University of Köln, August 1980.

31. Godfried T. Toussaint. Bibliography on estimation of misclassification. *IEEE Transactions on Information Theory*, 20:472–479, 1974.

32. Godfried T. Toussaint. On the divergence between two distributions and the probability of misclassification of several decision rules. In *Proceedings of the Second International Joint Conference on Pattern Recognition*, pages 27–34, Copenhagen, 1974.

33. Godfried. T. Toussaint. Algorithms for computing relative neighbourhood graph. *Electronics Letters*, 16(22):860, 1980.

34. Godfried T. Toussaint. Pattern recognition and geometrical complexity. In *Fifth International Conference on Pattern Recognition*, pages 1324–1347, Miami, December 1980.

35. Godfried T. Toussaint. The relative neighbourhood graph of a finite planar set. *Pattern Recognition*, 12:261–268, 1980.

36. Godfried T. Toussaint. Proximity graphs for nearest neighbor decision rules: recent progress. In *Interface-2002, 34th Symposium on Computing and Statistics*, Ritz-Carlton Hotel, Montreal, 2002.

37. G. T. Toussaint, B. K. Bhattacharya, and R. S. Poulsen. The application of Voronoi diagrams to nonparametric decision rules. In *Computer Science and Statistics: The Interface*, pages 97–108, Atlanta, 1985.

38. G. T. Toussaint and R. S. Poulsen. Some new algorithms and software implementation methods for pattern recognition research. In *Proc. IEEE Int. Computer Software Applications Conf.*, pages 55–63, Chicago, 1979.

39. V. N. Vapnik. *Statistical Learning Theory*. Wiley, New York, 1998.

40. Gordon Wilfong. Nearest neighbor problems. In *Proc. 7th Annual ACM Symposium on Computational Geometry*, pages 224–233, 1991.

41. D. Randall Wilson and Tony R. Martinez. Instance pruning techniques. In D. Fisher, editor, *Machine Learning: Proceedings of the Fourteenth International Conference*, pages 404–411. Morgan Kaufmann Publishers, San Francisco, CA, 1997.

42. D. Randall Wilson and Tony R. Martinez. Reduction techniques for instance-based learning algorithms. *Machine Learning*, 38:257–286, 2000.

43. Charles T. Zahn. Graph theoretical methods for detecting and describing gestalt clusters. *IEEE Transactions on Computers*, 20:68–86, 1971.

Author Index

Lecture Notes in Computer Science

For information about Vols. 1–2834
please contact your bookseller or Springer-Verlag

Vol. 2873: J. Lawry, J. Shanahan, A. Ralescu (Eds.), Modelling with Words. XIII, 229 pages. 2003. (Subseries LNAI)

Vol. 2874: C. Priami (Ed.), Global Computing. Proceedings, 2003. XIX, 255 pages. 2003.

Vol. 2875: E. Aarts, R. Collier, E. van Loenen, B. de Ruyter (Eds.), Ambient Intelligence. Proceedings, 2003. XI, 432 pages. 2003.

Vol. 2876: M. Schroeder, G. Wagner (Eds.), Rules and Rule Markup Languages for the Semantic Web. Proceedings, 2003. VII, 173 pages. 2003.

Vol. 2877: T. Böhme, G. Heyer, H. Unger (Eds.), Innovative Internet Community Systems. Proceedings, 2003. VIII, 263 pages. 2003.

Vol. 2878: R.E. Ellis, T.M. Peters (Eds.), Medical Image Computing and Computer-Assisted Intervention - MICCAI 2003. Part I. Proceedings, 2003. XXXIII, 819 pages. 2003.

Vol. 2879: R.E. Ellis, T.M. Peters (Eds.), Medical Image Computing and Computer-Assisted Intervention - MICCAI 2003. Part II. Proceedings, 2003. XXXIV, 1003 pages. 2003.

Vol. 2880: H.L. Bodlaender (Ed.), Graph-Theoretic Concepts in Computer Science. Proceedings, 2003. XI, 386 pages. 2003.

Vol. 2881: E. Horlait, T. Magedanz, R.H. Glitho (Eds.), Mobile Agents for Telecommunication Applications. Proceedings, 2003. IX, 297 pages. 2003.

Vol. 2882: D. Veit, Matchmaking in Electronic Markets. XV, 180 pages. 2003. (Subseries LNAI)

Vol. 2883: J. Schaeffer, M. Müller, Y. Björnsson (Eds.), Computers and Games. Proceedings, 2002. XI, 431 pages. 2003.

Vol. 2884: E. Najm, U. Nestmann, P. Stevens (Eds.), Formal Methods for Open Object-Based Distributed Systems. Proceedings, 2003. X, 293 pages. 2003.

Vol. 2885: J.S. Dong, J. Woodcock (Eds.), Formal Methods and Software Engineering. Proceedings, 2003. XI, 683 pages. 2003.

Vol. 2886: I. Nyström, G. Sanniti di Baja, S. Svensson (Eds.), Discrete Geometry for Computer Imagery. Proceedings, 2003. XII, 556 pages. 2003.

Vol. 2887: T. Johansson (Ed.), Fast Software Encryption. Proceedings, 2003. IX, 397 pages. 2003.

Vol. 2888: R. Meersman, Zahir Tari, D.C. Schmidt et al. (Eds.), On The Move to Meaningful Internet Systems 2003: CoopIS, DOA, and ODBASE. Proceedings, 2003. XXI, 1546 pages. 2003.

Vol. 2889: Robert Meersman, Zahir Tari et al. (Eds.), On The Move to Meaningful Internet Systems 2003: OTM 2003 Workshops. Proceedings, 2003. XXI, 1096 pages. 2003.

Vol. 2891: J. Lee, M. Barley (Eds.), Intelligent Agents and Multi-Agent Systems. Proceedings, 2003. X, 215 pages. 2003. (Subseries LNAI)

Vol. 2892: F. Dau, The Logic System of Concept Graphs with Negation. XI, 213 pages. 2003. (Subseries LNAI)

Vol. 2893: J.-B. Stefani, I. Demeure, D. Hagimont (Eds.), Distributed Applications and Interoperable Systems. Proceedings, 2003. XIII, 311 pages. 2003.

Vol. 2894: C.S. Laih (Ed.), Advances in Cryptology - ASIACRYPT 2003. Proceedings, 2003. XIII, 543 pages. 2003.

Vol. 2895: A. Ohori (Ed.), Programming Languages and Systems. Proceedings, 2003. XIII, 427 pages. 2003.

Vol. 2896: V.A. Saraswat (Ed.), Advances in Computing Science – ASIAN 2003. Proceedings, 2003. VIII, 305 pages. 2003.

Vol. 2897: O. Balet, G. Subsol, P. Torguet (Eds.), Virtual Storytelling. Proceedings, 2003. XI, 240 pages. 2003.

Vol. 2898: K.G. Paterson (Ed.), Cryptography and Coding. Proceedings, 2003. IX, 385 pages. 2003.

Vol. 2899: G. Ventre, R. Canonico (Eds.), Interactive Multimedia on Next Generation Networks. Proceedings, 2003. XIV, 420 pages. 2003.

Vol. 2901: F. Bry, N. Henze, J. Maluszyński (Eds.), Principles and Practice of Semantic Web Reasoning. Proceedings, 2003. X, 209 pages. 2003.

Vol. 2902: F. Moura Pires, S. Abreu (Eds.), Progress in Artificial Intelligence. Proceedings, 2003. XV, 504 pages. 2003. (Subseries LNAI).

Vol. 2903: T.D. Gedeon, L.C.C. Fung (Eds.), AI 2003: Advances in Artificial Intelligence. Proceedings, 2003. XVI, 1075 pages. 2003. (Subseries LNAI).

Vol. 2904: T. Johansson, S. Maitra (Eds.), Progress in Cryptology – INDOCRYPT 2003. Proceedings, 2003. XI, 431 pages. 2003.

Vol. 2905: A. Sanfeliu, J. Ruiz-Shulcloper (Eds.), Progress in Pattern Recognition, Speech and Image Analysis. Proceedings, 2003. XVII, 693 pages. 2003.

Vol. 2906: T. Ibaraki, N. Katoh, H. Ono (Eds.), Algorithms and Computation. Proceedings, 2003. XVII, 748 pages. 2003.

Vol. 2910: M.E. Orlowska, S. Weerawarana, M.P. Papazoglou, J. Yang (Eds.), Service-Oriented Computing – ICSOC 2003. Proceedings, 2003. XIV, 576 pages. 2003.

Vol. 2911: T.M.T. Sembok, H.B. Zaman, H. Chen, S.R. Urs, S.H.Myaeng (Eds.), Digital Libraries: Technology and Management of Indigenous Knowledge for Global Access. Proceedings, 2003. XX, 703 pages. 2003.

Vol. 2913: T.M. Pinkston, V.K. Prasanna (Eds.), High Performance Computing – HiPC 2003. Proceedings, 2003. XX, 512 pages. 2003.

Vol. 2914: P.K. Pandya, J. Radhakrishnan (Eds.), FST TCS 2003: Foundations of Software Technology and Theoretical Computer Science. Proceedings, 2003. XIII, 446 pages. 2003.

Vol. 2916: C. Palamidessi (Ed.), Logic Programming. Proceedings, 2003. XII, 520 pages. 2003.

Vol. 2918: S.R. Das, S.K. Das (Eds.), Distributed Computing – IWDC 2003. Proceedings, 2003. XIV, 394 pages. 2003.

Vol. 2923: V. Lifschitz, I. Niemelä (Eds.), Logic Programming and Nonmonotonic Reasoning. Proceedings, 2004. IX, 365 pages. 2004. (Subseries LNAI).

Vol. 2927: D. Hales, B. Edmonds, E. Norling, J. Rouchier (Eds.), Multi-Agent-Based Simulation III. Proceedings, 2003. X, 209 pages. 2003. (Subseries LNAI).

Vol. 2929: H. de Swart, E. Orlowska, G. Schmidt, M. Roubens (Eds.), Theory and Applications of Relational Structures as Knowledge Instruments. Proceedings. VII, 273 pages. 2003.